Animal Production: Animal Welfare, Feeding and Sustainability

Animal Production: Animal Welfare, Feeding and Sustainability

Editor: Arden Barnes

Callisto Reference
www.callistoreference.com

Callisto Reference,
118-35 Queens Blvd., Suite 400,
Forest Hills, NY 11375, USA

Visit us on the World Wide Web at:
www.callistoreference.com

ISBN: 978-1-64116-814-4 (Hardback)

Cataloging-in-Publication Data

Animal production : animal welfare, feeding and sustainability / edited by Arden Barnes.
p. cm.
Includes bibliographical references and index.
ISBN 978-1-64116-814-4
1. Animal culture. 2. Livestock--Breeding. 3. Animal welfare. 4. Animal feeding. I. Barnes, Arden.
SF41 .A55 2023
636.009--dc23

Table of Contents

Preface

It is often said that books are a boon to mankind. They document every progress and pass on the knowledge from one generation to the other. They play a crucial role in our lives. Thus I was both excited and nervous while editing this book. I was pleased by the thought of being able to make a mark but I was also nervous to do it right because the future of students depends upon it. Hence, I took a few months to research further into the discipline, revise my knowledge and also explore some more aspects. Post this process, I begun with the editing of this book.

Animal production refers to the process of raising animals for profit. This includes housing, feeding, marketing and breeding. Proper nutrition provides the animals with the necessary energy to develop and breed, as well as the immunity to combat infections. Animal nutrition is essential for livestock production, as feeding has a direct effect on the production capacity, growth rate and overall health of animals. The reproductive capability of male animals is also dependent on nutrition. Animal health plays a crucial role in sustainable livestock farming because it balances the three essential components which include social acceptability, environmental responsibility and economic feasibility. Animal feed safety has an impact on animal productivity, health and welfare, which in turn affects the livelihood of farmers and safety of human food. This book provides significant information to help develop a good understanding of animal welfare, feeding and sustainability. It presents researches and studies performed on animal production by experts across the globe. The extensive content of this book provides the readers with a thorough understanding of the subject.

I thank my publisher with all my heart for considering me worthy of this unparalleled opportunity and for showing unwavering faith in my skills. I would also like to thank the editorial team who worked closely with me at every step and contributed immensely towards the successful completion of this book. Last but not the least, I wish to thank my friends and colleagues for their support.

Editor

1

Effects of Long-Term Supplementation with Brown Seaweeds and Polyphenols in Rabbit on Meat Quality Parameters

Raffaella Rossi [1,*], Francesco Vizzarri [2], Sabrina Ratti [1], Marisa Palazzo [3], Donato Casamassima [3] and Carlo Corino [1]

1 Dipartimento di Medicina Veterinaria, Università Degli Studi di Milano, Via dell'Università 6, 26900 Lodi, Italy; sabrinaratti@hotmail.com (S.R.); carlo.corino@unimi.it (C.C.)
2 Department of Agricultural and Environmental Science, Università di Bari Aldo Moro, Via G. Amendola 165/A, 70126 Bari, Italy; francesco.vizzarri@uniba.it
3 Department of Agricultural, Environmental and Food Sciences, Università Degli Studi del Molise, Via F. De Sanctis 1, 86100 Campobasso, Italy; m.palazzo@unimol.it (M.P.); casamassima.d@unimol.it (D.C.)
* Correspondence: raffaella.rossi@unimi.it

Simple Summary: Safe and natural feed supplement are required to enhance animal health and welfare, answering to the agri-food system and consumers' needs. The research for sustainable feed additives for enhancing rabbit health and meat quality parameters are needed. Plants received considerable consideration as safe feed supplements in animal nutrition. Recent studies indicated that seaweed can be considered a sustainable dietary supplement that improve animals' health and meat quality parameters. Moreover, plant polyphenols have been studied as antioxidants and cholesterol-lowering agents in meat. No data on the effects of long term dietary supplementation of a mixture composed of brown seaweed and plant polyphenols has been available. For this reason, we decided to test the effects of dietary natural supplements in rabbits, based on performance and meat quality parameters.

Abstract: The objective of the present study was to evaluate the effects of dam and offspring dietary supplementation with a natural feed additive on the growth performance and meat quality parameters of growing rabbits. The growing rabbits are selected from lactating does receiving a control diet (C) or diets supplemented with 0.3% (SP1) and 0.6% (SP2) of feed additive containing brown seaweeds (*Laminaria* spp.) and plant extracts. In the postweaning phase, the growing rabbits remained in the treatment group defined by their does and the trial lasted 42 days. The average daily feed intake and feed conversion ratio were improved in the rabbit fed 0.6% of the natural feed additive. The cholesterol content tended to be lower in *Longissimus lumborum* (LL) muscle and decrease in *Semimembranosus* (SM) muscle (in SP2 −41.36% than controls). The α tocopherol and retinol content were enhanced in both muscles of rabbit fed the natural mixture (SP1 and SP2 groups). An improvement of sensory attributes of texture was observed in both muscles from rabbit fed natural mixture. In conclusion, long term supplementation of both lactating does and offspring with the high dosage of brown seaweed and plant polyphenols improves growth performance and enhances meat nutritional and sensory parameters.

Keywords: brown seaweed; meat quality parameters; plant polyphenols; rabbit

1. Introduction

The EU is the second highest world producer of meat rabbits. The majority (83%) of production is in Spain (48.5 million rabbits slaughtered), followed by France (29 million rabbits slaughtered) and Italy (24.5 million rabbits slaughtered). Rabbit meat represents a traditional dietary practice in the Mediterranean area and a new source of consumption in China and Mexico [1–3].

Even if the global rabbit meat consumption per capita is low if compared with other meat (0.19 kg/per capita/year), in EU countries the consumption is about 0.51 kg/per capita/year (Spain 1.09 kg; Italy 0.91 kg; France 0.75 kg) [3].

In the last decades, the genetic selection has turned towards improvement of the offspring number, growth rate and muscle mass. This causes some problems related to global farm efficiency [4,5]. In rabbit farming several diseases occur; digestive disorders are the main problem, affecting antibiotics consumption [6].

Considering that the EU has been committed for years towards a rational use of antibiotics in livestock, nutrition can be one potential efficient strategy for improving rabbit health and meat quality. Additional research that explores the effects of dietary sustainable additives and alternative feed ingredients in rabbit are required.

Moreover, in recent years, the demand for healthy foods that respect the environment and animal welfare has increased [7]. For products of animal origin, the idea of healthy and safe is linked not only to the product nutritional composition but also to the diet, additives, and antibiotics that the animals have taken. In fact, a strong consumers' preference for natural food, free from synthetic additives and antibiotics was observed [8].

Several studies reported that dietary integration with plant extracts containing polyphenols in rabbit improves health, increases nutrient digestibility and growth performances and enhances meat quality parameters [9,10]. Plant extracts offer a considerable range of activities such as antioxidant, anti-inflammatory, antiviral and antimicrobial effects, positively influencing feed digestibility and microbial ecology [6].

Other studies showed that dietary brown seaweeds in livestock had a positive effect on health, growth performance, and meat quality due to their nutraceuticals properties and content of sulfated polysaccharides, phlorotannins, diterpenes, omega-3 polyunsaturated fatty acids, minerals and vitamins [11,12]. Moreover, brown seaweeds represent a renewable and sustainable feed ingredient for their high productivity if compared to other conventional ingredients.

Our previous study reported the effects of a mixture of brown seaweeds and plant polyphenols on the performance of lactating does [13]. Moreover, we also investigated the effect of the same natural supplementation in growing rabbits [14]. In literature, no previous study reported the effects of long term supplementation in rabbit with brown seaweed and plant polyphenols from lactating does to rabbit slaughter. Therefore, we wanted to evaluate if dietary treatments during the entire rabbit rearing period may positively affect the growing rabbit's productive performance and meat quality parameters.

2. Materials and Methods

2.1. *Animals and Dietary Treatments*

Rabbits were handled following the guidelines of EU Directive 2010/63/EU and national guidelines for the care and use of animals [15]. All experimental procedures involving animals were approved by the National Agricultural and Food Centre ethical committee (No. NPPC 18-10-2016).

The dietary supplementation began in the lactation phase in which does (n = 60) were divided into three experimental groups, homogeneous for body weight (4.83 ± 0.19 kg) and parity order (second), receiving a control diet (C) or diets supplemented with 0.3% (SP1) and 0.6% (SP2) of feed additive containing prebiotic polysaccharides from brown seaweeds (*Laminaria* spp.) and plant extracts containing phenolic acid, hydroxycinnamic acids, tannins, and flavonoids. In the postweaning period, all rabbits remained in the treatment group defined by their does. The trial lasted 42 days.

The dosages of the supplement were chosen based on an in vitro study on the minimal inhibitory concentration (MIC) against *Clostridium* spp., *Staphylococcus* spp., and *Escherichia coli* spp. [16].

No anticoccidials, antibiotics or other medications were included in the diet. The feed additive was included in the mashed diets, then the diets were pelleted.

At weaning (35 days), 144 New Zealand White rabbits, balanced for body weight and sex were housed at the National Agricultural and Food Centre, Nitra (Slovak Republic).

Rabbits were randomly allotted into three experimental groups (48 rabbits per treatment) and were housed in cages (2 rabbits/cage). The cages were equipped with a hopper for feed and an automatic nipple drinking system. The lighting cycle throughout the trial was 16 h of light and 8 h of dark. The building temperature was maintained within 18 ± 4 °C, with a relative humidity of 70 ± 5%.

The animals were fed ad libitum and had free access to water. Table 1 reports the ingredients and chemical composition of the experimental diets.

Table 1. Ingredients (g/kg) and chemical composition of experimental diets.

Ingredients	C	Diet [a] SP1	SP2
Maize	282	279	276
Alfalfa hay	305	305	305
Sunflower meal	135	135	135
Carob bean meal	90	90	90
Wheat	80	80	80
Oat	53	53	53
Cane molasses	20	20	20
Palm seed oil	8	8	8
Soybean oil	7	7	7
Calcium carbonate	7	7	7
Sodium Chloride	3	3	3
DL-Methionine (99%)	2.5	2.5	2.5
Vitamin and mineral premix [b]	2.5	2.5	2.5
Dicalcium phosphate	2	2	2
L-Lysine HCl (78.5%)	1.6	1.6	1.6
Choline (75%)	1.4	1.4	1.4
Dietary supplement	0	3	6
Chemical composition [c]			
Crude protein	184	183.6	183.5
Ether extract	37.5	35.5	35.5
Crude fiber	187	186.8	187
Ash	86	85.7	85.8
Nitrogen free extract	507	507.1	506.9
NDF	302.1	301.5	301.7
ADF	195.8	195.4	195.3
ADL	39.9	39.5	39.5

[a] C = control diet; SP1 = diet supplemented with 0.3% of brown seaweed and plant polyphenols; SP2 = diet supplemented with 0.6% of brown seaweed and plant polyphenols. [b] Supplied per kg diet: 13,500 I.U. vitamin A (trans-retinyl acetate); 800 I.U. vitamin D3 (cholecalciferol); 35 mg vitamin E (α-tocopherol min 91%), 35 mg copper (cupric sulphate pentahydrate). [c] Determined in triplicate.

The chemical composition was analyzed according to the methods of the Association of Analytical Chemists [17], following the recommendations of the European group on rabbit nutrition [18].

Table 2 reports the chemical composition, phenolic composition and carotenoid content of the feed supplement. The supplement phenolic compounds were determined using HPLC-UV-DAD [19]. The beta-carotene content of the feed supplement was performed according to Rakusa et al. [20].

Table 2. Chemical composition and polyphenols content of the dietary supplement.

Item	
	% DM
Dry matter	93.6 ± 5.05
Crude protein	7.2 ± 0.99
Ether extract	0.32 ± 0.01
Crude fiber	11.2 ± 1.02
Carbohydrates	49.6 ± 3.18
Ash	32.7 ± 1.38
Chemical compounds: [a]	mg/kg DM
β-Carotene	402 ± 30.89
Phenolic acid:	
Syringic acid	1059.8 ± 62.82
Hydroxycinnamic acids:	
Neochlorogenic acid	7979.2 ± 468.11
Rosmarinic acid	126.5 ± 8.67
Trans sinapic acid	105.5 ± 8.09
Chlorogenic acid	21.4 ± 3.65
Tannins:	
Ellagic acid	2440.9 ± 148.29
Rutin	272.4 ± 20.82
Flavonoids:	
Myricetin	53.9 ± 5.68

[a] Values are expressed as means (n = 4) ± standard deviation.

2.2. *Sampling*

The animals were weighed at 0 and 42 days of the experimental trial and were monitored daily to assess the health conditions. Daily feed intake was recorded (feed offered and refused weekly). The data were used to calculate the average daily gain (ADG), average daily feed intake (ADFI) and feed conversion ratio (FCR).

At the end of the trial the rabbits (77 days old) were weighted and after a fasting period of 6 h 1 male rabbit per cage (12 animals/group) were randomly selected and slaughtered. Rabbits were subjected to electrical stunning (100 V, 50 Hz, 2–3 s) and sacrificed by bleeding according to the guidelines established by the European Community (1099/2009/EC) for the protection of animals during slaughter [21].

Carcasses were chilled for 24 h at +4 °C and then dissected, according to the recommendations of the WRSA [22], by removing the skin, distal part of the limbs, genital organs, bladder and the gastrointestinal tract. Carcasses were then weighted, and the dressing percentages were determined. Samples of whole *Longissimus lumborum* (LL) from the 7th thoracic vertebra to the 7th lumbar vertebra and whole thighs were collected. The physical parameters were evaluated on freshly cut LL muscle and *Semimembranosus* (SM) muscle. The whole muscles were then sampled for chemical and sensory analysis, vacuum packed and stored at −20 °C. The chemical parameters were performed on LL and SM muscles and the sensory evaluation was assessed on the LL muscle and whole thighs.

2.3. *Physical and Chemical Parameters*

All the analyses were performed on LL muscle and *Semimembranosus* (SM) muscle. The pH and color parameters were measured 24 h after slaughter. The pH measurement was performed using a pH meter (HI98191 microcomputer; Hanna Instruments, Vila do Conde, Portugal), calibrated with a standard buffer of pH 4.0 and 7.0.

Meat color parameters were evaluated at 24 h after slaughter on the fresh cut surface of LL and SM muscles, after 20 min of blooming, using a chromameter (Chromameter CR 300 Minolta Ltd., Osaka, Japan) equipped with a D65 light source and a 0° viewing angle geometry according to the reflectance coordinates (L*, a*, b*). The instrument was calibrated using a white calibration plate (Calibration Plate CR-A43; Minolta Camera Co.). The color determination was assessed in triplicate.

The chemical composition of both muscles was determined according to the AOAC methods [17].

The cholesterol content was determined following the method of Du and Ahn [23], using a gas chromatographic method. The cholesterol was identified based on the retention time of the standard (Sigma Aldrich, St. Louis, CA, USA) and quantified with the Chrom Card Data System (version 1.17) software. All samples were analyzed in triplicate.

2.4. Vitamin Content

The α-tocopherol and retinol content in LL and SM muscle were determined using the modified procedure of Zaspel and Csallany [24]. The samples were analyzed using an HPLC system (Kontron Instruments, Milan, Italy) with an autosampler (HPLC autosampler 360, Kontron Instruments, Milan, Italy) with a loop of 20 μL, a high-pressure pump and a C18 column 5 μm, 150 × 4.6 mm (Phenomenex, Torrance, CA, USA). The mobile phase was acetonitrile and methanol (75:25 *v*/*v*) and the flow rate was 1 mL per min. The α-tocopherol and retinol were identified using a fluorimeter detector, comparing the retention time with the standards (Sigma Aldrich, St. Louis, CA, USA). The quantification was performed using the Geminyx system (version 1.91).

2.5. Sensory Analysis

The sample preparation was performed after thawing thighs and LL muscle for 24 h at 4 °C. The whole samples of LL and thighs were prepared in an uncovered stainless-steel dish in a conventional oven (REX, Milano, Italy) at 180 °C. A thermocouple (Pentronic AB, Gunnebobruk, Sweden) was used to register the sample core temperature and samples were removed at 79–80 °C. The roasted samples were then cut into 1.5 cm thick slices (Electrolux 50, 220–24, kW0.2) and the slices were warmed to 60 °C before the evaluation. A trained sensory panel, involving eight members familiar with descriptive analysis procedures [25], was employed. All evaluations were performed in a sensory laboratory equipped according to EN ISO 8598 recommendations [26]. The descriptors, definitions, and standards are reported in Palazzo et al. [27].

The sensory profile was assessed according to EN ISO 13,299 [25] and the panel evaluated the two samples (thigh and LL) on different days in triplicate. Within each session the design was balanced for order and carry-over effects [28]. During training and sampling, the panelists had access to unlimited water and unsalted crackers. They were requested to evaluate the intensity of each attribute by assigning a score between 1 (absence of the sensation) and 9 (extremely intense).

2.6. Statistical Analysis

Data were analyzed using SPSS software (SPSS, PC Statistics 25.0 SPSS Inc., IBM, Segrate, MI, Italy). Productive performance and slaughter and meat quality parameters were analyzed using one-way analysis of variance (ANOVA), with diet as the fixed effect. The sensory data were submitted to ANOVA with samples, judges, replicates, and their interactions as effects (EN ISO 13299, 2010). The significance of these effects was tested with F tests. Post-hoc pairwise contrasts were evaluated by Duncan's test. The cage was considered as the experimental unit for growth performance and the rabbit for the meat quality parameters. Data are reported as mean ± SEM. Differences among treatments were considered significant at $p < 0.05$.

3. Results

3.1. Productive Parameters and Carcass Characteristics

The growth performances and carcass parameters of rabbit fed experimental diets are reported in Table 3. The final weight of the rabbits did not differ among experimental groups ($p > 0.05$), even if the ADG tended to be higher ($p = 0.09$) in both the SP1 and SP2 groups. The ADFI result was lower

($p < 0.05$) in the SP1 group than the results in the other groups. The feed conversion ratio was positively affected ($p < 0.001$) by both the dosage of brown seaweed and plant polyphenols (SP1 and SP2).

Table 3. Growth performances of rabbits fed the control diet (C) and diet supplemented with 3 g/kg (SP1) and 6 g/kg (SP2) of brown seaweed and plant polyphenols.

Item	C	Diet SP1	SP2	p-Value
Initial weight, g	829.1 ± 0.21	838.7 ± 0.22	818.8 ± 0.17	0.686
Final weight, g	2657.2 ± 58.62	2757.9 ± 41.54	2864.1 ± 35.51	0.220
ADG, g/d	42.5 ± 1.16	45.7 ± 1.04	46.3 ± 0.73	0.090
ADFI, g/d	172.1 ± 2.45 ^{a}	148.8 ± 3.17 ^{b}	161.9 ± 1.66 ^{a}	0.014
FC	3.95 ± 0.04 ^{A}	3.25 ± 0.06 ^{B}	3.49± 0.03 ^{B}	0.001

ADG, average daily gain; ADFI, average daily feed intake; FC, feed conversion ratio. Data are reported as means ± standard error of the means (SEM); n = 48 ^{a,b} values in the same row are different at $p < 0.05$. ^{A,B} values in the same row are different at $p < 0.01$.

The slaughter weight (2791.3 ± 69.2 g C vs. 3001.2 ± 59.4 g SP1 vs. 2986.1 ± 42.4 g SP2; $p = 0.021$) and the carcass weight (1473.4 ± 38.6 g C vs. 1585.2 ± 33.4 g SP1 vs. 1610.4 ± 19.7 g SP2; $p = 0.010$) of the sampled rabbit (n = 12) resulted higher in groups fed brown seaweed and polyphenols (SP1 and SP2) than in the control. The dressing percentage resulted higher in SP2 group than the others (52.8 ± 0.37% C vs. 52.4 ± 0.32% SP1 vs. 53.9 ± 0.41% SP2; $p = 0.025$).

3.2. Meat Physical and Chemical Parameters

The chemical composition of LL and SM muscle are reported in Tables 4 and 5, respectively. The physical parameters in both muscles were not affected ($p > 0.05$) by long term supplementation of both lactating does and the offspring with low and high dosage of seaweed and polyphenols. The nutritional parameters moisture, crude protein, ether extract and ash were also unaffected by dietary treatments ($p > 0.05$). The cholesterol content in LL muscle tended to be lower ($p = 0.059$) in SP2 than SP1 and the control groups. The cholesterol content of the SM muscle resulted lower ($p < 0.01$) in the group that received the high dosage of seaweed and polyphenols (SP2) than SP1 and the control groups. The cholesterol content reduction was about −40.3% than control and −36.4% than SP1 groups.

Table 4. Physical and chemical parameters of *Longissimus lumborum* muscle of rabbits fed the control diet (C) and diet supplemented with 3 g/kg (SP1) and 6 g/kg (SP2) of brown seaweed and plant polyphenols.

Item	C	Diet SP1	SP2	p-Value
pH 24 h	5.85 ± 0.25	5.91± 0.21	5.88 ± 0.16	0.222
Color parameters				
L*	55.85 ± 0.39	56.30 ± 1.01	56.44 ± 0.81	0.873
a*	3.89 ± 0.71	3.99 ± 0.32	4.28 ± 0.42	0.911
b*	11.92 ± 0.67	11.00 ± 0.52	11.28 ± 0.26	0.484
Chemical composition:				
Moisture, %	73.23 ± 0.18	73.02 ± 0.21	72.84 ± 0.18	0.383
Crude Protein, %	24.33 ± 0.14	24.52 ± 0.17	24.65 ± 0.21	0.404
Ether extract, %	1.17 ± 0.06	1.05 ± 0.06	1.19 ± 0.10	0.547
Ash, %	1.22 ± 0.01	1.36 ± 0.08	1.39 ± 0.06	0.132
Cholesterol, mg/100 g	33.23 ± 2.98	28.18 ± 1.24	25.07 ± 1.18	0.059

Data are reported as means ± standard error of the means (SEM); n = 12.

Table 5. Physical and chemical parameters of *Semimembranosus* muscle of rabbits fed the control diet (C) and diet supplemented with 3 g/kg (SP1) and 6 g/kg (SP2) of brown seaweed and plant polyphenols.

Item	C	Diet SP1	SP2	*p*-Value
pH 24 h	5.77 ± 0.20	5.79± 0.18	5.82± 0.21	0.853
Color parameters				
L*	63.92 ± 1.59	63.84 ± 2.01	64.41 ± 1.94	0.951
a*	6.01 ± 0.64	5.94 ± 0.56	6.11 ± 0.74	0.894
b*	3.92 ± 0.32	4.02 ± 0.49	3.78 ± 0.56	0.831
Chemical composition:				
Moisture, %	74.29 ± 0.86	74.22 ± 0.78	74.17 ± 0.72	0.594
Crude Protein, %	23.12 ± 0.53	23.37 ± 0.73	23.61 ± 0.62	0.338
Ether extract, %	1.66 ± 0.46	1.92 ± 0.38	1.51 ± 0.43	0.958
Ash, %	1.21 ± 0.06	1.19 ± 0.09	1.15 ± 0.07	0.329
Cholesterol, mg/100 g	39.85 ± 4.48 ^{A}	37.39 ± 2.07 ^{A}	23.78 ± 1.17 ^{B}	0.002

Data are reported as means ± standard error of the means (SEM); n = 12. ^{A,B} values in the same row are different at $p < 0.01$.

3.3. Muscle Vitamin Content

Figure 1 shows the α-tocopherol (A) and retinol (B) content of both muscles in relation to dietary treatments. The α-tocopherol content resulted higher ($p < 0.001$) in the LL and SM muscles of rabbit fed both dosages of brown seaweed and polyphenols (SP1 and SP2) than in the control. The retinol content was higher ($p < 0.001$) in both muscles of rabbit fed brown seaweed and plant polyphenols (SP1 and SP2) than in the control group.

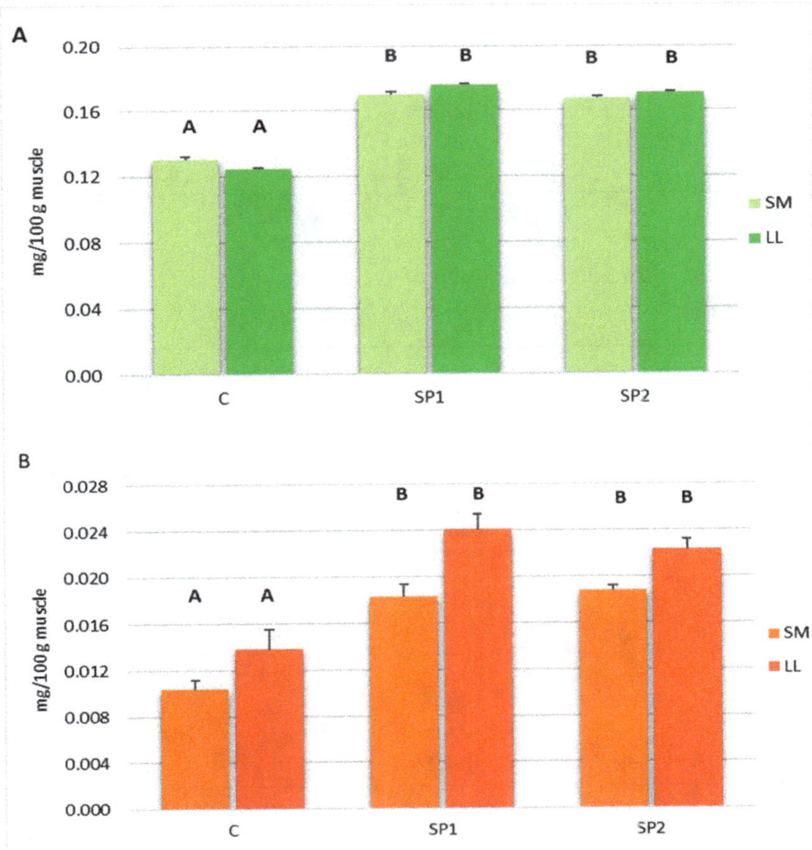

Figure 1. α-tocopherol (**A**) and retinol (**B**) content of *Longissimus lumborum* (LL) and *Semimembranosus* muscle (SM) of rabbits fed control diet (C) and diet supplemented with 3 g/kg (SP1) and 6 g/kg (SP2) of brown seaweed and plant polyphenols. Data are reported as means ± standard error of the means (SEM); n = 12. ^{A,B} values with different superscript letters are different at $p < 0.001$.

3.4. Sensory Profile

The F values for the measured parameters of the LL muscle and thigh sensory profile are reported in Tables 6 and 7, respectively. The data show that dietary supplementation with brown seaweed and plant polyphenols mixture affected ($p < 0.05$) the flavor and texture of the LL muscle. The F values for replicates and interactions were not affected ($p > 0.05$) for all the descriptors. In thigh, taste, flavor and texture were affected ($p < 0.05$) by dietary treatments. Judges presented differences ($p < 0.05$) for aroma, taste, flavor and texture as observed for LL muscle. The judges and the interaction of the treatments per judge revealed differences ($p < 0.05$) for aroma, taste, flavor and texture.

Table 6. The F value and statistical significance of treatments (C, SP1, SP2) and their interaction for sensory descriptors of *Longissumus lumborum* muscle, where the number of judges, n = 8, and the number of replicates, n = 3.

Descriptors	Treatments	Judges	F Value Replicates	T*J	T*R	J*R
Aroma						
Rabbit	2.27	4.52	0.26	1.68	1.55	0.85
Liver	0.05	1.73	1.37	0.70	1.75	1.03
Metallic	2.27	4.83 **	0.08	1.63	1.41	0.83
Taste						
Sweet	0.71	2.67 *	1.33	1.64	0.46	1.23
Salty	2.12	2.83 *	0.63	1.10	1.16	1.22
Flavor						
Rabbit	4.61 *	1.37	0.73	0.95	2.25	1.45
Liver	12.88 ***	5.56 ***	0.30	3.41 **	0.77	1.47
Metallic	5.18 *	5.83 ***	0.45	2.19 *	0.30	1.00
Texture						
Tender	8.53 **	7.80 ***	0.81	2.98 **	0.48	1.12
Juicy	20.31 ***	2.69 *	2.04	3.84 **	1.00	1.12
Stringy	3.56 *	3.26 *	1.04	4.83 ***	1.48	0.77

*** $p < 0.0001$; ** $p < 0.001$; * $p < 0.05$.

Table 7. The F value and statistical significance of treatments (C, SP1, SP2) and their interaction for sensory descriptors of thigh, where the number of judges, n = 8, and the number of replicates, n = 3.

Descriptors	Treatments	Judges	F Value Replicates	T*J	T*R	J*R
Aroma						
Rabbit	1.91	2.58 *	0.09	1.02	1.28	0.80
Liver	0.11	1.35	1.15	0.65	1.07	1.09
Metallic	035	5.68 **	0.89	2.07 *	0.56	1.10
Taste						
Sweet	3.97 *	0.70	2.90	1.29	2.42	0.81
Salty	0.36	4.06 **	1.02	2.18 *	0.37	0.71
Flavor						
Rabbit	6.51 **	3.26 *	1.22	1.26	0.58	2.42
Liver	2.66	2.46 *	1.66	1.47	1.42	1.43
Metallic	2.97 *	5.09 ***	0.95	1.84	1.13	1.17
Texture						
Tender	16.95 ***	4.44 **	0.93	5.23 ***	0.97	0.46
Juicy	12.92 ***	1.47	0.57	4.92 ***	0.86	0.71
Stringy	7.45 **	14.44 ***	0.39	9.57 ***	0.39	0.84

*** $p < 0.0001$; ** $p < 0.001$; * $p < 0.05$.

Figure 2 reports the least squares mean of the different attributes for the LL muscle (A) and thighs (B). In LL muscle the rabbit, liver and metallic flavors were affected by the high dosage of brown seaweed and plant polyphenols, resulting higher ($p < 0.05$) in the SP2 group than the SP1 group and the control groups. The texture parameters were also enhanced in muscle from rabbit fed the natural mixture. In particular, tenderness and juiciness resulted higher ($p < 0.05$) in both groups receiving brown seaweed and plant polyphenols (SP1 and SP2) than the controls. The stringiness resulted lower ($p < 0.05$) in the control and SP2 groups than results in the SP1 group. The attributes related to aroma and taste were unaffected ($p > 0.05$) by dietary treatments. In tights the rabbit and metallic flavors were affected by brown seaweed and plant polyphenols, resulting higher ($p < 0.05$) in both groups receiving the supplement (SP1 and SP2) than in the controls. The sweety taste was positively affected ($p < 0.05$) by the high dosage of natural supplement, in fact, thigh from SP2 group resulted sweetness higher than controls. Tenderness and juiciness resulted higher ($p < 0.05$) in groups receiving the high dosage of brown seaweed and plant polyphenols (SP2) than the SP1 group and the controls. The stringiness resulted lower ($p < 0.05$) in the control group than the SP1 group. The attributes related to aroma were unaffected ($p > 0.05$) by dietary treatments.

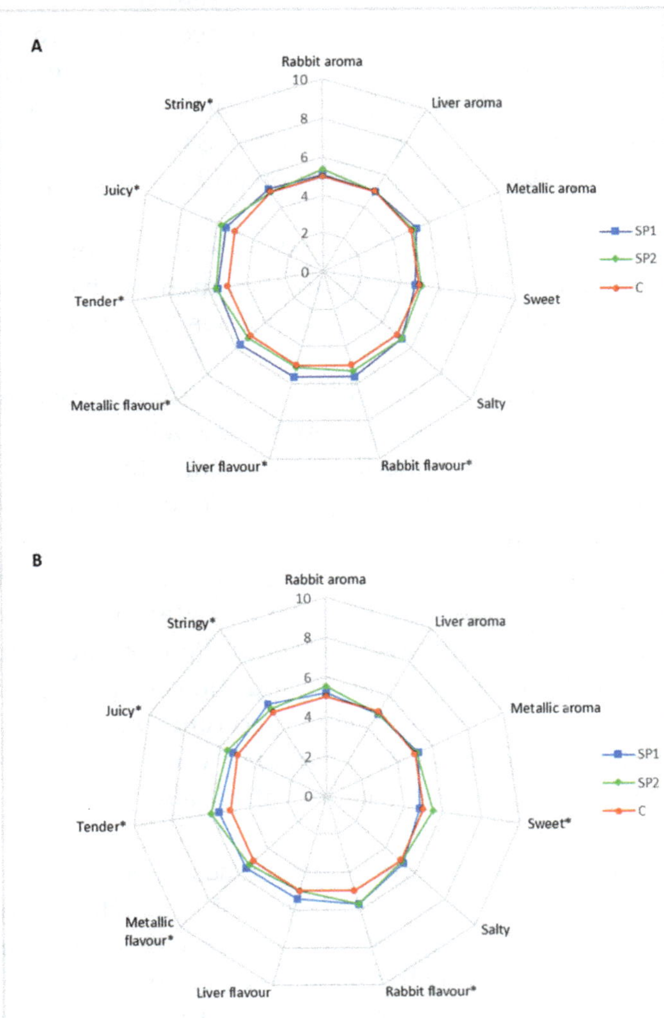

Figure 2. Spider plot of the sensory profile of *Longissimus lumborum* muscle (**A**) and thigh (**B**) of rabbit fed the control diet (C) and the diet supplemented with 3 g/kg (SP1) and 6 g/kg (SP2) of brown seaweed and plant polyphenols. * Significant at $p < 0.05$.

4. Discussion

In the recent years, several studies reported the effect of dietary supplementation with plant polyphenols in rabbit to improve growth performances and meat quality parameters [9,29]. However, there are only few studies on dietary supplementation with seaweed in rabbit. Previous study reported that dietary green seaweed *Ulva Lactuca* in does and male rabbit improve reproductive parameters [30,31]. Moreover, Dalle Zotte et al. [32] reported that a mixture containing dietary microalga *Spirulina* and *Thymus vulgaris*, did not affect growth performance and energy or nutrients digestibility.

To the best of our knowledge, the first study describing the effect of dietary brown seaweed and plant polyphenols in rabbit is our recent study limited to the growing phase [14], but the effects of long term supplementation of this mixture from lactating does to rabbit slaughter has been not yet studied. The present data showed that ADG tended to be higher in rabbit fed the natural mixture and the feed conversion ratio was enhanced in rabbit fed the dietary supplement, suggesting a better feed conversion due to the probiotic, antibacterial and antioxidant effects of the natural mixture used [11]. An enhancement of nutrient digestibility was also observed in several animal species fed seaweed [12]. The present data are in line with studies in rabbit showing that natural extract dietary supplementation positively affected rabbit's growth performances [6,10].

The data on carcass characteristics showed an enhancement in rabbit fed the natural dietary supplement. Previous studies reported no effects of dietary supplementation with polyphenols on carcass characteristics in pigs and rabbit [9,33]. Other studies in pigs reported an enhancement of carcass characteristics due to polyphenols supplementation [34,35]. These different results are probably due to the length and dosage of dietary supplementation and the active principles content of the plant extract used.

The present data show that dietary supplementation with polyphenols and brown seaweed did not affect pH and color parameters in both muscles, in agreement with our previous study in rabbit fed the same mixture [14]. The data fall in the means of the values reported in other studies for LL and SM muscles [36,37].

Likewise, the chemical composition of the LL and SM muscles was not affected by the dietary treatments, except for the cholesterol content. Comparing with our study in rabbit fed the same mixture [14], the dietary supplementation of both the does and the offspring reduces the cholesterol content in both muscles, and this reduction compared to controls was relevant in the SP2 groups (−24.5% and −40.3% in LD and SM muscle, respectively).

Previous studies reported a lowering cholesterol effect of seaweed in livestock and humans. Al-Harthi and El-Deek [38] reported that dietary supplementation with 3–6% of brown seaweed in laying hens improved egg quality, and decreased yolk cholesterol content. In humans, seaweed intake was positively linked to a modulation of blood glucose and cholesterol [39]. This mechanism is probably linked to the activity of the seaweed polysaccharide fucoidan that decreases cholesterol absorption and increases its excretion, modulating reverse cholesterol transport-related protein expression [40]. A cholesterol lowering effect was also observed in rabbit and hares fed plant extract containing polyphenols [41,42].

Our results show that the dietary mixture of brown seaweed and plant polyphenols enhances the content of α tocopherol and retinol in both LL and SM muscles. Seaweeds show antioxidant properties due to the phenols, carotenoid fucoxanthin, tannins, phlorotannins and polysaccharides content. Furthermore, seaweeds are an excellent source of vitamins and, above all, brown seaweeds contain a large amount of α tocopherol and carotenoid [11]. In fact, previous studies in pork reported that seaweed dietary supplementation enhances oxidative stability of LD muscle due to higher muscle α tocopherol content [43,44].

Moreover, plant extract containing polyphenols increased α tocopherol and retinol levels in several tissues, reducing the oxidative markers [45]. In fact, an earlier study in pork, rabbit and hares reported that natural extract containing polyphenols enhanced muscle α tocopherol and retinol [27,46,47].

The data on sensory profile revealed that the dietary mixture of brown seaweed and plant polyphenols affected the flavor in both LL muscle and thigh, even if the differences are hardly perceived. The texture attributes related to tenderness, juiciness and stringiness was improved in in both LL muscle and thigh from rabbit fed the natural mixture. The judges and the interaction of the treatments per judges revealed differences for aroma, taste, flavor and texture, but this is common in sensory evaluations, due to the different use of the scale [48]. Even if there is selection and training of judges, some variability always remains, but the interaction treatments per judges was not significant for most descriptors, indicating low variability among the panel members.

In literature no previous study reported the effects of dietary brown seaweed supplementation on meat sensory quality parameters while several studies reported the effects of dietary plant polyphenols on this parameter. Dietary plant polyphenols from *Lippia* spp. improve the texture parameters related to tenderness and juiciness in donkey and horse *Longissimus dorsi* muscle and it is probably related to the protection against the oxidation process [27]. Moreover, Zhao et al. [49] also reported an improvement of tenderness of lamb meat due to dietary supplementation with wine grape pomace, a rich source of polyphenols. Probably, the protection of the proteolytic enzymes μ-calpain and m-calpain from the oxidative process, increases their functionality, enhancing meat tenderness [50,51].

5. Conclusions

Our results show that dietary supplementation with a high dosage of brown seaweeds and plant polyphenols of both lactating does and the offspring have a positive effect on the average daily feed intake and feed conversion ratio, without affecting other meat physical parameters.

Nutritional characteristics of both LL and SM muscle were positively affected by the natural extract, with a reduction of the cholesterol content and an increase of α tocopherol and retinol content. Moreover, an improvement of the sensory quality perceived by consumers in terms of texture was observed in both LL muscle and thigh from rabbit fed the high dosage of the natural mixture.

From these data, we can conclude that the long term supplementation of both lactating does and the offspring with a high dosage of brown seaweeds and plant polyphenols is a valid approach to boost rabbits' performance and enhance meat nutritional and sensory parameters.

Author Contributions: Conceptualization, C.C. and D.C.; investigation, F.V., R.R., S.R. and M.P.; data curation, F.V. and R.R.; writing—original draft preparation, R.R.; writing—review and editing, C.C., F.V., R.R., S.R. and M.P.; project administration, supervision C.C. and D.C. All authors have read and agreed to the published version of the manuscript.

Acknowledgments: The authors would like to thank the technical staff of the National Agricultural and Food Centre, Nitra (Slovak Republic) for the availability of rabbit, and technical support during the whole experimental period.

References

1. EFSA Panel on Animal Health and Welfare (AHAW). Health and welfare of rabbit farmed in different production system. *EFSA J.* **2020**, *18*, e05944.
2. Petracci, M.; Cavani, C. Rabbit meat processing: Historical perspective to future directions. *World Rabbit Sci.* **2013**, *21*, 217–226. [CrossRef]
3. Szendrő, K.; Szabó-Szentgróti, E.; Szigeti, O. Consumers' Attitude to Consumption of Rabbit Meat in Eight Countries Depending on the Production Method and Its Purchase Form. *Foods* **2020**, *9*, 654. [CrossRef]
4. Gondret, F.; Larzul, C.; Combes, S.; De Rochambeau, H. Carcass composition, bone mechanical properties, and meat quality traits in relation to growth rate in rabbits. *J. Anim. Sci.* **2005**, *83*, 1526–1535. [CrossRef]
5. Hernandez, P.; Arino, B.; Grimal, A.; Blasco, A. Comparison of carcass and meat characteristics of three rabbit lines selected for litter size or growth rate. *Meat Sci.* **2006**, *73*, 645–650. [CrossRef] [PubMed]
6. Dalle Zotte, A.; Celia, C.; Szendrő, Z.S. Herbs and spices inclusion as feed stuff or additive in growing rabbit diets and as additive in rabbit meat: A review. *Livest. Sci.* **2016**, *189*, 82–90. [CrossRef]

7. Petrescu, D.C.; Petrescu-Mag, R.M. Consumer behaviour related to rabbit meat as functional food. *World Rabbit Sci.* **2018**, *26*, 321–333. [CrossRef]
8. Román Luis, S.; Sánchez-Siles, M.; Siegrist, M. The importance of food naturalness for consumers: Results of a systematic review. *Trends Food Sci. Technol.* **2017**, *67*, 44–57. [CrossRef]
9. Palazzo, M.; Vizzarri, F.; Arvay, J.; D'Alessandro, A.G.; Martemucci, G.; Casamassima, D.; Ratti, S.; Corino, C.; Rossi, R. Dietary effect of dried bay leaves (*Laurus nobilis*) meal on selected productive performances and on quality meat traits in growing rabbits. *Livest. Sci.* **2020**, *242*, 104301. [CrossRef]
10. Miteva, D.; Velikov, K.; Ivanova, S.; Dimov, K. Production of rabbit meat with functional properties. *Agro Life Sci. J.* **2020**, *9*, 221–228.
11. Corino, C.; Modina, S.C.; Di Giancamillo, A.; Chiapparini, S.; Rossi, R. Seaweeds in Pig Nutrition. *Animals* **2019**, *9*, 1126. [CrossRef] [PubMed]
12. Morais, T.; Inácio, A.; Coutinho, T.; Ministro, M.; Cotas, J.; Pereira, L.; Bahcevandziev, K. Seaweed Potential in the Animal Feed: A Review. *J. Mar. Sci. Eng.* **2020**, *8*, 559. [CrossRef]
13. Vizzarri, F.; Chiapparini, S.; Corino, C.; Casamassima, D.; Palazzo, M.; Parkanyi, V.; Ondruska, L.; Rossi, R. Dietary supplementation with natural extracts mixture: Effects on reproductive performances, blood biochemical and antioxidant parameters in rabbit does. *Ann. Anim. Sci.* **2019**. [CrossRef]
14. Rossi, R.; Vizzarri, F.; Chiapparini, S.; Ratti, S.; Casamassima, D.; Palazzo, M.; Corino, C. Effects of dietary levels of brown seaweeds and plant polyphenols on growth and meat quality parameters in growing rabbit. *Meat Sci.* **2020**, *161*, 107987. [CrossRef] [PubMed]
15. European Community. Council Directive 2010/63/EU of 22 September 2010 on the Protection of Animals Used for Scientific Purposes. 2010. Available online: Eurlex.europa.eu/LexUriServ/LexUriServ.do?uri=OJ:L:2010:276:0033:0079:EN:PDF (accessed on 30 September 2020).
16. Tosi, G.; (Experimental Zooprophylactic Institute of the Lombardy and Emilia, Forlì, Italy). Personal communication, 2012.
17. AOAC. *Official Methods of Analysis*, 17th ed.; Association of Official Analytical Chemists: Arlington, VA, USA, 2002.
18. European Group on Rabbit Nutrition (EGRN). Technical note: Attempts to harmonize chemical analyses of feeds and faeces for rabbit feeding evaluation. *World Rabbit Sci.* **2001**, *9*, 57–64.
19. Russo, R.; Pucci, L.; Giorgetti, L.; Árvay, J.; Vizzarri, F.; Longo, V.; Pozzo, L. Polyphenolic characterisation of plant mixture (Lisosan® reduction) and its hypocholesterolaemic effect in high fat diet-fed mice. *Nat. Prod. Res.* **2017**, *33*, 651–658. [CrossRef] [PubMed]
20. Rakusa, Z.T.; Srecnik, E.; Roskar, R. Novel HPLC-UV method for simultaneous determination of fat-soluble vitamins and coenzyme Q10 in medicines and supplements. *Acta Chim. Slov.* **2017**, *64*, 523–529. [CrossRef] [PubMed]
21. Council of the European Union. Council Regulation (EC) No. 1099/2009 of 24 September 2009 on the Protection of Animals at the Time of Killing. 2009. Available online: Eur-lex.europa.eu/LexUriServ/LexUriServ.do?uri=OJ:L:2009:303:0001:0030:EN:PDF (accessed on 30 September 2020).
22. Blasco, A.; Ouhayoun, J. Harmonization of criteria and terminology in rabbit meat research. Revised proposal. *World Rabbit Sci.* **1996**, *4*, 93–99. [CrossRef]
23. Du, M.; Ahn, D.U. Simultaneous analysis of tocopherols, cholesterol, and phytosterols using gas chromatography. *J. Food Sci.* **2002**, *67*, 1696–1700. [CrossRef]
24. Zaspel, B.J.; Csallany, A.S. Determination of alpha-tocopherol in tissues and plasma by high-performance liquid chromatography. *Anal. Biochem.* **1983**, *1*, 146–150. [CrossRef]
25. International Organization for Standardization. *Sensory Analysis-Methodology-General Guidance to Establish a Sensory Profile*; ISO 13299; ISO: Geneva, Switzerland, 2010.
26. International Organization for Standardization. *Sensory Analysis: General Guidance for Design of Test Rooms*; ISO 8598; ISO: Geneva, Switzerland, 1989.
27. Palazzo, M.; Vizzarri, F.; Nardoia, M.; Ratti, S.; Pastorelli, G.; Casamassima, D. Dietary *Lippia citriodora* extract in rabbit feeding: Effects on quality of carcass and meat. *Arch. Anim. Breed.* **2015**, *58*, 355–364. [CrossRef]
28. MacFie, H.J.; Bratchell, N.; Greenhoff, K.; Vallis, L.V. Designs to balance the effect of order of presentation and first-order carry-over effect in halls tests. *J. Sens. Stud.* **1989**, *4*, 129–148. [CrossRef]

29. Dabbou, S.; Gai, F.; Renna, M.; Rotolo, L.; Dabbou, S.; Kovitvadhi, C.L.A.; Brugiapaglia, A.; De Marco, M.; Helal, A.N.; Zoccarato, I.; et al. Inclusion of bilberry pomace in rabbit diets: Effects on carcass characteristics and meat quality. *Meat Sci.* **2017**, *124*, 77–83. [CrossRef] [PubMed]
30. El-banna, S.G.; Hassan, A.A.; Okab, A.B.; Koriem, A.A.; Ayoub, M.A. Effect of feeding diets supplemented with seaweed on growth performance and some blood hematological and biochemical characteristics of male Baladi rabbits. In Proceedings of the 4th International Conference on Rabbit Production in Hot Climates, Sharm Elsheikh, Egypt, 24–27 February 2005; pp. 373–382.
31. Okab, A.B.; Samara, E.M.; Abdoun, K.A.; Rafay, J.; Ondruska, L.; Parkanyi, V.; Pivko, J.; Ayoub, M.A.; Al-Haidary, A.A.; Riyadh, S.A.; et al. Effects of dietary seaweed (*Ulva lactuca*) supplementation on the reproductive performance of buck and doe rabbits. *J. Appl. Anim. Res.* **2013**, *41*, 347–355. [CrossRef]
32. Dalle Zotte, A.; Sartori, A.; Bohatir, P.; Remignon, H.; Ricci, R. Effect of dietary supplementation of Spirulina (*Arthrospira platensis*) and thyme (*Thymus vulgaris*) on growth performance, apparent digestibility and health status of companion dwarf rabbits. *Livest. Sci.* **2013**, *152*, 182–191. [CrossRef]
33. Korniewicz, D.; Rózanski, H.; Usydus, Z.; Dobrzanski, Z.; Korniewicz, A.; Kaczmarek, P.; Frankiewicz, A.; Szulc, K. Efficiency of plant extracts (herbiplant cs) in pigs fattening. *Pol. J. Food Nutr. Sci.* **2007**, *57*, 309–315.
34. Rossi, R.; Stella, S.; Ratti, S.; Maghin, F.; Tirloni, E.; Corino, C. Effects of antioxidant mixtures in the diet of finishing pigs on the oxidative status and shelf life of Longissimus dorsi muscle packaged under modified atmosphere. *J. Anim. Sci.* **2017**, *95*, 4986–4997. [CrossRef]
35. Dávila-Ramírez, J.L.; Munguía-Acosta, L.L.; Morales-Coronado, J.G.; García-Salinas, A.D.; González-Ríos, H.; Celaya-Michel, H.; Sosa-Castañeda, J.; Sánchez-Villalba, E.; Anaya-Islas, J.; Barrera-Silva, M.A. Addition of a mixture of plant extracts to diets for growing-finishing pigs on growth performance, blood metabolites, carcass traits, organ weight as a percentage of live weight, quality and sensorial analysis of meat. *Animals* **2020**, *10*, 1229. [CrossRef]
36. Kozioł, K.; Maj, D.; Bieniek, J. Change in the color and pH of rabbit meat in the aging process. *Med. Weter.* **2015**, *71*, 104–108.
37. Tůmová, E.; Bízková, Z.; Skřivanová, V.; Chodová, D.; Martinec, M.; Volek, Z. Comparisons of carcass and meat quality among rabbit breeds of different sizes, and hybrid rabbits. *Livest. Sci.* **2014**, *165*, 8–14. [CrossRef]
38. Al-Harthi, M.A.; El-Deek, A.A. Effect of different dietary concentrations of brown marine algae (*Sargassum dentifebium*) prepared by different ways on plasma and yolk lipid profiles, yolk total carotene and lutein plus zeaxanthin of laying hens. *Ital. J. Anim. Sci.* **2012**, *11*, e64. [CrossRef]
39. Brown, E.M.; Allsopp, P.J.; Magee, P.J.; Gill, C.I.; Nitecki, S.; Strain, C.R.; McSorley, E.M. Seaweed and human health. *Nutr. Rev.* **2014**, *72*, 205–216. [CrossRef] [PubMed]
40. Yang, Z.; Liu, G.; Wang, Y.; Yin, J.; Wang, J.; Xia, B.; Li, T.; Yang, X.; Hou, P.; Hu, S.; et al. Fucoidan A2 from the Brown Seaweed Ascophyllum nodosum Lowers Lipid by Improving Reverse Cholesterol Transport in C57BL/6J Mice Fed a High-Fat Diet. *J. Agric. Food Chem.* **2019**, *67*, 5782–5791. [CrossRef] [PubMed]
41. Vizzarri, F.; Nardoia, M.; Palazzo, M. Effect of dietary Lippia citriodora extract on productive performance and meat quality parameters in hares (*Lepus europaeus* Pall.). *Arch. Tierzucht* **2014**, *57*, 1–7. [CrossRef]
42. Vizzarri, F.; Palazzo, M.; D'Alessandro, A.G.; Casamassima, D. Productive performance and meat quality traits in growing rabbit following the dietary supplementation of Lippia citriodora, Raphanus sativus and *Solanum lycopersicum* extracts. *Livest. Sci.* **2017**, *200*, 53–59. [CrossRef]
43. Rajauria, G.; Draper, J.; McDonnel, M.; O'Doherty, J.V. Effect of dietary seaweed extracts, galacto oligosaccharide and vitamin E supplementation on meat quality parameters in finisher pigs. *Innov. Food Sci. Emerg. Technol.* **2016**, *37*, 269–275. [CrossRef]
44. Moroney, N.C.; O'Grady, M.N.; Robertson, R.C.; Stanton, C.; O'Doherty, J.V.; Kerry, J.P. Influence of level and duration of feeding polysaccharide (laminarin and fucoidan) extracts from brown seaweed (*Laminaria digitata*) on quality indices of fresh pork. *Meat Sci.* **2015**, *99*, 132–141. [CrossRef]
45. Corbi, G.; Conti, V.; Komici, K.; Manzo, V.; Filippelli, A.; Palazzo, M.; Vizzari, F.; Davinelli, S.; Di Costanzo, A.; Scapagnini, G.; et al. Phenolic Plant Extracts Induce Sirt1 Activity and Increase Antioxidant Levels in the Rabbit's Heart and Liver. *Oxid. Med. Cell. Longev.* **2018**, *2018*, 2731289. [CrossRef]
46. Rossi, R.; Pastorelli, G.; Cannata, S.; Tavaniello, S.; Maiorano, G.; Corino, C. Effect of long term dietary supplementation with plant extract on carcass characteristics meat quality and oxidative stability in pork. *Meat Sci.* **2013**, *95*, 542–548. [CrossRef]

47. Palazzo, M.; Schiavitto, M.; Cinone, M.; Vizzarri, F. Rabbit metabolic response and selected meat quality traits. Evaluation of dietary PLX® 23 and LycoBeads® feed supplement. *J. Anim. Physiol. Anim. Nutr.* **2019**, *103*, 383–394. [CrossRef]
48. Lea, P.; Rødbotten, M.; Næs, T. *Analysis of Variance for Sensory Data*; Chirchester John Wiley & Sons Ltd.: New York, NY, USA, 1997.
49. Zhao, J.X.; Li, Q.; Zhang, R.X.; Liu, W.Z.; Ren, Y.S.; Zhang, C.X.; Zhang, J.X. Effect of dietary grape pomace on growth performance, meat quality and antioxidant activity in ram lambs. *Anim. Feed Sci. Technol.* **2018**, *236*, 76–85. [CrossRef]
50. Morán, L.; Andrés, S.; Bodas, R.; Prieto, N.; Giráldez, F.J. Meat texture and antioxidant status are improved when carnosic acid is included in the diet of fattening lambs. *Meat Sci.* **2012**, *91*, 430–434. [CrossRef] [PubMed]
51. Huff–Lonergan, E.; Lonergan, S.M. Mechanisms of water–holding capacity of meat: The role of postmortem biochemical and structural changes. *Meat Sci.* **2005**, *71*, 194–204. [CrossRef] [PubMed]

2

Dietary Application of Tannins as a Potential Mitigation Strategy for Current Challenges in Poultry Production

Janghan Choi and Woo Kyun Kim *

Department of Poultry Science, University of Georgia, Athens, GA 30602, USA; choij@uga.edu
* Correspondence: wkkim@uga.edu

Simple Summary: There are diverse challenges in the poultry production industry that decrease the productivity and efficiency of poultry production, impair animal welfare, and pose issues to public health. Furthermore, the use of antibiotic growth promoters (AGP) in feed, which have been used to improve the growth performance and gut health of chickens, has been restricted in many countries. Tannins, polyphenolic compounds that precipitate proteins, are considered as alternatives for AGP in feed and provide solutions to mitigate challenges in poultry production due to their antimicrobial, antioxidant, anti-inflammatory and gut health promoting effects. However, because high dosages of tannins have antinutritional effects when fed to poultry, determining appropriate dosages of supplemental tannins is critical for their potential implementation as a solution for the challenges faced in poultry production.

Abstract: The poultry industry has an important role in producing sources of protein for the world, and the size of global poultry production continues to increase annually. However, the poultry industry is confronting diverse challenges including bacterial infection (salmonellosis), coccidiosis, oxidative stress, including that caused by heat stress, welfare issues such as food pad dermatitis (FPD) and nitrogen and greenhouse gasses emissions that cumulatively cause food safety issues, reduce the efficacy of poultry production, impair animal welfare, and induce environmental issues. Furthermore, restrictions on the use of AGP have exacerbated several of these negative effects. Tannins, polyphenolic compounds that possess a protein precipitation capacity, have been considered as antinutritional factors in the past because high dosages of tannins can decrease feed intake and negatively affect nutrient digestibility and absorption. However, tannins have been shown to have antimicrobial, antioxidant and anti-inflammatory properties, and as such, have gained interest as promising bioactive compounds to help alleviate the challenges of AGP removal in the poultry industry. In addition, the beneficial effects of tannins can be enhanced by several strategies including heat processing, combining tannins with other bioactive compounds, and encapsulation. As a result, supplementation of tannins alone or in conjunction with the above strategies could be an effective approach to decrease the need of AGP and otherwise improve poultry production efficiency.

Keywords: tannins; chickens; gut health; feed additives; antimicrobials; antioxidants

---

1. Introduction

Poultry products including meat and eggs account for a significant part of global food production and constitute a protein staple throughout the world [1]. The United States Department of Agriculture (USDA) reported in 2016 that global egg production was approximately 70 million tons, and poultry meat production reached more than 100 million tons, which accounted for more than one-third of global meat production [2]. Moreover, global poultry production continues to increase annually [1]. However, there are many challenges in the poultry industry including: bacterial infection (salmonellosis); parasitic infection (coccidiosis); oxidative stress, including that caused by heat stress; welfare issues such as food pad dermatitis (FPD); and nitrogen and greenhouse gas emissions which can cause severe economic losses, threaten food safety and public health, impair animal welfare, and induce environmental pollution [2–5].

Antibiotic growth promoters (AGP) have been supplemented to chicken diets to improve growth performance and gut health, predominantly due to their antimicrobial effects and immunomodulatory functions in chickens [6,7]. However, because of the increased public concern about the transmission of antibiotic-resistant bacteria from poultry products, the use of AGP in poultry production has been banned or restricted in many countries [8–10]. In addition, some producers in the U.S. poultry industry have opted to entirely remove the use of antibiotics and instead raise chickens using "no antibiotics ever (NAE)" or "raised without antibiotics (RWA)" approaches [11,12]. As a consequence, the efficiency of poultry production has decreased due to increases in various bacterial and parasitic infections and reductions in the growth rate of chickens [13]. In addition, because there is no "magic bullet" that can replace AGP, some poultry producers are still using antibiotics in the U.S. and in many other countries, and the use of antibiotics for livestock animals in the world is expected to increase, possibly owing to population growth which is associated with a greater demand for livestock products in middle-income countries [14]. Therefore, it is essential to find alternatives to AGP, which must be cost-effective, eco-friendly and have antimicrobial and growth-promoting effects, without causing side effects (e.g., generation of resistant bacteria) to the animals and humans [15].

Tannins, defined as polyphenolic compounds that can precipitate proteins, are secondary metabolites, which are found in plants, seeds, bark, wood leaves and fruit skins and serve as plant defense mechanisms against predation [16]. High concentrations of tannins have been shown to have antinutritional effects in monogastric animals because tannins can decrease feed intake, nutrient digestibility and growth performance of chickens [17,18]. However, recently in poultry production, tannins have garnered a great deal of attention as an alternative for AGP because of their antimicrobial, antioxidants and anti-inflammation properties [19–21]. In addition, many tannins are considered sustainable feed additives, as they derive from byproducts of plant-based agriculture and industry. For example, chestnut tannins, which are already sometimes supplemented to poultry, are obtained by the distillation of wood that is used in the building industry [22]. However, the effects of tannins on the growth performance and gut ecosystem of the chickens are still inconsistent and their mode of action is unclear. Therefore, it is important to understand the chemical properties and biological effects of tannins to maximize the use of supplemental tannins in chickens. This review is mainly focused on the classification and bioavailability of tannins, the effects of distinct tannins on mitigating the challenges facing poultry production, and strategies to enhance the effects of tannins.

2. Classification and Bioavailability of Tannins

Tannins, defined as polyphenolic compounds that have a protein precipitation capacity, exist in several different types with various molecular weights [23]. Plant tannins are classified into hydrolysable tannins (HT) with tannin derivatives (e.g., gallic acid and ellagic acid) and condensed tannins (CT) [24] (Figure 1). Additionally, phlorotannins (PT) are a third class of tannin unique to brown algae. Tannins have different bioavailabilities (absorbability), and their level of bioavailability varies depending on several factors, including the derivatives of each tannin (e.g., gallic acid and ellagic acid), their affinity to protein, molecular structure and molecular weight. The bioavailability of tannins is an important trait for their functionality and should be considered for tackling different issues in poultry production. Tannins with a low bioavailability potentially have better antimicrobial effects in chickens, whereas highly bioavailable tannins would be more beneficial as antioxidant and anti-inflammatory agents.

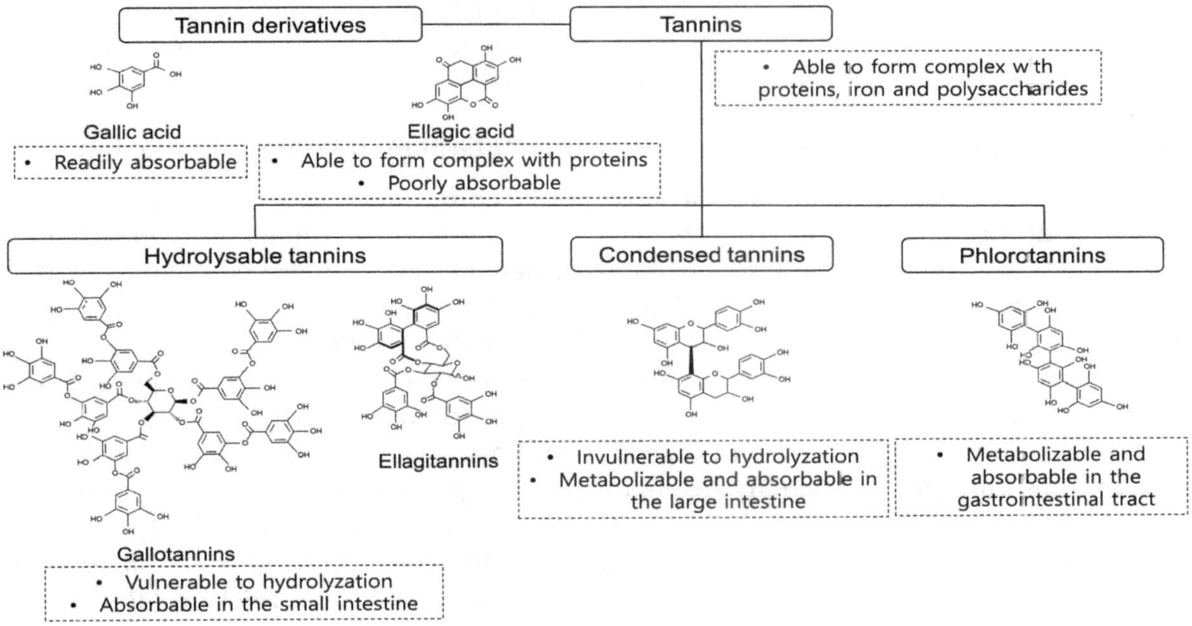

Figure 1. Classification and examples of molecular structure of tannins. Hydrolysable tannins and condensed tannins belong to plant tannins, and phlorotannins are found in brown algae.

The HT including predominately gallotannins and ellagitannins (molecular weight 500 to 3000 Da) contain a polyol (normally glucose) as a central core, which is esterified with phenolic groups (e.g., gallic acid and ellagic acid) [25]. Under certain conditions (e.g., thermal processing, esterification, and acid or base treatment), HT can be hydrolyzed to yield free gallic acid and ellagic acid. Gallotannins produce one molecule of sugar and 9 to 10 molecules of gallic acid when hydrolyzed, whereas ellagitannins yield one molecule of sugar and several molecules of gallic acid and ellagic acid [26]. The unhydrolyzed HT are partially absorbable in the small intestine [27]. Gallic acid is readily absorbable into the blood stream [28], while ellagic acid has low bioavailability due to strong affinity for proteins and poor absorption [29]. Hence, HT include gallotannins and ellagitannins and have different bioavailability depends on the components and structure.

The CT are defined as oligomeric or polymeric flavonoids containing flavan-3-olunits such as catechin, epicatechin, gallocatechin and epigallocatechin [24]. The CT exist in fruits (e.g., berries, pears and apples), forage legumes (e.g., lentils, black-eyed peas, chickpeas and red kidney beans), nuts, red and green grapes (and their juice and wine) [30–32]. The CT have higher molecular weight (1000—20,000 Da) and more complex structure compared to HT [24]. Unlike HT, CT are not vulnerable to hydrolyzation, which may imply low bioavailability in the gastrointestinal tract (GIT) of the chickens. This is because high molecular weight of tannins are hardly absorbable in the intestine [33]. Kahle et al. [34] reported that around 90% procyanidins (CT) were recovered in the distal ileum, which imply that most of CT can reach to the large intestine. The CT in the large intestine can be metabolized and absorbed by epithelial cells [35]. The CT have low bioavailability compared to HT and can be delivered to the large intestine of the chicken.

The PT, algae-derived polyphenols, have molecular sizes ranging 126 Da to 650 kDa, but the majority of them are 10 and 100 kDa [36,37]. By using in-vitro models, Corona et al. [38] demonstrated that PT can be metabolized and absorbed in the upper GIT potentially due to low molecular weight of PT, and limited amounts of biologically active PT were delivered to the colon. In addition, the author also demonstrated that higher molecular weight of PT are less vulnerable to metabolization and absorption in the upper GIT [38]. Nwosu et al. [39] showed that PT can be metabolized and absorbed by the colon cells (e.g., Caco-2). Therefore, PT can be easily metabolized and absorbed in the GIT tracts due to their low molecular weights. However, more future in-vivo experiments are required to study bioavailability of PT in chickens.

Taken together, tannins are classified into HT, CT and PT, and their bioavailability are depending on their components, structure and molecular weight. It is important to understand specific chemical properties each tannin prior to application in poultry diets.

3. Traditional Viewpoints on Tannins as Antinutritional Factors in Poultry Production

While tannins were known as beneficial bioactive compounds in ruminants, they have been considered as antinutritional factors in poultry diets [16]. In the (GIT) of ruminants, tannins bind proteins under rumen pH (pH 5.5–7.0), inhibiting the microbial degradation of dietary proteins. Upon reaching the abomasum (pH 2.5–3.5), the non-covalent linkages between proteins and tannins are broken, and free protein can be absorbed by the host in the distal small intestine (pH 7.5) [40]. As such, tannins are known to increase ruminant's protein utilization and decrease gas emission [41,42]. However, in chickens, tannins were considered as phytotoxins predominately due to their protein binding properties, which impair dietary protein digestion and decrease activity of digestive enzymes [43]. Moreover, tannins bind proline-rich, hydrophobic salivary proteins of chickens, forming complexes that are responsible for an astringent taste of the feed, and in turn lead to low palatability and decreased feed intake of chickens [44]. In addition, a high concentration of tannic acid (25 g/kg) showed toxicity to chickens by showing liver proteolytic activity in broiler chickens [45]. Lee et al. [46] reported that tannic acid impaired growth performance, hematological indices and plasma iron status in weaning pigs potentially because tannic acid can form complex with iron. However, these toxic properties of tannins were shown when tannins were included in the diets more than 7.5 g/kg (based on tannic acid, considered as the standard of HT), and many current studies proved that appropriate amounts of tannins (based on tannic acid; standard of tannins) ranging from to 0.5 g/kg to 5 g/kg in poultry could improve grow rate and gut health due to their potential antimicrobial, antioxidants and anti-inflammatory functions (Table 1). Therefore, while tannins were considered as anti-nutritional factors in the past, tannins at appropriate dosages have potentials to improve growth performance and gut health of chickens.

Table 1. Effects of different dosages of tannic acid on the broiler chickens.

Dosages of Tannic Acid	Outcomes	References
0.5 g/kg	Improved growth performance and immune system in broiler chickens challenged with *Eimeria* spp.	[47]
2 g/kg	Improved foot pad dermatitis of the chickens without affecting growth performance	[48]
5 g/kg	Improved growth performance and decreased lipid oxidation.	[49]
5 g/kg	Increased growth performance, modulated cecal microbial metabolites and decreased cecal pH.	[50]
7.5 and 15 g/kg	Decreased growth performance and did not inhibit the growth of *Salmonella typhimurium* in broiler chickens challenged with *S. typhimurium*.	[51]
10 g/kg	Attenuated fatty acid profile of breast but decreased growth performance in the heat stressed chickens.	[52]
25 g/kg	Decreased growth performance and showed liver toxicity by inducing liver proteolytic activity	[45]

4. Challenges in Poultry Production and Potential Solution by Using Tannins

4.1. Effects of Tannins on Bacterial Infection (Salmonellosis)

There are many pathogenic bacteria in chickens and poultry products (meat and eggs), including *Salmonella Typhimurium*, *Escherichia coli* O157:H7, *Campylobacter jejuni*, *Clostridium perfringens*, enterohemorrhagic *Escherichia. coli* (EHEC), *Listeria monocytogenes*, *Arcobacter butzleri*, *Mycobacterium avium* subsp. *Paratuberculosis* and *Aeromonas hydrophila* [53]. These pathogenic bacteria not only impair gut health and growth rate of chickens, but are also a public health threat as foodborne diseases in humans [54]. Salmonellosis, one of the main food-borne diseases from poultry products, is induced by *S. Typhimurium* and *S. Enteritis* and causes, illness, morbidity and mortality in humans [55].

Diverse in-vitro studies showed that tannins and their derivatives showed bacteriostatic (inhibiting bacterial growth) and bactericidal (killing bacteria) effects on *Salmonella* spp. and other pathogens, as shown in Table 2. Potential mechanisms of antibacterial effects of tannins include (1) direct interactions with components in the cell wall to alter morphology of the cell wall and to increase membrane permeability of bacteria [24]; (2) decreasing activities of microbial enzymes [56]; and (3) depriving essential nutrients such as proteins and minerals (e.g., iron) for pathogenic bacteria [57–59]. In addition to bacteriostatic and bactericidal effects of tannins, many in-vitro studies reported that sub-lethal dosages of tannins also restricted pathogenicity of *Salmonella* spp. and other pathogens by inhibiting motility [60], quorum sensing [61] and biofilm formation [62] of pathogenic bacteria.

Table 2. In-vitro antimicrobial effects tannins against diverse pathogenic bacteria.

Tannins Sources	Strains	Results and Conclusions	References
Ellagitannins from Chestnut wood Gallotannins from Tara and Sumach (Gall nuts) Condensed tannins from Quebaracho and *Calliandra calothyrsus* Flavanol gallates from Tea and *Acacia nilotica*	*Salmonella Typhimurium*	All of the tannins inhibited the growth of *S. Typhimurium.*	[63]
Tannic acid Gallic acid	*S. Typhimurium*	Tannic acid and gallic acid had bactericidal effects and gallic acid had higher bactericidal effects than tannic acid	[64]
Condensed tannins extracted from tree leaves viz. babool (*Acacia nilotica*), jamun (*Eugenia jambolana*), peepal (*Ficus religiosa*), subabul (*Leucaenia leucocephala*) and guajava (*Psidium guajava*)	*Escherichia coli* *Staphylococcus aureus* *S. enteritis Enterococcus faecalis*	All of the selected five condensed tannins inhibited the growth of the four pathogenic bacteria	[65]
Tannin extracts from *Cytinus hypocistis* and *C. ruber*	*S. aureus* *S. epidermidis* *E. faecium* *Pseudomonas aeruginosa* *Klebsiella pneumoniae*	Tannin extracts from *C. hypocistis* and *C. ruber* showed antibacterial and antibiofilm activities against gram positive and negative human pathogens.	[66]
Chestnut tannins (80% hydrolysable tannins) Quebracho tannins (75% condensed tannins)	*Clostridium perfringens*	Both hydrolysable and condensed showed antimicrobial effects against *C. perfringens* and neutralized its cytotoxicity.	[67]

In addition to their antibacterial effects, tannins can control systemic infection of *Salmonella* spp. by beneficially modulated components of gut ecosystem. *Salmonella* spp. can enter blood circulation via paracellular and transcellular pathways and use immune cells to enter enterocytes to be distributed in the internal organs and muscles in chickens (Figure 2). Many studies reported that tannins altered expression and functionality of tight junction proteins [68], mucus [69,70], and immune cells [71,72] of chickens. Nevertheless, supplemental tannins have not typically translated to antimicrobial effects against *Salmonella* spp.in in-vivo studies. For example, while Van Parys et al. [73] reported that whereas HT extract of sweet chestnut woods showed strong antimicrobial effects (minimum inhibitory concentration: 25–50 μg/mL and minimum inhibitory concentration: 100 μg/mL) against *S. Typhimurium* isolated from pigs under *in-vitro* conditions, the inclusion of 3000 mg/kg HT in the pig feed did not reduce the *Salmonella* spp. concentration in feces, intestine and internal organs of pigs inoculated 10^7 colony forming units (CFU)/mL of *S. Typhimurium* strain at four days post-inoculum. Similarly, Kubena et al. [51] reported that tannic acid (7.5 or 15 g/kg in the feed) did not modulate the salmonella concentration in cecal content of broiler chickens inoculated with 10^4 CFU of *S. typhimurium.* Potential explanations for such findings include (1) tannins were degraded by host or microbial enzymes or absorbed before they reached to the lower intestine where most of the *Salmonella* spp. and other pathogenic bacteria propagate; (2) tannins formed complexes with components of feedstuffs (polysaccharides and proteins) or endogenous proteins, which inhibits antimicrobial effects of tannins; and (3) experimental factors including the low dosages of tannins, the high dosages of salmonella inoculum or short period of the experiments were possibly obstacles to diminish antimicrobial effects of tannins in in-vivo models [73]. In contrast, Jamroz et al. [74] reported that 1000 mg/kg of sweet chestnut tannins reduced the number of *E. coli* and coliforms bacteria in the small intestine on 28 d; however, growth performance of the tannin-fed birds was decreased in this study.

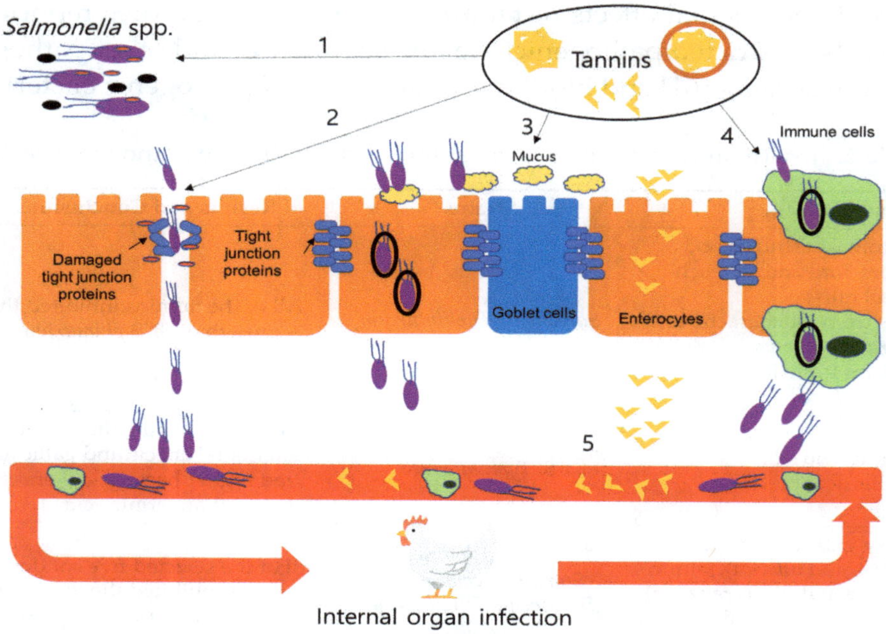

Figure 2. Systemic infection routes of *Salmonella* spp. and potential mechanisms of antibacterial actions of tannins in an in-vivo chicken model. Chickens can orally ingest *Salmonella* spp. from feed or their environment, and subsequently escape the small or large intestine [75]. (1) Tannins and tannin derivatives (tannins, hydrolyzed tannins, and tannin-protein complexes) are known to inhibit the growth of *Salmonella* spp. in the intestine and decrease quorum sensing of the bacteria [61]. Lipopolysaccharides (LPS) of *Salmonella* spp. can impair intestinal barrier function, which allows *Salmonella* spp. to pass the paracellular pathways of the intestine [76,77]. (2) Tannins and tannin derivatives may improve gut barrier integrity by neutralizing LPS or decreasing expression of cytokines, which can impair tight junction proteins [78]. (3) and (4) *Salmonella* spp. also utilized mucus and immune cells to invade epithelial cells [71]. Tannins and their derivates potentially modulate expression and morphology of immune cells and mucus. *Salmonella* spp. can invade epithelial cells in diverse pathways and enter blood circulation and finally colonize in internal organs (e.g., liver, kidney, spleen, etc.) and muscles in chickens [79]. (5) Tannins also can enter the blood circulation and potentially show antimicrobial effects and modulate immune cells to attenuate internal organ infection by *Salmonella* spp. in chickens [80,81].

While many in-vitro studies showed that tannins showed antimicrobial effects against *Salmonella* spp. and other pathogens, the optimal antimicrobial dosages in chickens have not yet been fully determined. Therefore, dose-specific antimicrobial effects of tannins in chickens, as well as their mode of actions, warrant further investigation.

4.2. *Effects of Tannins on Coccidiosis*

Coccidiosis, which is a parasitic disease induced by protozoa of the family *Eimeridae*, is one of the most prevalent and detrimental enteric diseases in poultry production [82]. The nine identified *Eimeria* species in chickens includes *E. acervulina*, *E. brunetti*, *E. maxima*, *E. necatrix*, *E. praecox*, *E. mitis*,

E. tenella, *E. mivati* and *E. hagani* to date [83,84]. *Eimeria* spp. infect and multiply within the mucosal epithelial layers in the different parts of the GIT through the fecal-oral route [85]. Coccidiosis can result in reduced growth rate and gut barrier integrity and induce inflammation, diarrhea, hemorrhaging, and even mortality in broiler chickens, negatively influencing the efficacy of poultry production and welfare [86,87]. The negative effects of coccidiosis on gut health of chickens are closely related to oxidative stress, as *Eimeria* infections cause lipid peroxidation and excessive production of reactive oxygen species (ROS) in chickens [88]. Furthermore, coccidiosis is closely associated with an enteric infectious disease, necrotic enteritis, which is predominately induced by *C. perfringens* with the presence of *Eimeria* spp. [89,90]. To cope with coccidiosis for the poultry industry, prophylactic coccidiostats and anticoccidial drugs have been supplemented in poultry diets [91]. However, resistance for all currently available drugs have been documented, making it imperative to discover novel drug alternatives that induce limited resistance and effectively control coccidiosis in broiler chickens [92]. To find alternatives for prophylactic coccidiostats and anticoccidial drugs, bioactive compounds including prebiotics [93], plant extracts [94], organic acids [95], essential oils [96], lipids (fatty acids) [97], minerals (e.g., zinc) [98] and nitro compounds [99] have been studied in chickens.

Tannins are known to have anticoccidial effects because tannins can form complexes with parasitic enzymes and metal ions, which are essential for *Eimeria* spp. and can stimulate immune system against of the chickens [100–102]. Tonda et al. [47] reported that the dietary supplementation of 500 mg/kg of gallnut tannic acid extract reduced total oocyst number in excreta, and 500 mg/kg of tannic acid or gallnut tannic acid extract decreased intestinal lesion scores in broilers infected with *Eimeria* spp. Furthermore, the authors showed that gallnut tannic acid extract enhanced feed conversion ratio of cocci-vaccinated birds, which possibly implies that gallnut tannic acid extract improved protective immunity following coccidiosis vaccination [47]. Supplemental chestnut HT and quebracho CT tannins reduced *Eimeria* spp. oocyst shedding and parasitic bacterial diarrhea and attenuated negative effects of coccidiosis via immunomodulating and anti-inflammatory effects in rabbits [103]. In broiler chickens, Kaleem et al. [104] showed that administration of *Emblica officinalis* derived tannins improved growth performance and showed immunostimulatory properties and enhanced protective immune system. The beneficial effects of tannins on gut health of chickens infected with *Eimeria* spp. are closely associated with antioxidant properties of tannins, which can restore an *Eimeria*-damaged gastrointestinal [105]. A study by [106] demonstrated that supplementation of grape seed proanthocyanidin extract, rich in CT, enhanced growth performance and attenuated clinical symptoms, potentially by improving antioxidant capacity in chickens infected with *Eimeria tenella*. In contrast, Mansoori and Modirsanei [107] showed that supplemental tannic acid (10 g/kg) numerically increased D-xylose absorption in chickens vaccinated against coccidiosis followed by challenging with *Eimeria* spp., however, supplemental tannic acid increased the total number of oocysts in excreta, indicating that high dosages of tannins can attenuate the efficiency of anticoccidial vaccines and impair appropriate development of immune system against coccidiosis in chickens. The discrepancy may be attributed to different sources and concentrations of tannins and dissimilar experimental conditions (e.g., challenge dosages of *Eimeria* spp.).

Although many studies reported the potential benefits of supplemental tannins in broiler chickens infected *Eimeria* spp., more comprehensive studies are required (1) to study the mechanisms under anticoccidial effects of tannins in chickens; (2) to elucidate mechanisms of the beneficial effects of tannins on the gut health of chickens infected with *Eimeria* spp.; (3) to investigate the effects of tannins in a necrotic enteritis challenge model by using inoculum of *Eimeria* spp. and *C. perfringens*; and (4) to find appropriate concentrations and types of tannins against coccidiosis in chickens.

4.3. *Effects of Tannins on Oxidative Stress, Including that Caused by Heat Stress*

Heat stress is one of the major obstacles in the poultry industry because heat stress negatively impacts growth performance, gut health, meat quality, and welfare of chickens [108,109]. A potential reason for negative effects of heat stress on chickens is closely associated with excessive production of

ROS via accelerated metabolic reactions due to mitochondrial respiration [110–112]. Under normal conditions, enzymatic and non-enzymatic antioxidants can neutralize ROS and maintain the balance between oxidants and antioxidants [113,114] (Figure 3). However, if there is an imbalance between oxidants and antioxidants in chickens, excessively produced ROS can impair gut health and induce inflammation, which results in decreased growth performance of chickens [115]. Antioxidants including plant extracts [116,117], L-carnitine [118], vitamin C [119,120], vitamin E [121], and selenium [122] have been studied to alleviate heat stress and oxidative stress.

Tannins are believed to relieve or attenuate effects of oxidative stress, including that caused by heat stress by scavenging ROS and modulating enzymatic antioxidants in animals (Figure 3) [116]. The reducing power of tannins including proanthocyanidins, catechins, epicatechin, and procyanidin from grade seeds is approximately 20 times higher than vitamin E and 50 times higher than vitamin C [123]. A study from Sahin et al. [124] reported that 200 or 400 mg/kg of epigallocatechin-3-gallate (the ester of epigallocatechin and gallic acid) from green tea relieved oxidative stress by controlling the hepatic nuclear transcription factors such as nuclear factor κ-light-chain-enhancer of activated B cells (NF-κB) and nuclear factor (erythroid-derived 2)-like 2 (Nrf2) in heat stressed quails. Furthermore, 10 mg/kg supplementation of tannic acid improved fatty acid profile (decreased monosaturated fatty acids) in breast muscle of chickens under heat stress [52]. Ramnath and Rekha [125] showed that supplementation of *Brahma Rasayana* containing various sorts of tannins enhanced activities of enzymatic antioxidants including superoxide dismutase (SOD), glutathione peroxidase (GPx), glutathione reductase (GR), and reduced glutathione (GSH) in blood of chickens raised at a cold temperature. Moreover, the inclusion of grape (*Vitis vinifera*) pomace, rich in CT, enhanced antioxidant enzyme activities (GPx and SOD) and intestinal morphology, and increased relative weight of bursa of Fabricius and thymus in the heat-stressed broiler chickens [126]. In a diquat-induced mouse model (oxidative stress model), tannic acid improved intestinal morphology, activated the antioxidative pathway by reducing protein expression of Kelch like-ECH-associated protein 1 (KEAP1) and enhancing protein expression of nuclear factor erythroid 2-related factor 2 (NRF2), as well as modulated intestinal barrier function in the jejunum. Thus, supplementing appropriate concentrations of tannins potentially would be an effective strategy to attenuate oxidative stress in heat-stressed birds.

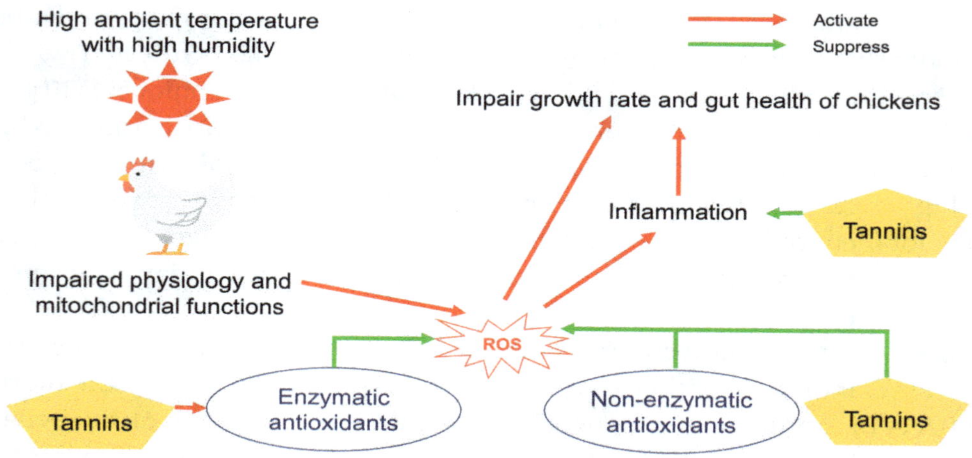

Figure 3. Potential mechanisms for the alleviating effects of tannins in heat-stressed chickens. High ambient temperature with high humidity increase core body temperature of chickens, and this can result in impaired physiology and mitochondrial functions, which lead to excessive reactive oxygen species (ROS) [127]. Tannins, polyphenolic compounds, directly scavenge ROS like non-enzymatic antioxidants [128], and also, tannins can increase activities of antioxidant enzymes such as superoxide dismutase (SOD), glutathione peroxidase (GPX), and glutathione reductase (GR) in broiler chickens. The ROS cause inflammation that can impair growth rate and gut health of chickens and tannins also have anti-inflammatory function [124].

4.4. *Effects of Tannins on Food Pad Dermatitis (FPD)*

Foot pad dermatitis (FPD) is defined as a condition that induces lesions on the plantar surface of the footpads in growing chickens [129]. The FPD causes severe economic losses in poultry production because paws are the third most crucial economic part in broiler chickens, and FPD can impair growth rate, gut health, and welfare of chickens [129,130]. Thøfner et al. [131] showed that FPD and systemic bacterial infections are closely correlated because pathogens can invade to the chickens through damaged epithelium on the foot pads [132]. Litter moisture and litter quality are the most two crucial features that cause FPD among varied factors such as bedding materials and depth, drinkers and nutrient deficiencies [129,133,134]. In addition, increased excreta viscosity has been known to increase the occurrence and severity of FPD in chickens by modulating litter moisture and litter quality [135].

Tannins can relieve the incidence and severity of FPD by enhancing fecal consistency (e.g., fecal dry matter contents) and litter quality [16]. Cengiz et al. [48] reported that 2000 mg/kg supplementation of tannic acid reduced the incidence and severity of FPD in broiler chickens without affecting growth performance, litter quality and intestinal viscosity of chickens. Moreover, 700 mg/kg and 2000 mg/kg of tannin-rich sweet chestnut wood extract increased fecal dry matter contents in chickens [136]. In addition, antimicrobial, antioxidant and anti-inflammatory properties of tannins probably helped to attenuate the incidence and severity of FPD in broiler chickens because oxidative stress and inflammation can exacerbate the severity and consequences of FPD in chickens [137,138]. Thus, supplementation of tannins has potential to reduce severity and incidence of FPD in broiler chickens by enhancing fecal consistency and litter quality.

4.5. *Effects of Tannins on Nitrogen Excretion and Emissions of Noxious and Greenhouse Gases*

Reactive nitrogen species (ammonia, nitrous oxide, and other oxides of nitrogen) and sulfur-containing compounds (hydrogen sulfide and sulfur dioxide) are produced in poultry production facilities and cause environment pollution and greenhouse gases such as carbon monoxide, carbon dioxide and methane [139–142]. Factors affecting the production of nitrogen compounds and detrimental gases include types of feedstuffs, manure conditions, and housing accessories (bedding and heating materials) [143,144]. Nitrogen and greenhouse gases are produced in the livestock animals due to enteric fermentation and manure fermentation, and ruminants contribute most of the livestock greenhouse gas production because of high fermentation rate in the rumen [140]. Although poultry production is not a major contributor of noxious or greenhouse gases, the continued growth of poultry production coupled with higher intensity of excretion and emission per unit compared to other species, more attention is needed to find strategies to mitigate the production of nitrogen excretion and emissions of noxious and greenhouse gases [145].

It is well-established that tannins improve N utilization and decrease methane production in ruminant animals [25,146]. The potential mechanisms underlying effects of tannins on mitigation of emissions of nitrogen and greenhouse gases are inhibiting growth of methanogens, reducing protozoal-associated methane production, and decreasing fiber fermentation [25]. Although chickens (monogastric animals) have distinct GIT from ruminants, tannins may alter gut health and microbiota and improve N utilization in chickens, thereby potentially reducing nitrogen and methane emissions. Ahmed and Yang [147] reported that supplementation of by-products of *Punica granatum* (fruit), which contains HT such as ellagitannins, punicalagin, punicalin and pedunculagin, decreases the emission of ammonia and methanethiol from excreta of broiler chickens. Moreover, a study by Bostami et al. [139] demonstrated that supplementation of fermented pomegranate byproducts, containing ellagitannins, reduced gas emission (ammonia and hydrogen sulfide) from excreta in broiler chickens, potentially via reducing microbial activity and pH of excreta. However, 2000 mg/kg of tannic acid from chestnut wood did not affect ammonia volatilization in broiler chickens. Hence, more studies are required to identify suitable sorts and dosages of tannins to reduce nitrogen excretion and emissions of noxious and greenhouse gases during poultry production.

4.6. Effects of Tannins on Growth Performance, Immune System, Gut Microbiota, Gut Ecosystem and in Chickens Raised under General Conditions

While some authors showed that low concentrations (0.5 g/kg to 5 g/kg) of tannic acid improved growth performance (shown in Table 1), the others reported that supplementation of different sources of tannins at low dosages (0.5 to 5 g/kg) did not affect growth performance [48,136] and even showed negative effects on growth performance of the birds [74]. These differences may be a result of different tannin sources, supplementation period, or specific experiment conditions (e.g., genetics of chickens, temperature and abundance of pathogens in the living conditions).

However, supplemental tannins at appropriate dosages can improve immune system, gut ecosystem and gut microbiota of chickens raised under general conditions. Ramah et al. [148] reported that whereas the high dosage of tannic acid (30 g/kg diet) have negative impacts on immune system by decreasing relative (cluster of differentiation) CD4+, CD8+, CD4+CD8+ and $\gamma\delta$+ cell populations in thymus, spleen, and cecal tonsils and by reducing cytokine mRNA expression in spleen cells, the low dosage of tannic acid (0.5 g/kg) enhanced CD4+CD8+ subpopulations and $\gamma\delta$+ cells in spleen and CD4+CD8+ subpopulations and B cells in cecal tonsils and increasing mRNA expression of IFN-γ in broiler chickens. A study by Karaffová et al. [149], reported that tannins were beneficial to maintain components of mucosal immunity of chickens via upregulating immunoglobulins A and mucin 2. Erlejman et al. [150] also demonstrated that CT can combine with receptors of tumor necrosis factor-α (pro-inflammatory cytokines) to inhibit inflammation, which implies that tannins directly modulate immune system without eliciting antimicrobial and antioxidant properties.

Diaz Carrasco et al. [151] reported that a blend of chestnut (HT) and quebracho tannins (CT) modulated diversity cecal microbiota of chickens and decreased genus *Bacteroides* and increased certain members of order Clostridiales predominately in the families Ruminococcaceae and Lachnospiraceae. A study by Koo and Nyachoti [152] demonstrated that tannic acid positively modulated microbial metabolites in pigs fed oxidized oil. Selective antimicrobial effects of tannins would beneficially modulate microbiome of animals [153]. Viveros et al. [154] also suggested that tannins probably can have prebiotic effects via stimulating the proliferation of the beneficial bacteria.

Microbiome modulating effects of tannins may partially explain the gut health promoting effects of the chickens because microbiota of the chicken is closely associated with the gut ecosystem of chickens [155]. Moreover, an in-vitro study by Brus et al. [19] reported that chestnut tannins stimulated proliferation of enterocytes and enhanced antioxidative properties of the chicken small intestinal epithelial cells. Bilić-Šobot et al. [156] demonstrated that HT decreased production of cell debris in the large intestine of pigs, which leaded to decreased production of skatole, which belongs to the indole family. Together, supplemental tannins have the potential to improve growth performance, gut microbiota and gut ecosystem in broiler chickens, even in the absence of challenge models.

5. Strategies to Maximize the Effects of Supplemental Tannins in the Chickens

5.1. Heat Process on Tannins

Some in-vitro studies showed that heat processed HT had better antimicrobial and antioxidant properties than unprocessed HT [143,144]. This would be because heat processing could partially hydrolyze tannic acid and release gallic acid molecules, and these newly produced gallic acid and galloyl groups had enhanced antimicrobial and antioxidant effects compared to the fresh tannic acid [157]. González et al. [145] also reported that thermal process of *hamamelis virginiana* containing gallotannins and CT improved efficacy of antioxidant properties for inhibiting lipid oxidation. However, because CT are hardly hydrolyzed, enhanced antioxidant of heat-processed *hamamelis virginiana* probably due to hydrolyzation of gallotannins in *hamamelis virginiana* rather than hydrolyzation of CT. Thus, while in-vitro studies found that heat process of tannins could improve their functional properties (e.g., antioxidant and antimicrobial effects) compared to unprocessed HT, it is unknown yet whether heat-processed tannins have more beneficial effects on animal models.

5.2. Co-Supplementation of Tannins with other Bioactive Compounds

Supplementation of tannins with other bioactive compounds could be more beneficial to chickens than supplementing tannins alone for several reasons: (1) complexed form with proteins or polysaccharides of tannins inhibit tannins to form a complex with endogenous and dietary proteins and metal ions; (2) distinct properties of bioactive compounds can show synergistic effects to antimicrobial effects against both gram negative and positive bacteria; (3) different bioactive compounds affect gut health in different ways, which can lead to synergistic effects in animals; and (4) by providing more than two bioactive compounds, pathogenic bacteria are hard to generate resistant system against diverse bioactive compounds. Table 3 shows that tannins have potentials to show synergistic effects with other bioactive compounds.

Probiotics are living microorganisms which beneficially affect the host animals by enhancing animal's intestinal microbial balance [158]. Probiotics may have different mode of actions from tannins to inhibit the growth of pathogenic bacteria and to improve gut health of chickens. Probiotics can improve gut integrity by modulating immune system and maintaining microflora of chickens and tannins, while tannins can show antioxidant and anti-inflammatory properties [159]. However, one of the concerns of using probiotics with tannins could be that tannins may show antimicrobial effects against probiotics. However, Khalil [160] showed that gallic acid and catechin polyphenols did not inhibit the growth of *Streptococcus thermophilus* (probiotics), and Pacheco-Ordaz et al. [153] reported that catechin, gallic, vanillic, ferulic and protocatechuic acids selectively inhibit the growth of pathogenic bacteria without decreasing viability of probiotics. More studies are required to establish synergistic effects and mechanisms of tannins and probiotics in *in-vivo* chicken models.

Table 3. Effects of tannins with other bioactive compounds on the chickens.

Tannins	Other Bioactive Compounds	Outcomes	References
100 mg/kg tannic acid extract	Probiotics (1×10^4 spores/kg *Bacillus coagulans*)	Improved feed conversion ratio of coccidiosis vaccinated broilers.	[47]
240 mg/kg tannic acid	Organic acids (420 mg/kg lactic, 480 mg/kg butyric acid and 480 mg/kg acetic acid)	Decreased *S. enteritis* horizontal transmission in broiler chickens	[161]
Chestnut tannins	Saturated short medium chain fatty acids (C4:0 to C12:0)	Showed strong antimicrobial effects in in-vitro conditions and did not affect growth performance and meat quality of in-vivo chicken models.	[22]

Organic acids, known as strong antimicrobials, are organic compounds with acidic properties. Tannins inhibit the growth of pathogens predominately by inhibiting activities of microbial enzymes and modulating bacterial membrane, but organic acids penetrate bacterial cell wall, and bacteria have to spend a lot of energy to pump out hydrogen molecules, which causes bacterial death [162,163]. Furthermore, organic acids are known to improve intestinal morphology and gut barrier integrity by being energy sources for epithelial cells, which may imply that organic acids with tannins can show synergistic effects [164–166]. Thus, combination of tannins and organic acids can have synergistically increased antimicrobial effects and gut health promoting effects due to different mode of actions.

5.3. Supplementation of Combined or Encapsulated Form of Tannins

If tannins are combined with proteins, polysaccharides and ions before being included in the chickens feed, the tannins in complexes would not bind dietary and endogenous proteins and metal ions in chickens. The tannin complexes would be loose in the high pH (>7.0) in the intestine of chickens, and proteins in the tannin complexes can be degraded by digestive enzymes in the small intestine of chickens [40,167]. However, Lee et al. [46] showed that supplementation of albumin-tannin complexes still decreased growth performance and negatively modulated microbiota, hematological indices and plasma iron status of weaning piglets. The delivery of tannin-protein or polysaccharide complexes in the GIT of chickens, and effects of diverse dosages of supplemental tannin complexes on growth rate and gut health of chickens should be further investigated.

Encapsulation techniques, which offer a physical barrier for bioactive compounds and separate the core material from the environment until their release, have obtained a lot of attention in the livestock industry because encapsulation can maximize the efficacy of feed additives that have stability, cost and environmental issues. Encapsulation has been applied to various vulnerable feed additives such as essential oils [96,168,169], probiotics [170], organic acids [171], bacteriophages [172], zinc [173] and exogenous enzymes [174]. Diverse materials including proteins [96,175], lipids [171], carbohydrates (starch) [169], and polysaccharides [172,176] have been used to encapsulate bioactive compounds.

Encapsulation techniques can be applied to decrease side effects and maximize benefits of tannins in chickens. Encapsulation of tannins can depress the protein binding capacity of tannins, which decreases feed intake by making astringent taste and digestibility of proteins and induces dietary and endogenous protein losses [177]. In addition, more tannins can be delivered to small or large intestine where many pathogens propagate by decreasing bioavailability for absorption in the upper GIT of chickens. Adejoro et al. [178] showed that lipid-encapsulated acacia tannin extracts reduced methane production and enhanced neutral detergent fiber digestibility in sheep. A study by Wang et al. [179] reported that microencapsulated tannic acid improved intestinal morphology in duodenum, increased expression of ileal nutrient transporters (sodium-dependent neutral amino acid transporter; B^0AT1 and peptide transporter 1; PepT1) and modulate microbiota without affecting growth performance of weaned piglets even though ileal maltase activity and gene expression of jejunal sodium-dependent glucose transporter 1 (SGLT1) was reduced. Future studies are required to develop effective encapsulated tannins and to determine appropriate dosages of encapsulated tannins for chickens.

6. Conclusions

There are various kinds of tannins, which have different bioavailability in chickens. Depends on the sorts and dosages of tannins, tannins can elicit detrimental effects on growth performance and gut ecosystem of the broiler chickens or can beneficially modulate growth performance and gut ecosystem of the broiler chickens. In addition, heat processing, co-supplementation with other bioactive compounds and encapsulation potentially enhance the beneficial biological effects of tannins. In conclusion, supplementation of tannins alone or in conjunction with those strategies have large potential to alleviate challenges, replace AGP and improve production efficiency in poultry productions.

Author Contributions: Conceptualization, J.C. and W.K.K.; supervision, W.K.K.; investigation, W.K.K.; Writing—Original draft preparation, J.C.; Writing—Reviewing and editing, W.K.K. and J.C. All authors have read and agreed to the published version of the manuscript.

Acknowledgments: We appreciate Brett Marshall for his help in the manuscript preparation.

References

1. Mottet, A.; Tempio, G. Global poultry production: Current state and future outlook and challenges. *World. Poult. Sci. J.* **2017**, *73*, 245–256. [CrossRef]
2. Pawar, S.S.; Basavaraj, S.; Dhansing, L.V.; Kurade, N.P.; Sahebrao, K.A.; Vitthal, N.A.; Pandit, B.M.; Kumar, B.S. Assessing and Mitigating the Impact of Heat Stress in Poultry. *Adv. Anim. Veter. Sci.* **2016**, *4*, 332–341. [CrossRef]
3. Nkukwana, T. Global poultry production: Current impact and future outlook on the South African poultry industry. *South. Afr. J. Anim. Sci.* **2019**, *48*, 869. [CrossRef]

4. Akil, L.; Ahmad, H.A. Quantitative Risk Assessment Model of Human Salmonellosis Resulting from Consumption of Broiler Chicken. *Diseases* **2019**, *7*. [CrossRef]
5. Hunter, J.M.; Anders, S.A.; Crowe, T.; Korver, D.R.; Bench, C.J. Practical assessment and management of foot pad dermatitis in commercial broiler chickens: A Field Study. *J. Appl. Poult. Res.* **2017**, *26*, 593–604. [CrossRef]
6. Hassan, Y.I.; Lahaye, L.; Gong, M.M.; Peng, J.; Gong, J.; Liu, S.; Gay, C.G.; Yang, C. Innovative drugs, chemicals, and enzymes within the animal production chain. *Veter. Res.* **2018**, *49*, 71. [CrossRef] [PubMed]
7. Kumar, S.; Chen, C.; Indugu, N.; Werlang, G.O.; Singh, M.; Kim, W.K.; Thippareddi, H. Effect of antibiotic withdrawal in feed on chicken gut microbial dynamics, immunity, growth performance and prevalence of foodborne pathogens. *PLoS ONE* **2018**, *13*, e0192450. [CrossRef]
8. Muhammad, J.; Khan, S.; Su, J.Q.; Hesham, A.E.-L.; Ditta, A.; Nawab, J.; Ali, A. Antibiotics in poultry manure and their associated health issues: A systematic review. *J. Soils Sediments* **2020**, *20*, 486–497. [CrossRef]
9. Brown, K.; Uwiera, R.R.; Kalmokoff, M.L.; Brooks, S.P.; Inglis, G.D. Antimicrobial growth promoter use in livestock: A requirement to understand their modes of action to develop effective alternatives. *Int. J. Antimicrob. Agents* **2017**, *49*, 12–24. [CrossRef]
10. Hu, Y.J.; Cowling, B.J. Reducing antibiotic use in livestock, China. *Bull. World Health Organ.* **2020**, *98*, 360. [CrossRef]
11. Karavolias, J.; Salois, M.J.; Baker, K.T.; Watkins, K. Raised without antibiotics: impact on animal welfare and implications for food policy. *Transl. Anim. Sci.* **2018**, *2*, 337–348. [CrossRef]
12. Singer, R.S.; Porter, L.J.; Thomson, D.U.; Gage, M.; Beaudoin, A.; Wishnie, J.K. Raising Animals Without Antibiotics: U.S. Producer and Veter.inarian Experiences and Opinions. *Front. Veter. Sci.* **2019**, *6*. [CrossRef]
13. Echeng, G.; Ehao, H.; Exie, S.; Ewang, X.; Edai, M.; Ehuang, L.; Yuan, Z. Antibiotic alternatives: The substitution of antibiotics in animal husbandry? *Front. Microbiol.* **2014**, *5*. [CrossRef]
14. Van Boeckel, T.P.; Brower, C.; Gilbert, M.; Grenfell, B.T.; Levin, S.A.; Robinson, T.P.; Teillant, A.; Laxminarayan, R. Global trends in antimicrobial use in food animals. *Proc. Natl. Acad. Sci. USA* **2015**, *112*, 5649–5654. [CrossRef]
15. Yang, C.; Chowdhury, M.A.K.; Hou, Y.; Gong, J. Phytogenic Compounds as Alternatives to In-Feed Antibiotics: Potentials and Challenges in Application. *Pathogens* **2015**, *4*, 137–156. [CrossRef]
16. Redondo, L.M.; Chacana, P.A.; Dominguez, J.E.; Miyakawa, M.E.D.F. Perspectives in the use of tannins as alternative to antimicrobial growth promoter factors in poultry. *Front. Microbiol.* **2014**, *5*. [CrossRef]
17. Trevino, J.; Ortiz, L.; Centeno, C. Effect of tannins from faba beans (Vicia faba) on the digestion of starch by growing chicks. *Anim. Feed. Sci. Technol.* **1992**, *37*, 345–349. [CrossRef]
18. Garcia, R.; Mendes, A.; Sartori, J.; Paz, I.C.D.L.A.; Takahashi, S.; Pelícia, K.; Komiyama, C.; Quinteiro, R. Digestibility of feeds containing sorghum, with and without tannin, for broiler chickens submitted to three room temperatures. *Brazil. J. Poultry Sci.* **2004**, *6*, 55–60. [CrossRef]
19. Brus, M.; Gradišnik, L.; Trapečar, M.; Škorjanc, D.; Frangež, R. Beneficial effects of water-soluble chestnut (Castanea sativa Mill.) tannin extract on chicken small intestinal epithelial cell culture. *Poult. Sci.* **2018**, *97*, 1271–1282. [CrossRef]
20. Amirmohammadi, M.; Khajoenia, S.; Bahmani, M.; Rafieian-Kopaei, M.; Eftekhari, Z.; Qorbani, M. In vivo evaluation of antiparasitic effects of Artemisia abrotanum and Salvia officinalis extracts on Syphacia obvelata, Aspiculoris tetrapetra and Hymenolepis nana parasites. *Asian Pac. J. Trop. Dis.* **2014**, *4*, S250–S254. [CrossRef]
21. Daglia, M. Polyphenols as antimicrobial agents. *Curr. Opin. Biotechnol.* **2012**, *23*, 174–181. [CrossRef]
22. Mannelli, F.; Team, H.; Tosi, G.; Secci, G.; Daghio, M.; Massi, P.; Fiorentini, L.; Galigani, I.; Lancini, S.; Rapaccini, S.; et al. Effect of Chestnut Tannins and Short Chain Fatty Acids as Anti-Microbials and as Feeding Supplements in Broilers Rearing and Meat Quality. *Animals* **2019**, *9*. [CrossRef]
23. Haslam, E. *Plant. Polyphenols: Vegetable Tannins Revisited. Australia (UK)*; Cambridge University Press: Cambridge, UK, 1989.
24. Huang, Q.; Liu, X.; Zhao, G.; Hu, T.; Wang, Y. Potential and challenges of tannins as an alternative to in-feed antibiotics for farm animal production. *Anim. Nutr.* **2018**, *4*, 137–150. [CrossRef]
25. Patra, A.K.; Saxena, J. Exploitation of dietary tannins to improve rumen metabolism and ruminant nutrition. *J. Sci. Food Agric.* **2011**, *91*, 24–37. [CrossRef]

26. Lamy, E.; Pinheiro, C.; Rodrigues, L.; Capela-Silva, F.; Lopes, O.; Tavares, S.; Gaspar, R. *Determinants of Tannin-Rich Food and Beverage Consumption: Oral Perception vs. Psychosocial Aspects*; Universidade De Évora: Évora, Portugal, 2016.
27. Kamiloglu, S.; Capanoglu, E.; Bilen, F.D.; Gonzales, G.B.; Grootaert, C.; Van de Wiele, T.; Van Camp, J. Bioaccessibility of polyphenols from plant-processing byproducts of black carrot (Daucus carota L.). *J. Agric. Food Chem.* **2016**, *64*, 2450–2458. [CrossRef]
28. Manach, C.; Williamson, G.; Morand, C.; Scalbert, A.; Rémésy, C. Bioavailability and bioefficacy of polyphenols in humans. I. Review of 97 bioavailability studies. *Am. J. Clin. Nutr.* **2005**, *81*, 230S–242S. [CrossRef] [PubMed]
29. Seeram, N.P.; Lee, R.; Heber, D. Bioavailability of ellagic acid in human plasma after consumption of ellagitannins from pomegranate (Punica granatum L.) juice. *Clin. Chim. Acta* **2004**, *348*, 63–68. [CrossRef]
30. Martinez, K.B.; Mackert, J.D.; McIntosh, M.K. Polyphenols and intestinal health. In *Nutrition and Functional Foods for Healthy Aging*; Elsevier: Amsterdam, The Netherlands, 2017; pp. 191–210. [CrossRef]
31. Selma, M.V.; Espin, J.C.; Tomas-Barberan, F.A. Interaction between phenolics and gut microbiota: Role in human health. *J. Agric. Food Chem.* **2009**, *57*, 6485–6501. [CrossRef]
32. Costain, D.J.; Guha, A.K.; Liwski, R.S.; Lee, T.D. Murine hypodense eosinophils induce tumour cell apoptosis by a granzyme B-dependent mechanism. *Cancer Immunol. Immunother.* **2001**, *50*, 293–299. [CrossRef]
33. Serrano, J.; Puupponen-Pimiä, R.; Dauer, A.; Aura, A.M.; Saura-Calixto, F. Tannins: Current knowledge of food sources, intake, bioavailability and biological effects. *Mol. Nutr. Food Res.* **2009**, *53*, S310–S329. [CrossRef]
34. Kahle, K.; Huemmer, W.; Kempf, M.; Scheppach, W.; Erk, T.; Richling, E. Polyphenols are intensively metabolized in the human gastrointestinal tract after apple juice consumption. *J. Agric. Food Chem.* **2007**, *55*, 10605–10614. [CrossRef]
35. Bawadi, H.; Bansode, R.; Trappey, A., II; Truax, R.; Losso, J. Inhibition of Caco-2 colon, MCF-7 and Hs578T breast, and DU 145 prostatic cancer cell proliferation by water-soluble black bean condensed tannins. *Cancer Lett.* **2005**, *218*, 153–162. [CrossRef]
36. Ragan, M.; Glombitza, K. *Handbook of Physiological Methods*; Cambridge University Press: Cambridge, UK, 1986.
37. Kirke, D.; Smyth, T.; Rai, D.; Kenny, O.; Stengel, D. The chemical and antioxidant stability of isolated low molecular weight phlorotannins. *Food Chem.* **2017**, *221*, 1104–1112. [CrossRef]
38. Corona, G.; Coman, M.M.; Guo, Y.; Hotchkiss, S.; Gill, C.; Yaqoob, P.; Spencer, J.P.; Rowland, I. Effect of simulated gastrointestinal digestion and fermentation on polyphenolic content and bioactivity of brown seaweed phlorotannin-rich extracts. *Mol. Nutr. Food Res.* **2017**, *61*, 1700223. [CrossRef]
39. Nwosu, F.; Morris, J.; Lund, V.A.; Stewart, D.; Ross, H.A.; McDougall, G.J. Anti-proliferative and potential anti-diabetic effects of phenolic-rich extracts from edible marine algae. *Food Chem.* **2011**, *126*, 1006–1012. [CrossRef]
40. Barry, T.; Manley, T.; Duncan, S. The role of condensed tannins in the nutritional value of Lotus pedunculatus for sheep:* 4. Sites of carbohydrate and protein digestion as influenced by dietary reactive tannin concentration. *Br. J. Nutr.* **1986**, *55*, 123–137. [CrossRef]
41. Berard, N.C.; Wang, Y.; Wittenberg, K.; Krause, D.; Coulman, B.; McAllister, T.; Ominski, K. Condensed tannin concentrations found in vegetative and mature forage legumes grown in western Canada. *Can. J. Plant Sci.* **2011**, *91*, 669–675. [CrossRef]
42. Goel, G.; Makkar, H.P. Methane mitigation from ruminants using tannins and saponins. *Trop. Anim. Health Prod.* **2012**, *44*, 729–739. [CrossRef]
43. Medugu, C.; Saleh, B.; Igwebuike, J.; Ndirmbita, R. Strategies to improve the utilization of tannin-rich feed materials by poultry. *Int. J. Poult. Sci.* **2012**, *11*, 417. [CrossRef]
44. Butler, L.G.; Riedl, D.J.; Lebryk, D.; Blytt, H. Interaction of proteins with sorghum tannin: Mechanism, specificity and significance. *J. Am. Oil Chem. Soc.* **1984**, *61*, 916–920. [CrossRef]
45. Marzo, F.; Urdaneta, E.; Santidrian, S. Liver proteolytic activity in tannic acid-fed birds. *Poult. Sci.* **2002**, *81*, 92–94. [CrossRef]
46. Lee, S.; Shinde, P.; Choi, J.; Kwon, I.; Lee, J.; Pak, S.; Cho, W.; Chae, B. Effects of tannic acid supplementation on growth performance, blood hematology, iron status and faecal microflora in weanling pigs. *Livest. Sci.* **2010**, *131*, 281–286. [CrossRef]

47. Tonda, R.; Rubach, J.; Lumpkins, B.; Mathis, G.; Poss, M. Effects of tannic acid extract on performance and intestinal health of broiler chickens following coccidiosis vaccination and/or a mixed-species Eimeria challenge. *Poult. Sci.* **2018**, *97*, 3031–3042. [CrossRef]
48. Cengiz, Ö.; Köksal, B.H.; Tatlı, O.; Sevim, Ö.; Ahsan, U.; Bilgili, S.F.; Önol, A.G. Effect of dietary tannic acid supplementation in corn-or barley-based diets on growth performance, intestinal viscosity, litter quality, and incidence and severity of footpad dermatitis in broiler chickens. *Livest. Sci.* **2017**, *202*, 52–57. [CrossRef]
49. Starčević, K.; Krstulović, L.; Brozić, D.; Maurić, M.; Stojević, Z.; Mikulec, Ž.; Bajić, M.; Mašek, T. Production performance, meat composition and oxidative susceptibility in broiler chicken fed with different phenolic compounds. *J. Sci. Food Agric.* **2015**, *95*, 1172–1178. [CrossRef]
50. Mašek, T.; Starčević, K.; Mikulec, Ž. The influence of the addition of thymol, tannic acid or gallic acid to broiler diet on growth performance, serum malondialdehyde value and cecal fermentation. *Yeast* **2014**, *5*, 5.0. [CrossRef]
51. Kubena, L.; Byrd, J.; Young, C.; Corrier, D. Effects of tannic acid on cecal volatile fatty acids and susceptibility to Salmonella typhimurium colonization in broiler chicks. *Poult. Sci.* **2001**, *80*, 1293–1298. [CrossRef]
52. Ebrahim, R.; Liang, J.B.; Jahromi, M.F.; Shokryazdan, P.; Ebrahimi, M.; Li Chen, W.; Goh, Y.M. Effects of tannic acid on performance and fatty acid composition of breast muscle in broiler chickens under heat stress. *Ital. J. Anim. Sci.* **2015**, *14*, 3956. [CrossRef]
53. Mor-Mur, M.; Yuste, J. Emerging bacterial pathogens in meat and poultry: An overview. *Food Bioproc. Tech.* **2010**, *3*, 24. [CrossRef]
54. Adhikari, P.; Cosby, D.E.; Cox, N.A.; Franca, M.S.; Williams, S.M.; Gogal, R.M., Jr.; Ritz, C.W.; Kim, W.K. Effect of dietary fructooligosaccharide supplementation on internal organs Salmonella colonization, immune response, ileal morphology, and ileal immunohistochemistry in laying hens challenged with Salmonella enteritidis. *Poult. Sci.* **2018**, *97*, 2525–2533. [CrossRef]
55. Thung, T.; Mahyudin, N.; Basri, D.; Radzi, C.W.M.; Nakaguchi, Y.; Nishibuchi, M.; Radu, S. Prevalence and antibiotic resistance of Salmonella Enteritidis and Salmonella Typhimurium in raw chicken meat at retail markets in Malaysia. *Poult. Sci.* **2016**, *95*, 1888–1893. [CrossRef]
56. Wu-Yuan, C.; Chen, C.; Wu, R. Gallotannins inhibit growth, water-insoluble glucan synthesis, and aggregation of mutans streptococci. *J. Dent. Res.* **1988**, *67*, 51–55. [CrossRef]
57. Delimont, N.M.; Rosenkranz, S.K.; Haub, M.D.; Lindshield, B.L. Salivary proline-rich protein may reduce tannin-iron chelation: A systematic narrative review. *Nutr. Metab.* **2017**, *14*, 47. [CrossRef]
58. Karamać, M. Chelation of Cu (II), Zn (II), and Fe (II) by tannin constituents of selected edible nuts. *Int. J. Mol. Sci.* **2009**, *10*, 5485–5497. [CrossRef]
59. Tan, Z. Roles of Iron Uptake in the Survival, Colonization and Virulence of Salmonella in Animal Guts. Master's Thesis, University of Manitoba, Winnipeg, MB, Canada, 2019.
60. Mahadwar, G.; Chauhan, K.; Bhagavathy, G.; Murphy, C.; Smith, A.; Bhagwat, A. Swarm motility of S almonella enterica serovar T yphimurium is inhibited by compounds from fruit peel extracts. *Lett. Appl. Microbiol.* **2015**, *60*, 334–340. [CrossRef]
61. Sivasankar, C.; Jha, N.K.; Ghosh, R.; Shetty, P.H. Anti quorum sensing and anti virulence activity of tannic acid and it's potential to breach resistance in Salmonella enterica Typhi/Paratyphi A clinical isolates. *Microb. Pathog.* **2020**, *138*, 103813. [CrossRef]
62. Yang, Q.; Wang, L.; Gao, J.; Liu, X.; Feng, Y.; Wu, Q.; Baloch, A.B.; Cui, L.; Xia, X. Tannin-rich fraction from pomegranate rind inhibits quorum sensing in Chromobacterium violaceum and biofilm formation in Escherichia coli. *Foodborne Pathog. Dis.* **2016**, *13*, 28–35. [CrossRef]
63. Costabile, A.; Sanghi, S.; Martín-Pelaez, S.; Mueller-Harvey, I.; Gibson, G.R.; Rastall, R.A.; Klinder, A. Inhibition of Salmonella Typhimurium by tannins in vitro. *J. Food. Agric. Environ.* **2011**, *9*, 119–124.
64. Reyes, A.W.B.; Hong, T.G.; Hop, H.T.; Arayan, L.T.; Huy, T.X.N.; Min, W.; Lee, H.J.; Lee, K.S.; Kim, S. The in vitro and in vivo protective effects of tannin derivatives against Salmonella enterica serovar Typhimurium infection. *Microb. Pathog.* **2017**, *109*, 86–93. [CrossRef]
65. Daing, M.I.; Pathak, A.; Bhat, M.A.; Zargar, M.A. Antioxidant and antibacterial potential of condensed tannins containing tree leaves extract. *Vet. Pract.* **2017**, *18*, 118–121.
66. Maisetta, G.; Batoni, G.; Caboni, P.; Esin, S.; Rinaldi, A.C.; Zucca, P. Tannin profile, antioxidant properties, and antimicrobial activity of extracts from two Mediterranean species of parasitic plant Cytinus. *BMC Complement Altern. Med.* **2019**, *19*, 82. [CrossRef]

67. Elizondo, A.M.; Mercado, E.C.; Rabinovitz, B.C.; Fernandez-Miyakawa, M.E. Effect of tannins on the in vitro growth of Clostridium perfringens. *Vet. Microbiol.* **2010**, *145*, 308–314. [CrossRef]
68. Liu, H.; Li, K.; Zhao, J.; Deng, W. Effects of chestnut tannins on intestinal morphology, barrier function, pro-inflammatory cytokine expression, microflora and antioxidant capacity in heat-stressed broilers. *J. Anim. Physiol. Anim. Nutr.* **2018**, *102*, 717–726. [CrossRef]
69. Kamali Sangani, A.; Masoudi, A.A.; Hosseini, S.A. The effects of herbal plants on Mucin 2 gene expression and performance in ascetic broilers. *Iran. J. Vet. Med.* **2014**, *8*, 47–52. [CrossRef]
70. Moghaddam, H.S.; Moghaddam, H.N.; Kermanshahi, H.; Mosavi, A.H.; Raji, A. The effect of threonine on mucin2 gene expression, intestinal histology and performance of broiler chicken. *Ital. J. Anim. Sci.* **2011**, *10*, e14. [CrossRef]
71. Urdaneta, V.; Casadesús, J. Interactions between bacteria and bile salts in the gastrointestinal and hepatobiliary tracts. *Front. Med.* **2017**, *4*, 163. [CrossRef]
72. Li, X.; Bleumink-Pluym, N.M.; Luijkx, Y.M.; Wubbolts, R.W.; van Putten, J.P.; Strijbis, K. MUC1 is a receptor for the Salmonella SiiE adhesin that enables apical invasion into enterocytes. *PLoS Pathog.* **2019**, *15*, e1007566. [CrossRef]
73. Van Parys, A.; Boyen, F.; Dewulf, J.; Haesebrouck, F.; Pasmans, F. The use of tannins to control Salmonella typhimurium infections in pigs. *Zoonoses Public Health* **2010**, *57*, 423–428. [CrossRef]
74. Jamroz, D.; Wiliczkiewicz, A.; Skorupińska, J.; Orda, J.; Kuryszko, J.; Tschirch, H. Effect of sweet chestnut tannin (SCT) on the performance, microbial status of intestine and histological characteristics of intestine wall in chickens. *Br. Poult. Sci.* **2009**, *50*, 687–699. [CrossRef]
75. Adhikari, P.; Cosby, D.; Cox, N.; Kim, W. Colonization of mature laying hens with Salmonella Enteritidis by oral or intracloacal inoculation. *J. Appl. Poult. Res.* **2017**, *26*, 286–294. [CrossRef]
76. Chen, Y.; Zhang, H.; Cheng, Y.; Li, Y.; Wen, C.; Zhou, Y. Dietary l-threonine supplementation attenuates lipopolysaccharide-induced inflammatory responses and intestinal barrier damage of broiler chickens at an early age. *Br. J. Nutr.* **2018**, *119*, 1254–1262. [CrossRef]
77. Omonijo, F.A.; Liu, S.; Hui, Q.; Zhang, H.; Lahaye, L.; Bodin, J.-C.; Gong, J.; Nyachoti, M.; Yang, C. Thymol improves barrier function and attenuates inflammatory responses in porcine intestinal epithelial cells during lipopolysaccharide (LPS)-induced inflammation. *J. Agric. Food Chem.* **2018**, *67*, 615–624. [CrossRef]
78. van Ampting, M.T.; Schonewille, A.J.; Vink, C.; Brummer, R.J.M.; van der Meer, R.; Bovee-Oudenhoven, I.M. Damage to the intestinal epithelial barrier by antibiotic pretreatment of salmonella-infected rats is lessened by dietary calcium or tannic acid. *J. Nutr.* **2010**, *140*, 2167–2172. [CrossRef]
79. Adhikari, P.; Lee, C.; Cosby, D.; Cox, N.; Kim, W. Effect of probiotics on fecal excretion, colonization in internal organs and immune gene expression in the ileum of laying hens challenged with SalmonellaEnteritidis. *Poult. Sci.* **2019**, *98*, 1235–1242. [CrossRef]
80. Sieniawska, E.; Baj, T. Tannins. In *Pharmacognosy*; Elsevier: Amsterdam, The Netherlands, 2017; pp. 199–232.
81. Konishi, Y.; Kobayashi, S.; Shimizu, M. Transepithelial transport of p-coumaric acid and gallic acid in Caco-2 cell monolayers. *Biosci. Biotechnol. Biochem.* **2003**, *67*, 2317–2324. [CrossRef]
82. Grilli, G.; Borgonovo, F.; Tullo, E.; Fontana, I.; Guarino, M.; Ferrante, V. A pilot study to detect coccidiosis in poultry farms at early stage from air analysis. *Biosyst. Eng.* **2018**, *173*, 64–70. [CrossRef]
83. Chapman, H. Milestones in avian coccidiosis research: A review. *Poult. Sci.* **2014**, *93*, 501–511. [CrossRef]
84. Clark, E.L.; Macdonald, S.E.; Thenmozhi, V.; Kundu, K.; Garg, R.; Kumar, S.; Ayoade, S.; Fornace, K.M.; Jatau, I.D.; Moftah, A. Cryptic Eimeria genotypes are common across the southern but not northern hemisphere. *Int. J. Parasitol.* **2016**, *46*, 537–544. [CrossRef]
85. Li, C.; Yan, X.; Lillehoj, H.S.; Oh, S.; Liu, L.; Sun, Z.; Gu, C.; Lee, Y.; Xianyu, Z.; Zhao, H. Eimeria maxima-induced transcriptional changes in the cecal mucosa of broiler chickens. *Parasites Vectors* **2019**, *12*, 285. [CrossRef]
86. Yin, G.; Lin, Q.; Wei, W.; Qin, M.; Liu, X.; Suo, X.; Huang, Z. Protective immunity against Eimeria tenella infection in chickens induced by immunization with a recombinant C-terminal derivative of EtIMP1. *Vet. Immunol. Immunopathol.* **2014**, *162*, 117–121. [CrossRef]
87. Teng, P.-Y.; Yadav, S.; de Souza Castro, F.L.; Tompkins, Y.H.; Fuller, A.L.; Kim, W.K. Graded Eimeria challenge linearly regulated growth performance, dynamic change of gastrointestinal permeability, apparent ileal digestibility, intestinal morphology, and tight junctions of broiler chickens. *Poult. Sci.* **2020**, *99*, 4203–4216. [CrossRef]

88. Abbas, R.; Iqbal, Z.; Mansoor, M. Role of natural antioxidants for the control of coccidiosis in poultry. *Pak. Vet. J.* **2013**, *33*, 401.
89. Prescott, J.F.; Smyth, J.A.; Shojadoost, B.; Vince, A. Experimental reproduction of necrotic enteritis in chickens: A review. *Avian Dis.* **2016**, *45*, 317–322. [CrossRef]
90. Wu, S.-B.; Stanley, D.; Rodgers, N.; Swick, R.A.; Moore, R.J. Two necrotic enteritis predisposing factors, dietary fishmeal and Eimeria infection, induce large changes in the caecal microbiota of broiler chickens. *Vet. Microbiol.* **2014**, *169*, 188–197. [CrossRef]
91. Peek, H.; Landman, W. Coccidiosis in poultry: Anticoccidial products, vaccines and other prevention strategies. *Vet. Q.* **2011**, *31*, 143–161. [CrossRef]
92. Noack, S.; Chapman, H.D.; Selzer, P.M. Anticoccidial drugs of the livestock industry. *Parasitol. Res.* **2019**, *118*, 2009–2026. [CrossRef]
93. Levine, R.; Horst, G.; Tonda, R.; Lumpkins, B.; Mathis, G. Evaluation of the effects of feeding dried algae containing beta-1, 3-glucan on broilers challenged with Eimeria. *Poult. Sci.* **2018**, *97*, 3494–3500. [CrossRef]
94. Christaki, E.; Florou-Paneri, P.; Giannenas, I.; Papazahariadou, M.; Botsoglou, N.A.; Spais, A.B. Effect of a mixture of herbal extracts on broiler chickens infected with Eimeria tenella. *Anim. Res.* **2004**, *53*, 137–144. [CrossRef]
95. Abbas, R.Z.; Munawar, S.H.; Manzoor, Z.; Iqbal, Z.; Khan, M.N.; Saleemi, M.K.; Zia, M.A.; Yousaf, A. Anticoccidial effects of acetic acid on performance and pathogenic parameters in broiler chickens challenged with Eimeria tenella. *Pesqui. Vet. Bras.* **2011**, *31*, 99–103. [CrossRef]
96. Yang, C.; Kennes, Y.M.; Lepp, D.; Yin, X.; Wang, Q.; Yu, H.; Yang, C.; Gong, J.; Diarra, M.S. Effects of encapsulated cinnamaldehyde and citral on the performance and cecal microbiota of broilers vaccinated or not vaccinated against coccidiosis. *Poult. Sci.* **2020**, *99*, 936–948. [CrossRef]
97. Bortoluzzi, C.; Rothrock, M.J.; Vieira, B.S.; Mallo, J.J.; Puyalto, M.; Hofacre, C.; Applegate, T.J. Supplementation of protected sodium butyrate alone or in combination with essential oils modulated the cecal microbiota of broiler chickens challenged with coccidia and Clostridium perfringens. *Front. Sustain. Food Syst.* **2018**, *2*, 72. [CrossRef]
98. He, B.; Bortoluzzi, C.; King, W.; Graugnard, D.; Dawson, K.; Applegate, T. Zinc source influences the gene expression of zinc transporters in jejunum and cecal tonsils during broiler challenge with Eimeria maximaand Clostridium perfringens. *Poult. Sci.* **2019**, *98*, 1146–1152. [CrossRef]
99. Teng, P.-Y.; Fuller, A.L.; Kim, W.K. Evaluation of nitro compounds as feed additives in diets of Eimeria-challenged broilers in vitro and in vivo. *Poult. Sci.* **2020**, *99*, 1320–1325. [CrossRef]
100. Min, B.; Hart, S. Tannins for suppression of internal parasites. *J. Anim. Sci.* **2003**, *81*, E102–E109. [CrossRef]
101. Scalbert, A. Antimicrobial properties of tannins. *Phytochemistry* **1991**, *30*, 3875–3883. [CrossRef]
102. Chung, K.-T.; Lu, Z.; Chou, M. Mechanism of inhibition of tannic acid and related compounds on the growth of intestinal bacteria. *Food Chem. Toxicol.* **1998**, *36*, 1053–1060. [CrossRef]
103. Parisi, F.; Mancini, S.; Mazzei, M.; Forzan, M.; Turchi, B.; Perrucci, S.; Poli, A.; Paci, G. Effect of dietary supplementation of a mix of chestnut and quebracho tannins on intestinal morphology, bacterial load, eimeria spp oocyst excretion and immune response after vaccination in rabbits. *Am. J. Anim.Vet. Sci.* **2018**. [CrossRef]
104. Kaleem, Q.M.; Akhtar, M.; Awais, M.M.; Saleem, M.; Zafar, M.; Iqbal, Z.; Muhammad, F.; Anwar, M.I. Studies on Emblica officinalis derived tannins for their immunostimulatory and protective activities against coccidiosis in industrial broiler chickens. *Sci. World J.* **2014**, *2014*. [CrossRef]
105. Mishra, B.; Jha, R. Oxidative stress in the poultry gut: Potential challenges and interventions. *Front. Vet. Sci.* **2019**, *6*, 60. [CrossRef]
106. Wang, M.; Suo, X.; Gu, J.; Zhang, W.; Fang, Q.; Wang, X. Influence of grape seed proanthocyanidin extract in broiler chickens: Effect on chicken coccidiosis and antioxidant status. *Poult. Sci.* **2008**, *87*, 2273–2280. [CrossRef]
107. Mansoori, B.; Modirsanei, M. Effects of dietary tannic acid and vaccination on the course of coccidiosis in experimentally challenged broiler chicken. *Vet. Parasitol.* **2012**, *187*, 119–122. [CrossRef]
108. Attia, Y.A.; Al-Harthi, M.A.; El-Shafey, A.S.; Rehab, Y.A.; Kim, W.K. Enhancing tolerance of broiler chickens to heat stress by supplementation with vitamin E, vitamin C and/or probiotics. *Ann. Anim. Sci.* **2017**, *17*, 1155–1169. [CrossRef]

109. Quinteiro-Filho, W.; Ribeiro, A.; Ferraz-de-Paula, V.; Pinheiro, M.; Sakai, M.; Sá, L.; Ferreira, A.; Palermo-Neto, J. Heat stress impairs performance parameters, induces intestinal injury, and decreases macrophage activity in broiler chickens. *Poult. Sci.* **2010**, *89*, 1905–1914. [CrossRef]
110. Laudicina, D.C.; Marnett, L.J. Enhancement of hydroperoxide-dependent lipid peroxidation in rat liver microsomes by ascorbic acid. *Arch. Biochem. Biophys.* **1990**, *278*, 73–80. [CrossRef]
111. Choi, J.; Li, W.; Schindell, B.; Ni, L.; Liu, S.; Zhao, X.; Gong, J.; Nyachoti, M.; Yang, C. Molecular cloning, tissue distribution and the expression of cystine/glutamate exchanger (xCT, SLC7A11) in different tissues during development in broiler chickens. *Anim. Nutr.* **2020**, *6*, 107–114. [CrossRef]
112. Lin, H.; Decuypere, E.; Buyse, J. Acute heat stress induces oxidative stress in broiler chickens. *Comp. Biochem. Physiol. Part A Mol. Integr. Physiol.* **2006**, *144*, 11–17. [CrossRef]
113. Lee, S.; Kim, D.; Keum, M.; Han, E.; An, B.; Chang, H.; Choi, Y.; Moon, B.; Lee, K. Effects of fumonisin B1 and mycotoxin binders on growth performance, tibia characteristics, gut physiology, and stress indicators in broiler chickens raised in different stocking densities. *Poult. Sci.* **2018**, *97*, 845–854. [CrossRef]
114. Lauridsen, C. From oxidative stress to inflammation: Redox balance and immune system. *Poult. Sci.* **2019**, *98*, 4240–4246. [CrossRef]
115. Nawab, A.; Ibtisham, F.; Li, G.; Kieser, B.; Wu, J.; Liu, W.; Zhao, Y.; Nawab, Y.; Li, K.; Xiao, M. Heat stress in poultry production: Mitigation strategies to overcome the future challenges facing the global poultry industry. *J. Therm. Biol.* **2018**, *78*, 131–139. [CrossRef]
116. Yang, R.; Hui, Q.; Jiang, Q.; Liu, S.; Zhang, H.; Wu, J.; Lin, F.; Yang, C. Effect of Manitoba-grown red-osier dogwood extracts on recovering Caco-2 cells from H2O2-induced oxidative damage. *Antioxidants* **2019**, *8*, 250. [CrossRef]
117. Jiang, Q.; Zhang, H.; Yang, R.; Hui, Q.; Chen, Y.; Mats, L.; Tsao, R.; Yang, C. Red-Osier Dogwood Extracts Prevent Inflammatory Responses in Caco-2 Cells and a Caco-2 BBe1/EA. hy926 Cell Co-Culture Model. *Antioxidants* **2019**, *8*, 428. [CrossRef]
118. Terruzzi, I.; Montesano, A.; Senesi, P.; Villa, I.; Ferraretto, A.; Bottani, M.; Vacante, F.; Spinello, A.; Bolamperti, S.; Luzi, L. L-Carnitine reduces oxidative stress and promotes cells differentiation and bone matrix proteins expression in human osteoblast-like cells. *Biomed Res. Int.* **2019**, *2019*. [CrossRef]
119. Sahin, K.; Sahin, N.; Kucuk, O. Effects of chromium, and ascorbic acid supplementation on growth, carcass traits, serum metabolites, and antioxidant status of broiler chickens reared at a high ambient temperature (32 C). *Nutr. Res.* **2003**, *23*, 225–238. [CrossRef]
120. Yun, S.-H.; Moon, Y.-S.; SoHn, S.-H.; Jang, I.-S. Effects of cyclic heat stress or vitamin C supplementation during cyclic heat stress on HSP70, inflammatory cytokines, and the antioxidant defense system in Sprague Dawley rats. *Exp. Anim.* **2012**, *61*, 543–553. [CrossRef]
121. Kumbhar, S.; Khan, A.Z.; Parveen, F.; Nizamani, Z.A.; Siyal, F.A.; Abd El-Hack, M.E.; Gan, F.; Liu, Y.; Hamid, M.; Nido, S.A. Impacts of selenium and vitamin E supplementation on mRNA of heat shock proteins, selenoproteins and antioxidants in broilers exposed to high temperature. *Amb Express* **2018**, *8*, 1–10. [CrossRef] [PubMed]
122. Shakeri, M.; Cottrell, J.J.; Wilkinson, S.; Le, H.H.; Suleria, H.A.; Warner, R.D.; Dunshea, F.R. Growth performance and characterization of meat quality of broiler chickens supplemented with betaine and antioxidants under cyclic heat stress. *Antioxidants* **2019**, *8*, 336. [CrossRef]
123. Shi, J.; Yu, J.; Pohorly, J.E.; Kakuda, Y. Polyphenolics in grape seeds—biochemistry and functionality. *J. Med. Food.* **2003**, *6*, 291–299. [CrossRef]
124. Sahin, K.; Orhan, C.; Tuzcu, M.; Ali, S.; Sahin, N.; Hayirli, A. Epigallocatechin-3-gallate prevents lipid peroxidation and enhances antioxidant defense system via modulating hepatic nuclear transcription factors in heat-stressed quails. *Poult. Sci.* **2010**, *89*, 2251–2258. [CrossRef]
125. Ramnath, V.; Rekha, P. Brahma Rasayana enhances in vivo antioxidant status in cold-stressed chickens (Gallus gallus domesticus). *Indian J. Pharmacol.* **2009**, *41*, 115. [CrossRef]
126. Hosseini-Vashan, S.J.; Safdari-Rostamabad, M.; Piray, A.H.; Sarir, H. The growth performance, plasma biochemistry indices, immune system, antioxidant status, and intestinal morphology of heat-stressed broiler chickens fed grape (Vitis vinifera) pomace. *Anim. Feed Sci. Technol.* **2020**, *259*, 114343. [CrossRef]
127. Mujahid, A.; Yoshiki, Y.; Akiba, Y.; Toyomizu, M. Superoxide radical production in chicken skeletal muscle induced by acute heat stress. *Poult. Sci.* **2005**, *84*, 307–314. [CrossRef]

128. Nakagawa, T.; Yokozawa, T. Direct scavenging of nitric oxide and superoxide by green tea. *Food Chem. Toxicol.* **2002**, *40*, 1745–1750. [CrossRef]
129. Shepherd, E.; Fairchild, B. Footpad dermatitis in poultry. *Poult. Sci.* **2010**, *89*, 2043–2051. [CrossRef]
130. Abd El-Wahab, A.; Radko, D.; Kamphues, J. High dietary levels of biotin and zinc to improve health of foot pads in broilers exposed experimentally to litter with critical moisture content. *Poult. Sci.* **2013**, *92*, 1774–1782. [CrossRef]
131. Thøfner, I.C.N.; Poulsen, L.L.; Bisgaard, M.; Christensen, H.; Olsen, R.H.; Christensen, J.P. Correlation between footpad lesions and systemic bacterial infections in broiler breeders. *Vet. Res.* **2019**, *50*, 38. [CrossRef] [PubMed]
132. Heidemann Olsen, R.; Christensen, H.; Kabell, S.; Bisgaard, M. Characterization of prevalent bacterial pathogens associated with pododermatitis in table egg layers. *Avian Dis.* **2018**, *47*, 281–285. [CrossRef] [PubMed]
133. El-Wahab, A.A.; Visscher, C.; Beineke, A.; Beyerbach, M.; Kamphues, J. Experimental studies on the effects of different litter moisture contents and exposure time to wet litter on development and severity of foot pad dermatitis in young fattening turkeys. *Arch. Geflugelkd.* **2012**, *76*, 55–62.
134. Mayne, R.; Else, R.; Hocking, P. High litter moisture alone is sufficient to cause footpad dermatitis in growing turkeys. *Br. Poult. Sci.* **2007**, *48*, 538–545. [CrossRef] [PubMed]
135. Cengız, Ö.; Hess, J.; Bilgili, S. Feed enzyme supplementation does not ameliorate foot pad dermatitis in broiler chickens fed on a corn-soyabean diet. *Br. Poult. Sci.* **2012**, *53*, 401–407. [CrossRef]
136. Rezar, V.; Salobir, J. Effects of tannin-rich sweet chestnut (Castanea sativa mill.) wood extract supplementation on nutrient utilisation and excreta dry matter content in broiler chickens. *Eur. Poult. Sci.* **2014**, *78*, 1–10. [CrossRef]
137. Salmond, G.G. Supplementation of Chelated Zn Cu and Mn in Broiler Diets to Improve Performance Skin Integrity and the Incidence and Severity of Footpad Dermatitis. Master's Thesis, University of Pretoria, Pretoria, South Africa, 2015.
138. Mayne, R.; Powell, F.; Else, R.; Kaiser, P.; Hocking, P. Foot pad dermatitis in growing turkeys is associated with cytokine and cellular changes indicative of an inflammatory immune response. *Avian Dis.* **2007**, *36*, 453–459. [CrossRef]
139. Bostami, A.; Ahmed, S.; Islam, M.; Mun, H.; Ko, S.; Kim, S.; Yang, C. Growth performance, fecal noxious gas emission and economic efficacy in broilers fed fermented pomegranate byproducts as residue of fruit industry. *Int. J. Adv. Res* **2015**, *3*, 102–114.
140. Wang, S.-Y.; Huang, D.-J. Assessment of greenhouse gas emissions from poultry enteric fermentation. *Asian-Australas J. Anim. Sci.* **2005**, *18*, 873–878. [CrossRef]
141. Wang, Y.; Huang, M.; Meng, Q. Effects of atmospheric hydrogen sulfide concentration on growth and meat quality in broiler chickens. *Poult. Sci.* **2011**, *90*, 2409–2414. [CrossRef]
142. McCrory, D.; Hobbs, P. Additives to reduce ammonia and odor emissions from livestock wastes: A review. *J. Environ. Qual.* **2001**, *30*, 345–355. [CrossRef] [PubMed]
143. Aneja, V.P.; Schlesinger, W.H.; Erisman, J.W.; Behera, S.N.; Sharma, M.; Battye, W. Reactive nitrogen emissions from crop and livestock farming in India. *Atmos. Environ.* **2012**, *47*, 92–103. [CrossRef]
144. Xin, H.; Gates, R.S.; Green, A.R.; Mitloehner, F.M.; Moore Jr, P.A.; Wathes, C.M. Environmental impacts and sustainability of egg production systems. *Poult. Sci.* **2011**, *90*, 263–277. [CrossRef] [PubMed]
145. Malomo, G.A.; Bolu, S.A.; Madugu, A.S.; Usman, Z.S. Nitrogen Emissions and Mitigation Strategies in Chicken Production. *Anim. Husband. Nutrit* **2018**, *43*, 43–62.
146. Aboagye, I.A.; Oba, M.; Koenig, K.M.; Zhao, G.Y.; Beauchemin, K.A. Use of gallic acid and hydrolyzable tannins to reduce methane emission and nitrogen excretion in beef cattle fed a diet containing alfalfa silage. *J. Anim. Sci.* **2019**, *97*, 2230–2244. [CrossRef]
147. Ahmed, S.T.; Yang, C.-J. Effects of dietary Punica granatum L. by-products on performance, immunity, intestinal and fecal microbiology, and odorous gas emissions from excreta in broilers. *J. Poult. Sci.* **2017**. [CrossRef]
148. Ramah, A.; Yasuda, M.; Ohashi, Y.; Urakawa, M.; Kida, T.; Yanagita, T.; Uemura, R.; Bakry, H.H.; Abdelaleem, N.M.; El-Shewy, E.A. Different doses of tannin reflect a double-edged impact on broiler chicken immunity. *Vet. Immunol. Immunopathol.* **2020**, *220*, 109991. [CrossRef]

149. Karaffová, V.; Bobíková, K.; Levkut, M.; Revajová, V.; Ševčíková, Z.; Levkut, M. The influence of Farmatan®and Flimabend®on the mucosal immunity of broiler chicken. *Poult. Sci.* **2019**, *98*, 1161–1166. [CrossRef]
150. Erlejman, A.G.; Jaggers, G.; Fraga, C.G.; Oteiza, P.I. TNFα-induced NF-κB activation and cell oxidant production are modulated by hexameric procyanidins in Caco-2 cells. *Arch. Biochem. Biophys.* **2008**, *476*, 186–195. [CrossRef]
151. Diaz Carrasco, J.M.; Redondo, E.A.; Pin Viso, N.D.; Redondo, L.M.; Farber, M.D.; Fernandez Miyakawa, M.E. Tannins and bacitracin differentially modulate gut microbiota of broiler chickens. *Biomed. Res. Int.* **2018**, *2018*. [CrossRef]
152. Koo, B.; Nyachoti, C.M. Effects of thermally oxidized canola oil and tannic acid supplementation on nutrient digestibility and microbial metabolites in finishing pigs. *J. Anim. Sci.* **2019**, *97*, 2468–2478. [CrossRef] [PubMed]
153. Pacheco-Ordaz, R.; Wall-Medrano, A.; Goñi, M.G.; Ramos-Clamont-Montfort, G.; Ayala-Zavala, J.F.; González-Aguilar, G. Effect of phenolic compounds on the growth of selected probiotic and pathogenic bacteria. *Lett. Appl. Microbiol.* **2018**, *66*, 25–31. [CrossRef] [PubMed]
154. Viveros, A.; Chamorro, S.; Pizarro, M.; Arija, I.; Centeno, C.; Brenes, A. Effects of dietary polyphenol-rich grape products on intestinal microflora and gut morphology in broiler chicks. *Poult. Sci.* **2011**, *90*, 566–578. [CrossRef] [PubMed]
155. Shang, Y.; Kumar, S.; Oakley, B.; Kim, W.K. Chicken gut microbiota: Importance and detection technology. *Front. Vet. Sci.* **2018**, *5*, 254. [CrossRef]
156. Bilić-Šobot, D.; Kubale, V.; Škrlep, M.; Čandek-Potokar, M.; Prevolnik Povše, M.; Fazarinc, G.; Škorjanc, D. Effect of hydrolysable tannins on intestinal morphology, proliferation and apoptosis in entire male pigs. *Arch. Anim. Nutr.* **2016**, *70*, 378–388. [CrossRef]
157. Kim, T.; Silva, J.; Kim, M.; Jung, Y. Enhanced antioxidant capacity and antimicrobial activity of tannic acid by thermal processing. *Food Chem.* **2010**, *118*, 740–746. [CrossRef]
158. Adhikari, P.A.; Kim, W.K. Overview of prebiotics and probiotics: Focus on performance, gut health and immunity–a review. *Ann. Anim. Sci.* **2017**, *17*, 949–966. [CrossRef]
159. Huyghebaert, G.; Ducatelle, R.; Van Immerseel, F. An update on alternatives to antimicrobial growth promoters for broilers. *Vet. J.* **2011**, *187*, 182–188. [CrossRef]
160. Khalil, R.K. Influence of gallic acid and catechin polyphenols on probiotic properties of Streptococcus thermophilus CHCC 3534 strain. *World J. Microbiol. Biotechnol.* **2010**, *26*, 2069–2079. [CrossRef]
161. Jarquin, R.; Nava, G.; Wolfenden, A.; Donoghue, A.; Hanning, I.; Higgins, S.; Hargis, B. The evaluation of organic acids and probiotic cultures to reduce Salmonella enteriditis horizontal transmission and crop infection in broiler chickens. *Int. J. Poult. Sci* **2007**, *6*, 182–186. [CrossRef]
162. Al-Harbi, R.; Shaaban, M.; Al-Wegaisi, R.; Moharram, F.; El-Rahman, O.A.; El-Messery, S. Antimicrobial Activity and Molecular Docking of Tannins from Pimenta dioica. *Lett. Drug Des. Discov.* **2018**, *15*, 508–515. [CrossRef]
163. Khan, S.H.; Iqbal, J. Recent advances in the role of organic acids in poultry nutrition. *J. Appl. Anim. Res.* **2016**, *44*, 359–369. [CrossRef]
164. De Lange, C.; Pluske, J.; Gong, J.; Nyachoti, C. Strategic use of feed ingredients and feed additives to stimulate gut health and development in young pigs. *Livest. Sci.* **2010**, *134*, 124–134. [CrossRef]
165. Chen, J.; Li, Y.; Yu, B.; Chen, D.; Mao, X.; Zheng, P.; Luo, J.; He, J. Dietary chlorogenic acid improves growth performance of weaned pigs through maintaining antioxidant capacity and intestinal digestion and absorption function. *J. Anim. Sci.* **2018**, *96*, 1108–1118. [CrossRef] [PubMed]
166. Choi, J.; Wang, L.; Liu, S.; Lu, P.; Zhao, X.; Liu, H.; Lahaye, L.; Santin, E.; Liu, S.; Nyachoti, M. Effects of a microencapsulated formula of organic acids and essential oils on nutrient absorption, immunity, gut barrier function, and abundance of enterotoxigenic Escherichia coli F4 in weaned piglets challenged with E. coli F4. *J. Anim. Sci.* **2020**, *98*, skaa259. [CrossRef] [PubMed]
167. Gracia, L.G.; de Castro, M.L. Development and validation of a flow-injection method for the determination of albumin tannate, the active component of a pharmaceutical preparation. *J. Pharm. Biomed. Anal.* **1997**, *15*, 447–452. [CrossRef]
168. Choi, J.; Wang, L.; Ammeter, E.; Lahaye, L.; Liu, S.; Nyachoti, M.; Yang, C. Evaluation of lipid matrix microencapsulation for intestinal delivery of thymol in weaned pigs. *Transl. Anim. Sci.* **2020**, *4*, 411–422. [CrossRef]

169. Omonijo, F.A.; Kim, S.; Guo, T.; Wang, Q.; Gong, J.; Lahaye, L.; Bodin, J.-C.; Nyachoti, M.; Liu, S.; Yang, C. Development of novel microparticles for effective delivery of thymol and lauric acid to pig intestinal tract. *J. Agric. Food Chem.* **2018**, *66*, 9608–9615. [CrossRef]
170. Liu, H.; Gong, J.; Chabot, D.; Miller, S.S.; Cui, S.W.; Ma, J.; Zhong, F.; Wang, Q. Protection of heat-sensitive probiotic bacteria during spray-drying by sodium caseinate stabilized fat particles. *Food Hydrocoll.* **2015**, *51*, 459–467. [CrossRef]
171. Choi, J. Lipid Matrix Microencapsulation for Effective Delivery of Essential Oils and Organic Acids to Improve gut Health in Weaned Piglets. Master's Thesis, University of Manitoba, Winnieg, MB, Canada, 2019.
172. Tang, Z.; Huang, X.; Baxi, S.; Chambers, J.R.; Sabour, P.M.; Wang, Q. Whey protein improves survival and release characteristics of bacteriophage Felix O1 encapsulated in alginate microspheres. *Food Res. Int.* **2013**, *52*, 460–466. [CrossRef]
173. Jang, I.; Kwon, C.H.; Ha, D.M.; Jung, D.Y.; Kang, S.Y.; Park, M.J.; Han, J.H.; Park, B.-C.; Lee, C.Y. Effects of a lipid-encapsulated zinc oxide supplement on growth performance and intestinal morphology and digestive enzyme activities in weanling pigs. *J. Anim. Sci. Technol.* **2014**, *56*, 29. [CrossRef] [PubMed]
174. Liu, L.J.; Zhu, J.; Wang, B.; Cheng, C.; Du, Y.J.; Wang, M.Q. In vitro stability evaluation of coated lipase. *Asian-Australas J. Anim. Sci.* **2017**, *30*, 192. [CrossRef] [PubMed]
175. Zhang, Y.; Gong, J.; Yu, H.; Guo, Q.; Defelice, C.; Hernandez, M.; Yin, Y.; Wang, Q. Alginate-whey protein dry powder optimized for target delivery of essential oils to the intestine of chickens. *Poult. Sci.* **2014**, *93*, 2514–2525. [CrossRef] [PubMed]
176. Trabelsi, I.; Ktari, N.; Slima, S.B.; Bouchaala, K.; Salah, R.B. Effects of supplementation with L. plantarum TN8 encapsulated in alginate-chitosan in broiler chickens. *Int. J. Biol. Macromol.* **2016**, *89*, 677–681. [CrossRef]
177. Fernández, H.; Catanese, F.; Puthod, G.; Distel, R.A.; Villalba, J.J. Depression of rumen ammonia and blood urea by quebracho tannin-containing supplements fed after high-nitrogen diets with no evidence of self-regulation of tannin intake by sheep. *Small Rumin. Res.* **2012**, *105*, 126–134. [CrossRef]
178. Adejoro, F.A.; Hassen, A.; Akanmu, A.M. Effect of lipid-encapsulated acacia tannin extract on feed intake, nutrient digestibility and methane emission in sheep. *Animals* **2019**, *9*, 863. [CrossRef]
179. Wang, M.; Huang, H.; Hu, Y.; Huang, J.; Yang, H.; Wang, L.; Chen, S.; Chen, C.; He, S. Effects of dietary microencapsulated tannic acid supplementation on the growth performance, intestinal morphology, and intestinal microbiota in weaning piglets. *J. Anim. Sci.* **2020**, *98*, skaa112. [CrossRef]

3

Influence of Different Time and Frequency of Multienzyme Application on the Efficiency of Broiler Chicken Rearing and Some Selected Metabolic Indicators

Youssef A. Attia [1,*], Mohammed A. Al-Harthi [1,*] and Ali S. El-Shafey [2]

1 Arid Land Agriculture Department, Faculty of Meteorology, Environment and Arid Land Agriculture, King Abdulaziz University, Jeddah 21589, Saudi Arabia

2 Animal and Poultry Production Department, Faculty of Agriculture, Damanhour University, Damanhour 22516, Egypt; alielshafey2015@gmail.com

* Correspondence: yaattia@kau.edu.sa (Y.A.A.); malharthi@kau.edu.sa (M.A.A.-H.)

Simple Summary: Enzymes are a useful, valuable and economic tool to improve feed utilization and thus animal performance. Costs of enzymes supplementation and time and application frequency of enzymes have received little attention other than dose and type of enzymes. This study showed that multienzymes offered intermittently during both early and late growth periods up to market age of broilers enhanced productive performance and economic profits and can substitute the daily administration with a considerable lowering of the supplementation cost.

Abstract: This study looks at the influence of time and/or frequency of multienzymes application on productivity, carcass characteristics, metabolic profile, and red blood cell characteristics of broiler chickens. Two hundred and eighty, one-day-old Arbor Acres broiler male chicks were randomly distributed into seven treatment groups. Each group consisted of eight replicates of five unsexed birds. The same basal diet was fed in a crumble form to all experimental groups: group one was the unsupplemented control that did not receive multienzymes supplementation. Additionally, multienzymes in water were supplemented in six groups in a factorial arrangement, including three times of application (starter time only which included days 1–21 of age, grower time only which included days 22–37 of age, and starter and grower time which included days 1–37 of age) and two application frequencies (continuously or intermittently). In the continuous application, the multienzymes were added to water over 24 h in a day, while in the intermittent frequency multienzymes were added to water for one day followed by a day off according to the time of application. Regardless of time and frequency of application, enzymes supplementation significantly increased growth rate, feed intake, European Production Index (EPI), protein digestibility, serum albumin, and high-density lipoprotein (HDL). Intermittent multienzymes application during days 1–21 of age or days 22–37 of age resulted in significantly greater growth, better feed conversion rate (FCR), and higher EPI of broilers during the whole rearing period than those under continuous multienzymes during different growth periods. Besides, intermittent multienzymes addition during days 1–37 of age improved FCR of broiler chicks compared to constant application. The intermittent addition of multienzymes during days 1–21 of age or 22–37 days of age and days 1–37 of age caused a significant increase in dry matter (DM) digestibility than the continuous application. The intermittent addition of multienzymes during days 1–21 of age significantly increased the digestibility of crude protein (CP), ether extract (EE), and crude fiber (CF) compared to continuous application. A similar trend was shown in the digestibility of CP and EE due to intermittent use during days 22–37 of age. Intermittent enzymes addition significantly increased high density lipoprotein (HDL) of groups receiving enzymes during days 22–37 of age compared to continuous application of enzymes. In conclusion, the use of multienzymes intermittently during days 1–21 of age and 22–37 days of age significantly increased growth, improved FCR, and raised EPI. Intermittent use can replace continuous multienzyme applications which can save 68.6% of the cost, even though further research is need from the cost-saving edge.

Keywords: broilers; application time and way; digestibility; carcass traits; metabolic profiles

1. Introduction

The digestive tract of chickens develops over time and is nearly mature by the fourth week of age, hence the digestibility of nutrients is weak and feed utilization is inadequate during this early age, resulting in increased costs of feeding and environmental pollution [1–4]. Nutrient digestibility and use of feeds were shown to improve over time and by enzyme supplementations [5,6]. Supplementation with enzymes in broiler diets increased the activity of the digestive enzymes [7,8] and improved the endogenous enzyme production, thus increasing the quantity of nutrients available for absorption [8–10]. Nonetheless, multienzymes impacts dietary composition, enzymes type, concentration and profiles, and strain and age of birds [11–14]. It is clear from the literature that the efficacy of enzymes decreased with the increasing age of chickens, and thus it is correlated with maturation of gut in terms of capacity, endogenous enzyme secretion, and ecology [3,12,13]. It was found that enzymes supplementation had more pronounced effects on broiler chickens than laying hens [2,14] and during the early stage of growth than later ones. The enzyme supplementation is effective for the total feeding cost because costs can be cut considerably if the efficiency of multienzymes can be determined according to the time of feeding [15]. Even so, until now and to the best of our knowledge, no studies that have addressed the relationship between time of growth and time and frequency of multienzymes administration on chickens' performance, nutrient digestibility, carcass traits, metabolic parameters, and blood constituents are available in the literature. Hence, the hypothesis of this research assumes that multienzymes improve the growth performance of broilers, and the intermittent application during the early growth period may replace the continuous use without adverse effects on growth performance, carcass traits and metabolic profits while enhancing economic profits.

2. Materials and Methods

2.1. Animals and Dietary Treatments

Two hundred and eighty, one-day-old Arbor Acres broiler chicks were randomly distributed into seven treatment groups. Each group consisted of eight replicates of five male chicks each. The same basal diet was offered in crumble form to all experimental groups. The group one (unsupplemented control group) was fed an un-supplemented diet. Further, there were six multienzyme-supplemented groups in a factorial arrangement that included three groups supplemented during different times of growth by two frequencies of administration. The times of growth were: the starter time only (days 1–21 of age), the grower time only (days 22–37 of age), and the starter plus grower time (days 1–37 of age). The administration of the multienzymes within each time of application were continuous or intermittent. In the continuous frequency, the multienzymes were added to the water over 24 h in a day. In the intermittent frequency, the multienzymes were added to the water for one day, followed by a day off. The diets in this experiment were formulated to meet the requirements of broiler chickens [16].

The multienzymes (Galzym®produced by Tex Biosciences (P) limited, Ashok Nagar, Chennai, India) used were a combination of a group of exogenous and fibrolytic enzymes consisting of: cellulase (100,000,000 units), xylanase (1,500,000 units), lipase (6500 units), alpha amylase (25,0000 units), protease (40,0000 units) and Pectinase (30,000 units). The recommended dose of enzymes was 0.333 mL/1 L water. The composition of the experimental diets is presented in Table 1.

Table 1. Ingredients and chemical-nutritional composition of the experimental diets (as feed basis) during different experimental periods.

Ingredients (g/kg)	Diets	
	Starter	Grower-Finisher
Maize	502.3	508.1
Rye	0	50
Soybean meal, 44% crude protein	328	244
Dicalcium phosphate	18.00	16.00
Limestone	10.00	10.00
NaCl	3.00	4.50
Full fat soybean meal	100	130
Vit+Min premix [1]	3.00	3.00
L-Lysine	1.00	1.90
DL-Methionine	2.00	2.50
Vegetable oil	22.70	20.00
Celite ™	10.0	10.0
Calculated and determined composition, (g/kg)		
Dry matter [2]	891	896
Crude protein, % [2]	223	205
Metabolisable energy (MJ/kg) [3]	12.50	12.64
Crude fat,% [2]	58	61
Crude fibre,% [2]	34.1	33.3
Nitrogen free extract [2]	513.4	532
Calcium [3]	8.72	8.13
Available phosphorus [3]	4.17	3.88
Methionine [3]	5.42	5.68
Methionine+cystine [3]	9.10	9.05
Lysine [3]	13.08	12.41
Ash [2]	62.5	64.7

[1] Vit+Min mix. provides per kilogram of the diets: Vit. A; 12,000 internation unit (IU), vit. E (dl-α-tocopheryl acetate) 20 mg, menadione 2.3 mg, Vit. D_3, 2200 international chick unit (ICU), riboflavin 5.5 mg, calcium pantothenate 12 mg, nicotinic acid 50 mg, Choline 250 mg, vit. B_{12} 10 µg, vit. B_6 3 mg, thiamine 3 mg, folic acid 1 mg, d-biotin 0.05 mg. Trace mineral (mg/kg of diets): Mn 80 Zn 60, Fe 35, Cu 8, Selenium 0.1 mg. [2] Analyzed values. [3] Calculated values.

All the experimental procedures were approved by Deanship of Scientific Research King Abdulaziz University, under proposal number G 170-155-1440H. The treatment of animals was according to the Royal Decree number M59 in 14/9/1431H, which recommends animal rights, welfare, and minimal stress and prevents any harm or suffering.

2.2. *Animal Housing and Management*

Broilers were housed in batteries of five chickens per cage (30 × 35 × 45 cm). Chicks were offered free access to feed and water during the experimental period. The housing temperature was 33 °C during the 1st week and declined gradually by 2 °C each week and was kept constant at 25 °C until slaughter. The average minimum and maximum outdoor and indoor relative humidity and temperature were 25.7 and 38.6%, 48.5 and 54.2%, 28.7 and 37.5 °C, and 27.8 and 34.6 °C, respectively. The light-dark cycle schedule was 23:1 h daily.

2.3. *Experimental Measurements*

Broilers in each replicate were weighed (g) at day 1, 21, and 37 of age, and the body weight gain (BWG) (g/chick) was calculated. Feed intake was recorded for each replicate (g/chick), and thereby FCR (g feed/g gain), and survival rate (SR, 100 − mortality rate) were calculated during days 1–21, 22–37, and 1–37 of age. Water intake was not measured due to a lack of precise research facility. European production index was for the index of economic evaluation, as reported by [12].

Apparent digestibility for crude protein (CP), dry matter (DM), ether extract (EE), crude fiber (CF), nitrogen-free extract (NFE) and ash during different times of growth using various methods of application was determined using Celite^{TM} (Celite Corp., Lompar, CA, USA) as a marker according to Dourado et al. [17]. The total collection method of excreta was followed for 5 days in each application time during days 15–20, 25–30, and 32–36 of age, respectively. In each time (1–21), (22–37) and (1–37) excreta and feed samples were constituted, and samples of each (25%) were collected, cleaned from feed residuals, scale and feathers, and then the DM, CP, EE, CF and ash of feeds and excreta were assayed according to [18] and expressed on dry matter basis. The NFE of feed and excreta was estimated by differences between the dry matter and sum of CP, EE, CF, and ash. The method adopted by [17] was used for analyses of acid-insoluble ash (AIA). The AIA recovery was estimated as g execrated/g intake × 100 using dry matter basis of excreta voided and feed intake data. It was found to range from 94% to 109%.

At 37 days of age, eight male broiler chicks per treatment group were slaughtered after 8 h fasting according to the Islamic method using a sharp knife and cutting into the jugular vein, carotid artery and windpipe, processed, and the weight of the carcass and inner organs was measured and estimated as a percentage of live body weight.

At 37 days of age, eight blood samples of 5 mL each per treatment group were collected and equally divided into two clean tubes with or without heparin. The tubes without heparin were centrifuged at 1500× g for 10 min at 4 °C, and then serum was separated and stored at −18 °C until analysis. The serum profiles were determined using commercial diagnostic kits (Diamond Diagnostics Company, Cairo, Egypt). Total protein (g/dl) [19], albumin (g/dl) [20], and globulin (g/dl) [21] were estimated. In addition, triglycerides [22], total cholesterol [23], high-density lipoprotein (HDL) [24]; and low-density lipoprotein (LDL) [25] were appraised.

Red blood cells [26], hemoglobin [27], packed cells volume (PCV) [28], mean cell volume (MCV), mean cell hemoglobin (MCH), and mean cell hemoglobin concentration (MCHC) [29] were manually measured.

2.4. Statistical Analysis

The statistical analysis software (SAS) was done using a completely randomized design and data were subjected to a two-way ANOVA procedure based on only time of supplementation and administration frequency without the unsupplemented control [30]. The statistical model included the effects of the time of multienzyme application (starter, grower-finisher and whole period), frequency of application (continuously vs. intermittently) and their interactions according to the subsequent model:

$$\text{Yijk} = \mu + \text{Di} + \text{AMj} + (\text{D} \times \text{AM}) + \text{eijk} \quad (1)$$

where: Y = the dependent variables; μ = general mean; D = time effect; AM = administration frequency effect; (D × AM) = interaction between the two factors; e = random error. Orthogonal contrast was done to test the effect of enzyme treatment regardless of time and frequency of application against the unsupplemented control. Before analysis, all percentages were subjected to logarithmic transformation (log10x+1) to normalize data distribution. The mean differences were determined using student Newman Keuls [30]. When p values found to be between 0.05 and 0.1 were considered as a tendency and thus indicate in the results "tend to be higher/lower."

3. Results

The effect of different times of multienzymes application in water continuously or intermittently on the production performance of broiler chickens is summarized in Table 2.

The contrast analyses showed that multienzymes supplementation significantly increased the growth of broilers during days 1–21 of age and days 1–37 of age compared to the unsupplemented control.

Both application time and application frequency had significant effects on the growth of broilers during 1–21, 22–37, and 1–37 days of age, but these effects were confounded by the significant interaction between the time and the frequency of application.

The interaction between application time and application frequency of multienzymes was significant for 1–21, 22–37, and days 1–37 of age. It was found that intermittent use during days 1–21 of age significantly improved the rate of growth of broilers during periods 1–21, 22–37, and 1–37 days of age compared with other continuous or intermittent groups. Moreover, intermittent enzyme application during days 22–37 of age substantially increased the growth of broilers during periods 1–37 days of age compared to continuous enzyme supplementation during periods 1–21 and 22–37 days of age, and permanent and intermittent use during periods 1–37 days of age. Enzyme application either as constant or intermittent during days 1–37 of age showed a similar growth rate during all tested periods.

The contrast analyses indicated that feed intake was significantly higher for chickens fed enzyme-supplemented diets compared to the un-supplemented ones during periods 1–21 and 1–37 days of age.

Feed intake during periods 1–21, 22–37, and 1–37 days of age was not significantly affected by the time of application and the interaction between time and frequency of enzyme application. However, feed intake during 22–37 and 1–37 days of age was significantly higher for broilers fed diets supplemented with enzymes continuously compared to those on the intermittent addition.

The contrast analyses for FCR showed no significant difference between the enzyme treatments and the un-supplemented group. The time of application significantly affected FCR during days 1–37 of age and application frequency influenced FCR during different tested periods, but these effects were confounded by the interaction between the time and frequency of application during days 1–21 and 1–37 of age.

In general, intermittent application during days 1–21 and 22–37 of age significantly improved FCR of broilers compared to continuous supplementation during periods 1–21 and 1–37 days of age. Besides, intermittent enzyme application during 1–37 days of age significantly improved FCR for broilers during days 1–37 compared to constant use.

There was no significant effect of time and/or frequency of enzyme application on the survival rate of broilers during days 1–37 of age.

The contrast analyses showed that enzyme application increased EPI compared to the unsupplemented control. The time and/or frequency of enzyme application significantly affected EPI. The intermittent enzyme application during days 1–21 of age significantly improved EPI of broilers compared to other groups. In addition, intermittent enzyme application during days 22–37 of age significantly enhanced EPI of broilers compared with other groups on continuous or intermittent frequency, except for intermittent frequency during days 1–21 of age. The intermittent application during 1–21 days of age had a greater effect on EPI than the same use during 22–37 days of age.

Results concerning the influence of the time and/or application frequency on nutrients digestibility of broilers are shown in Table 3. The contrast analyses demonstrated that enzyme supplementations significantly increased only crude protein digestibility compared to unsupplemented control, and had no effect on other digestibility traits.

Time and frequency of enzyme application had a significant effect on the digestibility of DM, CP, EE, and CF, but these factors were confounded by the interaction between the time and frequency of enzyme application. The interaction between application time and/or application frequency had a significant impact on most nutrient digestibility with the exception of the digestibility of NFE ($p = 0.834$) and ash ($p = 0.078$).

The interaction effect indicates that the intermittent use of enzymes during 1–21, 22–37, and days 1–37 of age improved the DM digestibility compared with continuous applications.

Table 2. Effect of multienzymes supplemented in water continuously or intermittently during the starter, grower, and the whole time on growth performance of broiler chicks.

Treatments/Period	Body Weight Gain (g)			Feed Intake (g)			Feed Conversion Rate (g/g)			Survival Rate (%)	Production Index
	1–21 d	22–37 d	1–37 d	1–21 d	22–37 d	1–37 d	1–21 d	22–37 d	1–37 d		
Effect of enzyme treatments vs unsupplemented control (contrast)											
Control	573 ^{b}	1113	1686 ^{b}	907 ^{b}	2119	3026 ^{b}	1.59	1.91	1.80	97.5	246 ^{b}
Enzymes	674 ^{a}	1139	1813 ^{a}	1031 ^{a}	2093	3125 ^{a}	1.53	1.84	1.72	99.6	286 ^{a}
Effect of administration time											
1–21	692 ^{a}	1158 ^{a}	1850 ^{a}	1031	2066	3096	1.50	1.79	1.68 ^{b}	100	301 ^{a}
22–37	676 ^{b}	1156 ^{a}	1831 ^{a}	1020	2106	3126	1.51	1.82	1.71 ^{ab}	98.8	287 ^{b}
1–37	654 ^{c}	1102 ^{b}	1755 ^{b}	1043	2108	3152	1.57	1.90	1.77 ^{a}	100	269 ^{c}
Effect of administration frequency											
Continuous	658 ^{b}	1117 ^{b}	1775 ^{b}	1035	2128 ^{a}	3164 ^{a}	1.58 ^{a}	1.91 ^{a}	1.78 ^{a}	99.2	267 ^{b}
Intermittent	690 ^{a}	1160 ^{a}	1850 ^{a}	1027	2058 ^{b}	3086 ^{b}	1.47 ^{b}	1.77 ^{b}	1.66 ^{b}	100	304 ^{a}
Interaction between time and frequency of application											
1–21 Con	653 ^{c}	1093 ^{d}	1746 ^{d}	1047	2107	3154	1.60 ^{a}	1.93	1.81 ^{a}	100	261 ^{c}
1–21 Int	731 ^{a}	1222 ^{a}	1953 ^{a}	1014	2024	3038	1.39 ^{e}	1.66	1.56 ^{d}	100	340 ^{a}
22–37 Con	667 ^{bc}	1143 ^{bc}	1810 ^{c}	1027	2143	3170	1.54 ^{c}	1.88	1.75 ^{b}	97.5	273 ^{c}
22–37 Int	685 ^{b}	1167 ^{b}	1852 ^{b}	1012	2069	3082	1.48 ^{d}	1.77	1.66 ^{c}	100	300 ^{b}
1–37 Con	652 ^{c}	1114 ^{cd}	1766 ^{d}	1031	2135	3167	1.58 ^{ab}	1.92	1.79 ^{a}	100	267 ^{c}
1–37 Int	654 ^{c}	1089 ^{d}	1743 ^{d}	1054	2081	3136	1.55 ^{bc}	1.88	1.75 ^{b}	100	270 ^{c}
RMSE	32.9	41.3	37.8	52.4	84.9	88.5	0.071	0.109	0.067	5.21	13.5
p value											
Control vs. enzymes	0.002	0.609	0.011	0.001	0.975	0.012	0.620	0.672	0.750	0.465	0.001
Time	0.048	0.009	0.001	0.616	0.462	0.391	0.088	0.112	0.018	0.552	0.001
Frequency	0.012	0.009	0.001	0.679	0.034	0.024	0.001	0.002	0.001	0.693	0.001
Interaction	0.041	0.001	0.001	0.492	0.927	0.562	0.017	0.066	0.005	0.348	0.001

D = day; Con = Continuous; Int = Intermittent; RMSE = Root mean square error; a–d, means with different superscripts in the same column in similar treatment groups are significantly different.

Table 3. Effect of multienzymes supplemented in water continuously or intermittently on apparent nutrient digestibility of broiler chicks.

Treatments/Period	Apparent Nutrient Digestibility, %					
	Dry Matter	Crude Protein	Ether Extract	Crude Fiber	Nitrogen Free Extract	Ash
			Unsupplemented control vs. enzymes treatment (contrast)			
Control	83.2	74.4^{b}	85.2	27.3	76.9	34.8
Enzymes	85.6	79.5^{a}	88.4	29.7	77.6	37.3
			Effect of administration time			
1–21	86.8^{a}	80.5^{a}	89.6^{a}	30.2^{a}	78.0	37.9
22–37	85.0^{b}	79.5^{b}	87.9^{b}	29.8^{a}	77.3	37.0
1–37	84.9^{b}	78.5^{c}	87.6^{b}	28.9^{b}	77.3	36.9
			Effect of administration frequency			
Continuous	84.2^{b}	78.5^{b}	87.2^{b}	28.8^{b}	77.1^{b}	36.8^{b}
Intermittent	86.9^{a}	80.5^{a}	89.5^{a}	30.5^{a}	78.0^{a}	37.8^{a}
			Interaction between time and frequency of application			
1–21 Con	83.9^{c}	78.5^{c}	86.9^{d}	28.3^{e}	76.8	36.7^{b}
1–21 Int	89.7^{a}	82.5^{a}	92.3^{a}	32.1^{a}	77.9	39.2^{a}
22–37 Con	84.3^{c}	78.7^{c}	87.1^{cd}	29.5^{bc}	77.4	36.6^{b}
22–37 Int	85.5^{b}	80.2^{b}	88.7^{b}	30.1^{b}	78.6	37.1^{b}
1–37 Con	84.3^{c}	78.2^{c}	87.6^{cd}	28.6^{de}	77.1	36.9^{b}
1–37 Int	85.7^{b}	78.8^{c}	87.7^{c}	29.3^{cd}	77.5	37.1^{b}
RMSE	2.61	1.76	2.74	1.52	2.42	1.74
			p value			
Control vs. enzymes	0.732	0.001	0.554	0.866	0.399	0.271
Time	0.045	0.002	0.049	0.036	0.565	0.116
Fequency	0.001	0.001	0.001	0.001	0.157	0.027
Interaction	0.011	0.011	0.011	0.002	0.834	0.078

Con = Continuous; Int = Intermittent; RMSE = Root means square error; a–e, means with different superscripts in the same column in similar treatment groups are significantly different.

The intermittent application of multienzymes during days 1–21 of age improved the digestibility of CP, EE, and CF compared to the other groups. The intermittent application of enzymes during days 22–37 of age enhanced the digestibility of CP and EE compared with the continuous and intermittent use during days 1–37 of age. The intermittent application during 1–21 days of age increased the digestibility of DM, CP, EE, and CF more than intermittent use during 22–37 days of age.

There were no significant differences between continuous and intermittent enzyme applications during days 1–37 of age on the digestibility of CP, EE, and CF.

Regardless of the time of application, intermittent supplementation increased ash digestibility compared to continuous supplementation.

The carcass characteristics and the inner organs of the broilers influenced by time of multienzymes application and/or the way of addition are shown in Table 4. The contrast analyses showed no significant effect of multienzymes compared to the unsupplemented control on all carcass traits and inner body organs.

Dressing, abdominal fat, gizzard, proventriculus, intestinal tract, liver, and heart were not significantly affected by the time and frequency, or by the interaction between time and frequency of application of enzymes.

The pancreas percentage was affected ($p = 0.051$) by the time of application, showing a trend of high value with enzymes supplemented during days 22–37 of age compared to the other periods. The effect of the application frequency approached significant ($p = 0.092$) for the heart percentage, showing the increase was proportional due to intermittent use.

The biochemical indices of blood serum of the broilers affected by the time of application and/or the application frequency are shown in Table 5.

The contrast analyses showed that enzyme treatments significantly increased the albumin, albumin/globulin ratio, total lipids, HDL, and HDL/LDL ratio in blood serum compared to the unsupplemented control, but decreased serum globulin. Differences between the enzymes' treatment and the unsupplemented control approached significant for total protein ($p < 0.054$).

There was no significant impact of time of application and/or frequency of use on blood serum proteins (total protein, albumin, globulin, albumin/globulin ratio). On the other hand, plasma total lipids were significantly affected by the application time, showing that total lipids were lower when enzymes were applied during days 22–37 of age than during the other times. The effect of time of application approached significant ($p < 0.063$) for serum triglycerides, with the highest being the 1–21 days group.

There was a significant interaction between application time and application frequency for serum HDL, LDL, and LDL/HDL ratio. Enzyme application intermittently during days 1–21 of age significantly decreased serum LDL compared to other groups, except for the continuous use during days 1–37 of age. On the other hand, the continuous offer of multienzymes during days 1–21 and 1–37 of age significantly increased the HDL and HDL/LDL ratio compared to the intermittent application, but the contrary was observed during days 22–37 of age.

Characteristics of red blood cells of broiler chickens are reported in Table 6. There were no significant effects due to enzyme treatments compared to the unsupplemented control when contrast analyses were considered for all red blood parameters.

The application time, application frequency, and the interaction between time and frequency of enzyme application did not significantly affect red blood cell traits, except for a significant increase in MCHC ($p < 0.039$) due to the continuous use compared to intermittent addition.

Table 4. Effect of multienzymes supplemented in water continuously or intermittently during the starter, grower, and the whole times on carcass characteristics and inner body organs of 37-day-old broiler chicks.

Treatments/Period	Dressing and Inner Organs,%							
	Dressing	Abdominal Fat	Gizzard	Proventriculus	Intestinal	Liver	Pancreas	Heart
Unsupplemented control vs enzymes treatment (contrast)								
Control	74.7	0.983	1.23	0.333	4.83	2.23	0.235	0.418
Enzymes	71.7	1.12	1.17	0.387	5.08	2.04	0.190	0.457
Effect of administration time								
1–21	74.1	1.078	1.16	0.379	5.12	2.14	0.186	0.434
22–37	70.5	1.170	1.21	0.388	5.15	2.10	0.213	0.476
1–37	70.3	1.140	1.12	0.394	4.96	1.88	0.171	0.461
Effect of administration frequency								
Continuous	71.6	1.199	1.15	0.411	5.04	2.07	0.193	0.439
Intermittent	71.7	1.054	1.18	0.363	5.11	2.01	0.187	0.475
Interaction between time and frequency of application								
1–21 Con	73.1	1.058	1.07	0.402	5.06	2.18	0.182	0.414
1–21 Int	75.1	1.098	1.25	0.356	5.18	2.10	0.190	0.454
22–37 Con	71.9	1.190	1.20	0.372	5.08	2.10	0.230	0.454
22–37 Int	68.7	1.150	1.22	0.404	5.22	2.10	0.196	0.498
1–37 Con	69.7	1.350	1.17	0.458	4.98	1.92	0.168	0.450
1–37 Int	71.3	0.877	1.07	0.330	4.94	1.84	0.174	0.472
RMSE	4.41	0.329	0.169	0.078	1.06	0.292	0.036	0.055
p value								
Control vs. enzymes	0.233	0.174	0.333	0.268	0.906	0.172	0.553	0.311
Time	0.122	0.821	0.510	0.911	0.911	0.122	0.051	0.244
Frequency	0.938	0.222	0.625	0.110	0.851	0.622	0.624	0.092
Interaction	0.341	0.219	0.184	0.093	0.979	0.941	0.369	0.893

Con = Continuous; Int = Intermittent; RMSE = Root means square error.

Table 5. Effect of multienzymes supplemented in water continuously or intermittently during the starter, grower, and the whole times on blood biochemical constituents of 37-day-old broiler chicks.

Treatments/Period	Blood Biochemical Constituents									
	Total Protein, g/dL	**Albumin, g/dL**	**Globulin, g/dL**	**A/G Ratio**	**Total Lipids, mg/dL**	**Triglycerides, mmol/L**	**Cholesterol, mmol/L**	**HDL, mmol/L**	**LDL, mol/L**	**LDL/HDL Ratio**
Unsupplemented control vs. enzyme treatments (contrast)										
Control	6.38	2.92^{b}	3.46^{a}	0.844^{b}	7.4^{b}	2.07	5.56	0.936^{b}	2.23	0.421^{b}
Enzymes	6.21	2.98^{a}	3.23^{b}	0.931^{a}	9.3^{a}	1.94	5.73	0.978^{a}	2.09	0.466^{a}
Effect of administration time										
1–21	6.24	3.03	3.21	0.953	9.7^{a}	1.95	5.73	0.993^{a}	2.12	0.469
22–37	6.17	3.03	3.15	0.975	8.5^{b}	1.92	5.73	0.996^{a}	2.09	0.475
1–37	6.21	2.87	3.34	0.864	9.7^{a}	1.94	5.73	0.944^{b}	2.07	0.456
Effect of administration frequency										
Continuous	6.18	2.99	3.18	0.951	9.6	1.93	5.66^{b}	0.998^{a}	2.12^{a}	0.471
Intermittent	6.23	2.96	3.28	0.910	9.0	1.94	5.79^{a}	0.957^{b}	2.07^{b}	0.461
Interaction between time and frequency of application										
1–21 Con	6.26	3.10	3.16	0.984	9.4	1.94	5.66	1.071^{a}	2.21^{a}	0.485^{a}
1–21 Int	6.22	2.96	3.16	0.923	10.0	1.97	5.81	0.915^{bc}	2.02^{c}	0.453^{bc}
22–37 Con	6.10	3.08	3.02	1.036	9.4	1.92	5.68	0.946^{bc}	2.09^{b}	0.452^{bc}
22–37 Int	6.24	2.98	3.28	0.915	7.6	1.92	5.78	1.045^{a}	2.10^{b}	0.497^{a}
1–37 Con	6.18	2.80	3.38	0.834	10.0	1.92	5.66	0.978^{b}	2.05^{bc}	0.476^{b}
1–37 Int	6.24	2.94	3.30	0.893	9.4	1.95	5.80	0.910^{c}	2.10^{b}	0.434^{c}
RMSE	0.146	0.206	0.251	0.131	1.13	0.030	0.062	0.047	0.051	0.024
p value										
Control vs. enzymes	0.054	0.048	0.009	0.025	0.001	0.731	0.896	0.005	0.351	0.029
Time	0.689	0.156	0.249	0.150	0.039	0.063	0.995	0.035	0.219	0.197
Frequency	0.327	0.661	0.318	0.395	0.159	0.188	0.001	0.023	0.025	0.281
Interaction	0.399	0.279	0.333	0.311	0.081	0.384	0.605	0.001	0.001	0.001

Con = Continuous; Int = Intermittent; HDL = High density lipoprotein; LDL = Low density lipoprotein; RMSE = Root mean square error; a–c, means with different superscripts in the same column in similar treatment groups are significantly different.

Table 6. Effect of multienzymes supplemented in water continuously or intermittently during the starter, grower and the complete times on red blood cell characteristics in broiler chicks.

Treatments/Period	Red Blood Cell Characteristics, %					
	RBC's (106/mm^3)	Hemoglobin (g/dL)	PCV (mL/1 mL)	MCV, μm^3/RBC	MHC, pg	MCHC, %
Unsupplemented control vs enzymes (contrast)						
Control	1.52	10.5	32.3	217	70.9	32.9
Enzymes	1.33	11.2	33.2	250	84.0	33.6
Effect of administration time						
1–21	1.34	11.4	33.5	252	84.8	33.7
22–35	1.31	11.1	33.0	253	84.91	33.6
1–35	1.35	11.1	33.1	246	82.47	33.5
Effect of administration frequency						
Continuous	1.31	11.3	33.2	253	86.4	34.1
Intermittent	1.35	11.1	33.2	246	81.6	33.1
Interaction between time and frequency of application						
1–21 Con	1.34	11.6	33.8	253	86.7	34.3
1–21 Int	1.34	11.1	33.2	250	82.9	33.1
22–35 Con	1.26	11.0	32.4	257	87.3	33.9
22–35 Int	1.36	11.2	33.6	248	82.5	33.3
1–35 Con	1.34	11.4	33.4	250	85.4	34.1
1–35 Int	1.36	10.8	32.8	242	79.6	32.9
RMSE	0.089	0.87	1.89	21.1	7.87	1.26
p value						
Control vs. enzymes	0.991	0.718	0.741	0.848	0.799	0.862
Time	0..589	0.838	0.824	0.756	0.742	0.948
Frequency	0.232	0.302	0.993	0.370	0.108	0.039
Interaction	0.429	0.501	0.482	0.916	0.959	0.825

Con = Continuous; Int = Intermittent; PCV = Packed cell volume; MCV = Mean cell volume; MCH = Mean hemogloblin concentration; MCHC = Mean cell hemogloblin concentration; RMSE = Root means square error.

4. Discussion

It is interesting to report that enzyme supplementation regardless of time and frequency of application enhanced growth performance and protein digestibility compared to unsupplemented controls. The increase in the growth of enzyme-supplemented groups was connected with increasing feed intake and digestibility of crude protein. These results are similar to the findings by [8,12,31–34]. The effect of the enzyme cocktail was attributed to improved digestibility and increased protein and energy availability for growth [32–40].

The intermittent enzymes application during days 1–21 of age resulted in better growth, improved FCR, and improved EPI compared to the continuous administration along the same period, but also to the constant or intermittent use during the other tested times (22–37 and days 1–37 of age). The positive effects of offering enzymes during the early growth phase (1–21 days of age) may be due to the supportive influence of exogenous enzymes on gut development and functions [3,13,15], and thus enhancing the digestibility of nutrients during the first stage of growth [1,10]. The gut of young chicks is under development and maturation in terms of enzyme secretion, capacity, and ecology during day 1–28 of age [1,5].

The results demonstrated that intermittent administration was adequate, which can save considerable cost (68.6%) of enzyme application. The high growth and better FCR of broilers on the intermittent use condition are tied to the increased digestibility of DM, CP, EE, and CF. Similar results concerning the positive impact of enzymes on nutrient digestibility were recently cited [6,8,13,31].

However, continuous use of enzymes throughout the experimental period (1–37 days of age) neither improved growth performance nor the digestibility of nutrients, and this may be due to the negative feedback mechanism of exogenous enzymes on pancreatic enzyme secretion [12,15].

It was found that the application of multienzymes during the period from 1 to 21 days of age yielded stronger effects than the use during the later ages and the whole period of growth on BWG and FCR than those supplemented during 22–37 or and days 1–37 of age. This may be because the digestive tract is under development up to four weeks of age. The enzyme applications during days 1–37 of age yield the least response. The continuous use during days 22–37 of age produced better growth during days 1–37 of age and FCR during days 1–21 and 1–37 of age than other permanent groups. The relationship between the time of application and growth performance of broilers needs, therefore, further research.

The positive impact of enzymes on animal productivity reported herein is in concert with those observed by [7,14,32]. These enhancements were attributed to the increasing activity of digestive enzymes [7,13], as estimated by indirect measurements, such as growth performance, or directly by digestibility of nutrients and eliminating the adverse effects of anti-nutritional factors [6,33,34]. Besides, multienzymes were found to increase the use of energy in sorghum and maize-soybean meal-based diets due to enhancing cereal cell walls and starch digestion [3,10]. Enzyme application may improve beneficial intestinal microbiota favoring *Lactobacillus* [14,31] and reduce environmental pollution [6,15] due to increased digestibility of nutrients (protein, fat and fiber, and minerals), while decreasing nutrients available in the hindgut for growth of harmful microbiota and increasing animal productive performance [9,13,15]. However, multienzymes' influence depends on enzyme type and concentrations, diet profile, form [4,6,31,35], and stage of growth [11,15].

Based on the present results, an enzyme blend supplemented in water during a different time of growth by various application frequencies at the recommended level did not affect most of the carcass and organ traits of broiler chickens. However, the pancreas percentage was increased due to enzyme supplementations during days 22–37 of age, suggesting a negative feedback mechanism of exogenous enzymes on the pancreas function [2,36]. There was also a decrease in proventriculus percentage when enzymes were added during the early time (1–21 days of age) and the growing time (22–37 days of age). In general, the absence of effects of enzymes on most of carcass and organ traits is constant with the findings by [7,37–39]. These authors indicated that application with enzymes did not affect carcass parameters and body organs of broilers and Japanese quail except for liver percent that was reduced with multienzymes application. Likewise, carcass traits were not influenced by dietary enzyme addition to broiler chicken diets [29,40,41].

In general, with few exceptions, our results showed that there were no significant effects from the application time and/or application frequency on the blood serum biochemistry and hematology. In other studies, dietary enzyme addition did not affect serum protein concentrations (alpha 1-, alpha 2-, beta and gamma-globulins and albumin) and albumin/globulin ratios [42–45].

The most pronounced effects found herein were a decrease in plasma cholesterol due to the continuous addition of enzymes and an increase in HDL with the intermittent applications during days 22–37 of age. There was also a decrease in LDL when enzymes were supplemented intermittently during days 1–21 of age. These beneficial effects due to multienzyme application on lipid metabolites require further investigation. However, enzymes had no significant effect on concentrations of triglyceride and cholesterol compared with animals fed on the same diet without the addition of enzymes [46]. Similar to the current results, different blood cholesterols and HDL/LDL ratios were increased in the unsupplemented control groups than enzyme-supplemented groups, showing the positive influence of enzymes on blood cholesterol [7,41,47].

5. Conclusions

Enzymes positively affect growth rate, feed intake, EPI, protein digestibility, serum albumin, total lipids, and HDL compared to the control group, irrespective of time and frequency of enzyme application. The application of multienzymes intermittently during 1–21 days of age and 22–35 days of age for broilers significantly increased growth, improved FCR, and enhanced EPI. This can replace continuous multienzymes applications while saving 68.6% of the cost, even though further research is needed to explore the beneficial economic impact of of time and frequency of enzyme application.

Author Contributions: Y.A.A., M.A.A.-H., A.S.E.-S. designed the experiment, and A.S.E.-S. carried the fieldwork and LAB analyses. Y.A.A. ran the statistical analyses. M.A.A.-H. prepared the first draft. All authors contributed to the final preparation and revision of the manuscript. All authors have read and agreed to the published version of the manuscript.

References

1. Noy, Y.; Sklan, D. Digestion and absorption in the young chick. *Poult. Sci.* **1995**, *74*, 366–373. [CrossRef] [PubMed]
2. Attia, Y.A.; Tag El-Din, A.E.; Zeweil, H.S.; Hussein, A.S.; Qota, E.M.; Arafat, M.A. The effect of application of enzyme on laying and reproductive performance in Japanese Quail hens fed nigella seed meal. *J. Poult. Sci.* **2008**, *45*, 110–115. [CrossRef]
3. Attia, Y.A.; El-Tahawy, W.S.; Abd El-Hamid, A.E.; Hassan, S.S.; Nizza, A.; El-Kelaway, M.I. Effect of phytase with or without multienzyme application on performance and nutrient digestibility of young broiler chicks fed mash or crumble diets. *Ital. J Anim. Sci.* **2012**, *11*, 303–308.
4. Fafiolu, A.O.; Oduguwa, O.O.; Jegede, A.V.; Tukura, C.C.; Olarotimi, I.D.; Teniola, A.A.; Alabi, J.O. Assessment of enzyme application on growth performance and apparent nutrient digestibility in diets containing undecorticated sunflower seed meal in layer chicks. *Poult. Sci.* **2015**, *94*, 1917–1922. [CrossRef]
5. Lee, S.A.; Apajalahti, J.; Vienola, K.; Gonzalez-Ortiz, G.; Fontes, C.M.G.A.; Bedford, M.R. Age and dietary xylanase application affects ileal sugar residues and short chain fatty acid concentration in the ileum and caecum of broiler chickens. *J. Anim. Feed Sci. Technol.* **2017**, *234*, 29–42. [CrossRef]
6. Attia, Y.A. The use of enzymes in poultry diets trends and approaches, a minimum Review. *Acta Sci. Nutr. Health* **2019**, *3*, 95–97.
7. Alagawany, M.; Attia, A.I.; Ibrahim, Z.A.; Mahmoud, R.A.; El-Sayed, S.A. The effectiveness of dietary sunflower meal and exogenous enzyme on growth, digestive enzymes, carcass traits, and blood chemistry of broilers. *Environ. Sci. Pol. Res.* **2017**, *24*, 12319–12327. [CrossRef]
8. Angel, C.R.; Saylor, W.; Vieira, S.L.; Ward, N. Effects of a monocomponent protease on performance and protein utilization in 7- to 22-day-old broiler chickens. *Poult. Sci.* **2011**, *90*, 2281–2286. [CrossRef]
9. Cozannet, P.; Kidd, M.T.; Neto, R.M.; Geraert, P.A. Next-generation non-starch polysaccharide-degrading, multi-carbohydrase complex rich in xylanase and arabinofuranosidase to enhance broiler feed digestibility. *Poult. Sci.* **2017**, *96*, 2743–2750. [CrossRef]
10. Williams, M.P.; O'Neil, H.V.M.; York, T.; Lee, J.T. Effects of nutrient variability in corn and xylanase inclusion on broiler performance, nutrient utilisation, and volatile fatty acid profiles. *J. Appl. Anim. Nutr.* **2018**, *6*, e1. [CrossRef]
11. Hong, D.; Burrows, H.; Adeola, O. Addition of enzyme to starter and grower diets for ducks. *Poult. Sci.* **2002**, *81*, 842–1849. [CrossRef]
12. Attia, Y.A.; El-Tahawy, W.S.; Abd El-Hamid, A.E.; Nizza, A.; Bovera, F.; Al-Harthi, M.A.; El-Kelway, M.I. Effect of feed form, pellet diameter and enzymes application on growth performance and nutrient digestibility of broiler during days 21–37 of age. *Arch. Tierzucht* **2014**, *57*, 1–11.
13. Aftab, U.; Bedford, M.R. The use of NSP enzymes in poultry nutrition: Myths and realities. *Worlds Poult. Sci. J.* **2018**, *74*, 277–286. [CrossRef]

14. Al-Harthi, M.A.; Attia, Y.A.; Al-sagan, A.; Elgandy, M.F. The effects of autoclaving or/and multi-enzymes complex application on performance, egg quality and profitability of laying hens fed whole Prosopis juliflora pods meal diet. *Eur. Poult. Sci.* **2018**, *82*. [CrossRef]
15. Choct, M. Enzymes for the feed industry: Past, present and future. *World's Poult. Sci. J.* **2006**, *62*, 5–16. [CrossRef]
16. National Research Council, NRC. *Nutrient Requirement of Poultry*; National Academy Press: Washington, DC, USA, 1994.
17. Dourado, L.R.B.; Siqueira, J.S.; Sakomura, N.K.; Pinheiro, S.R.F.; Marcato, S.M.; Fernandes, J.B.K.; Silva, J.H.V. Poultry feed metabolizable energy determination using total or partial excreta collection methods. *Rev. Bras. Cienc. Avic.* **2010**, *12*, 129–132. [CrossRef]
18. Association of Official Analytical Chemists, AOAC. *Official Methods of Analysis*, 18th ed.; Association of Official Analytical Chemists: Washington, DC, USA, 2004.
19. Henry, R.; Cannon, D.; Winkelman, J. *Clinical Chemistry, Principles and Techniques*, 2nd ed.; Harper and Row: New York, NY, USA, 1974.
20. Doumas, B.T.; Watson, W.A.; Biggs, H.G. Albumin standards and the measurement of serum albumin with bromcresol green. *Clin. Chim. Acta* **1971**, *31*, 21–30. [CrossRef]
21. Coles, E.H. *Veterinary Clinical Pathology*, 1st ed.; W.B. Saunder: Philadelphia, PA, USA, 1974.
22. Fossati, P.; Prencipe, L. Serum triglycerides determined colorimetrically with an enzyme that produces hydrogen peroxide. *Clin. Chem.* **1982**, *28*, 2077–2080. [CrossRef]
23. Stein, E.A. Quantitative enzymatic colorimetric determination of total cholesterol in serum or plasma. In *Textbook of Clinical Chemistry*; Tietz, N.W., Ed.; WB. Saunders: Philadelphia, PA, USA, 1986; pp. 879–886.
24. Lopez-Virella, M.F.; Stone, S.; Eills, S.; Collwel, J.A. Determination of HDL-cholesterol using enzymatic method. *Clin. Chem.* **1977**, *23*, 882–884.
25. Friedewald, W.T.; Levy, R.I.; Frederickson, D.S. Estimation of the concentration of low-density lipoprotein cholesterol in plasma without use of the preparative ultracentrifuge. *Clin. Chem.* **1972**, *18*, 499–502. [CrossRef] [PubMed]
26. Helper, O.E. *Manual of Clinical Laboratory Methods*, 4th ed.; Charles C Thomas Publishing Ltd.: Springfield, IL, USA, 1966.
27. Tietz, N.W. *Fundamentals of Clinical Chemistry*, 2nd ed.; Saunders Company Publishing: Philadelphia, PA, USA, 1982.
28. Wintrobe, M.M. *Clinical Hematology*; Lea & Febiger Publishing: Philadelphia, PA, USA, 1963.
29. Attia, Y.A.; El-Tahawy, W.S.; Abd El-Hamid, A.E.; Nizza, A.; El-Kelway, M.I.; Al-Harthi, M.A.; Bovera, F. Effect of feed form, pellet diameter and enzymes application on carcass characteristics, meat quality, blood plasma constituents and stress indicators of broilers. *Arch. Tierzucht* **2014**, *57*, 1–14.
30. SAS Institute. *SAS/STAT User's Guide Statistics*; SAS Institute INC: Cary, NC, USA, 2009.
31. Borda-Molina, D.; Zubar, T.; Siegert, W.; Camarinha-Silva, A.; Feuerstein, D.; Rodehutscord, M. Effects of protease and phytase supplements on small intestinal microbiota and amino acid digestibility in broiler chickens. *Poult. Sci.* **2019**, *98*, 2906–2918. [CrossRef]
32. Zeng, Q.; Huang, X.; Luo, Y.; Ding, X.; Bai, S.; Wang, J.; Xuan, Y.; Su, Z.; Liu, Y.; Zhang, K. Effects of a multi-enzyme complex on growth performance, nutrient utilization and bone mineralization of meat duck. *J. Anim. Sci. Biotechnol.* **2015**, 6–12. [CrossRef]
33. Kocher, A.; Hower, J.M.; Moran, C.A. A dualenzyme product containing protease in broiler diet: Efficacy and tolerance. *J. App. Anim. Nut.* **2015**, *3*, 1–14.
34. Abdel-Hafeez, H.M.; Saleh, E.S.E.; Tawfeek, S.S.; Youssef, I.M.I.; Abdel-Daim, A.S.A. Utilization of potato peels and sugar beet pulp with and without enzyme application in broiler chicken diets: Effects on performance, serum biochemical indices and carcass traits. *J. Anim. Phys. Anim. Nutr.* **2016**, *102*, 56–66. [CrossRef]
35. Nourmohammadi, R.; Hosseini, S.M.; Farhangfar, H.; Bashtani, M. Effect of citric acid and microbial phytase enzyme on ileal digestibility of some nutrients in broiler chicks fed corn-soybean meal diets. *Ital. J. Anim. Sci.* **2012**, *11*, 36–40. [CrossRef]
36. Attia, Y.A.; Qota, E.M.A.; Aggoor, F.A.M.; Kies, A.K. Value of rice bran, its maximal utilization and upgrading by phytase and other enzymes and diet- formulation based on available amino acid for broiler chicks. *Arch. Geflügelk.* **2003**, *67*, 157–166.

37. Mushtaq, T.; Sarwara, M.; Ahmad, G.; Mirza, M.A.; Ahmad, T.; Mushtaq, M.M.H.; Kamran, Z. Influence of sunflower meal based diets supplemented with exogenous enzyme and digestible lysine on performance, digestibility and carcass response of broiler chickens. *Anim. Feed Sci. Technol.* **2009**, *149*, 275–286. [CrossRef]
38. De Araujo, W.; Albino, L.F.T.; Rostagno, H.S.; Hannas, M.I.; Pessoa, G.B.S.; Messias, R.K.G.; Lelis, G.R.; Ribeiro, V., Jr. Sunflower meal and enzyme application of the diet of 21-to 42-d-old broilers. *Braz. J. Poult. Sci.* **2014**, *16*, 17–24. [CrossRef]
39. Rabie, M.H.; Abo El-Maaty, H.M.A. Growth performance of Japanese quail as affected by dietary protein level and enzyme application. *Asian J. Anim. Vet. Adv.* **2015**, *10*, 74–85. [CrossRef]
40. Dalólio, F.S.; Moreira, J.; Albino, L.F.T.; Vaz, D.P.; Pinheiro, S.R.F.; Valadares, L.R.; Pires, A.V. Exogenous enzymes in diets for broilers. *Rev. Bras. Saúde Prod. Anim.* **2016**, *17*, 149–161.
41. Al-Harthi, M.A. The effect of olive cake, with or without enzymes application, on growth performance, carcass characteristics, lymphoid organs and lipid metabolism of broiler chickens. *Rev. Bras. Cienc. Avic.* **2017**, *19*, 83–90. [CrossRef]
42. Mehri, M.; Adibmoradi, M.; Samie, A.; Shivazad, M. Effects of Mannanase on broiler performance, gut morphology and immune system. *Afr. J. Biotech.* **2010**, *9*, 6221–6228.
43. Gheisari, A.A.; Zavareh, M.S.; Toghyani, M.; Bahadoran, R. Application of incremental program, an effective way to optimize dilatory inclusion rate of guar meal in broiler chicks. *Livest. Sci.* **2011**, *140*, 117–123. [CrossRef]
44. El-Serwy, A.; Shoeib, M.S.; Fathey, I.A. Performance of broiler chicks fed mash or pelleted diets containing corn-with-cobs meal with or without enzyme application. *J. Anim. Poult. Prod.* **2012**, *3*, 137–155.
45. Dinani, O.P.; Tyagi, P.K.; Tyagi, P.K.; Mandal, A.B.; Jaiswal, S.K. Effect of feeding fermented guar meal vis-àvis toasted guar meal with or without enzyme application on immune response, caeca micro flora status and blood biochemical parameters of broiler quails international. *J. Pure Appl. Biosci.* **2017**, *5*, 624–630. [CrossRef]
46. El-Katcha, M.I.; Soltan, M.A.; El-Kaney, H.F.; Karwarie, E.R. Growth performance, blood parameters, immune response and carcass traits of broiler chicks fed on graded levels of wheat instead of corn without or with enzyme application. *Alex. J. Vet. Sci.* **2014**, *40*, 95–111.
47. Zarfar, A. Effect of hemicell enzyme on the performance, growth parameter, some blood factors and ileal digestibility of broiler chickens fed corn/soybean-based diets. *J. Cell Anim. Biol.* **2013**, *7*, 5–91.

4

Evaluation of Waste Mushroom Compost as a Feed Supplement and its Effects on the Fat Metabolism and Antioxidant Capacity of Broilers

Wen Yang Chuang [1], Chu Ling Liu [1], Chia Fen Tsai [1], Wei Chih Lin [1], Shen Chang Chang [2], Hsin Der Shih [3], Yi Ming Shy [4] and Tzu-Tai Lee [1,5,*]

1 Department of Animal Science, National Chung Hsing University, Taichung 402, Taiwan; xssaazxssaaz@yahoo.com.tw (W.Y.C.); s04610215@gmail.com (C.L.L.); bbella1123tw@gmail.com (C.F.T.); waynezi2227738@gmail.com (W.C.L.)

2 Kaohsiung Animal Propagation Station, Livestock Research Institute, Council of Agriculture, Tainan 71246, Taiwan; macawh@mail.tlri.gov.tw

3 Taiwan Agricultural Research Institute Council of Agriculture, Executive Yuan, Taichung City 41362, Taiwan; tedshih@tari.gov.tw

4 Hsinchu Branch, Taiwan Livestock Research Institute, Council of Agriculture, Executive Yuan, Tainan 71246, Taiwan; emshy@tlri.gov.tw

5 The iEGG and Animal Biotechnology Center, National Chung Hsing University, Taichung 402, Taiwan

* Correspondence: ttlee@dragon.nchu.edu.tw.

Simple Summary: Mushroom waste compost is the main byproduct when cultivating mushrooms. Due to its high mycelium content, mushroom waste compost may improve animal health by increasing antioxidant capacity. Furthermore, increasing evidence suggests that supplementing animal diets with fiber could improve body composition and health. The results showed that supplementation with mushroom waste compost accelerates adipolysis and enhances the antioxidant capacity of broilers. Among all treatment groups, broilers given dietary supplementation with 0.5% mushroom waste compost showed improved feed conversion rate and the highest adipose metabolism.

Abstract: *Pennisetum purpureum* Schum No. 2 waste mushroom compost (PWMC) is the main byproduct when cultivating *Pleurotus eryngii*. Due to the high mycelium levels in PWMC, it may have potential as a feed supplement for broilers. This study investigated the effects of PWMC supplementation on antioxidant capacity and adipose metabolism in broilers. In the study, 240 broilers were randomly allocated to one of four treatment groups: basal diet (control), 0.5%, 1%, or 2% PWMC supplementation. Each treatment group had 60 broilers, divided into three replicates. The results showed that supplementation with 0.5% PWMC decreased the feed conversion rate (FCR) from 1.36 to 1.28, compared to the control. Supplementation with 0.5% or 2% PWMC decreased glucose and triglyceride levels, compared to the control ($p < 0.0001$), the concentrations of adiponectin and oxytocin increased from 5948 to 5709, 11820, and 7938 ng/ mL; and 259 to 447, 873, and 963 pg/ mL, respectively. Toll-like receptor 4 was slightly increased in the 0.5% and 1% PWMC groups. Both interferon-γ (IFN-γ) and interleukin-1ß (IL-1ß) were significantly decreased, by about three to five times for IFN-γ ($p < 0.0001$) and 1.1 to 1.6 times for IL-1ß ($p = 0.0002$). All antioxidant-related mRNA, including nuclear factor erythroid 2–related factor 2 (Nrf-2) and superoxidase dismutase-1 (SOD-1), increased significantly following PWMC supplementation. Both claudin-1 and zonula occludens 1 increased, especially in the 2% PWMC group. Excitatory amino acid transporter 3 (EAAT3) significantly increased by about 5, 12, and 11 times in the 0.5%, 1%, and 2% PWMC groups. All adipolysis-related mRNA were induced in the PWMC treatment groups, further enhancing adipolysis. Overall, 0.5% PWMC supplementation was recommended due to its improving FCR, similar antioxidant capacity, and upregulated adipolysis.

Keywords: mushroom compost; fat metabolism; antioxidant; broilers

1. Introduction

Due to global climate change and rising crude oil prices, high-starch animal feed ingredients such as corn and cassava must be reconsidered in view of their value as raw materials for biomass energy [1]. Considering the cost of raising broilers, it is increasingly important to find cheap alternatives that can maintain animal growth and health. Crops are the main sources of raw feed materials, resulting in a large amount of high-fiber agricultural byproducts [2]. Traditionally, due to the limitations of animal digestion, high-fiber plant materials have not been considered for animal feed [3]. Unlike animals, mushrooms produce a large amount of cellulase and hemicellulase, and fiber is the main source of their growth [4]. As such, the culture medium used in mushroom cultivation is high in fiber and contains few minerals or crude proteins [4]. However, following cultivation, the medium is still high in crude fiber and may cause new environmental issues [5,6].

Due to its high-fiber characteristics, *Pennisetum* can be used as one of the raw materials or an adjustment material in the cultivation of mushrooms, reducing the environmental problems caused by the traditional use of wood chips [6]. In addition, mushroom waste compost produced by mushroom planting contains a lot of mycelium, which could improve animal antioxidant capacity and health [5]. Mahfuz et al. also indicated that although 2% mushroom waste compost addition would not increase the feed conversion rate or body weight of broilers, mushroom waste compost addition could decrease broilers' serum cholesterol content and improve immunomodulation [7].

In addition, previous researches pointed out that plant-based phenol-like compounds, such as epigallocatechin gallate (EGCG), could suppress fatty acid synthase and acetyl CoA carboxylase activities thereby decreasing the adipogenesis in broilers [8]. Lee et al. [9] also pointed out that the addition of mushroom waste compost could decrease crude fat in broiler meat and the results might be caused by the decrease of lipid peroxidation.

However, there is limited research on the reuse of mushroom waste compost or its effects when used as a supplement in broiler feed. This study therefore investigated the effect of mushroom waste compost on antioxidant capacity and how could it alter adipose metabolism in broilers.

2. Materials and Methods

2.1. The Collection and Characteristics of Pennisetum purpureum Schum No. 2 Waste Mushroom Compost (PWMC)

PWMC was the medium used to grow *Pleurotus eryngii*. It contained at least 70% *Pennisetum purpureum* Schum No. 2 (PP) and all mature *P. eryngii* were removed. The PWMC came from the Taiwan Agricultural Research Institute Council of Agriculture, Executive Yuan. Briefly, after removing *P. eryngii*, the medium with fresh residues had about 50% moisture and was immediately dried at 50 °C for 1 day before being crushed into powder. PWMC was stored at 4 °C until use.

The Folin–Ciocalteu method described by Tabart et al. [10] was used for the total phenol content analysis. Briefly, the sample was extracted by deionized water at 95 °C for 30 min and cooled down to room temperature before being used. Sample solution of 1 mL was mixed with 5 mL Folin–Ciocalteu reagent and 4 mL 7.5% sodium carbonate. The mixture was incubated in the room temperature for 30 min and measured the absorbance at 730 nm. The 0.01–0.1 mg/mL gallic acid was also measured by the methods described above for the standard curve. The presentation of total phenol content was described as gallic acid equivalent (GAE) mg/g sample.

PWMC was extracted by deionized water at 95 °C for 30 min and cooled down to room temperature, and filtrated by a 0.22 μm filter before used for the phenol-like compounds detection. HPLC (Hitachi, Kyoto, Japan) equipped with a pump (CM 5110), a column (C18-AR, 250 × 4.8 mm, maintained at

40 °C by the column oven (CM 5310)), an autosampler (L-2200) and a computer system with HPLC D-2000 Elite. The mobile phase conditions were (A) 0.05% *v/v* H_3PO_4 and (B) 3:2 *v/v* CH_3OH/CH_3CN solution, 1.0 mL/min, and UV detection (CM 5420) at 280nm. Also, 0.01~1.5 mg/ mL of gallocatechin (GC), epigallocatechin (EGC), catechin (CC), epicatechin (EC), gallic acid (GA), epigallocatechin gallate (EGCG), epicatechin gallate (ECG), catechin gallate (CG), and caffeic acid were measured by the methods described above for the standard curve.

The methods of total flavonoid content were described by Pourmorad et al. [11]. Briefly, a sample was extracted by methanol on ice for 30 min. The 1 mL sample solution, 3 mL methanol, 0.2 mL 10% $AlCl_3$, 0.2 mL 1M potassium acetate, and 5 mL distilled water were mixed and incubated at room temperature for 30 min. The absorbance of mixture was measured at 700 nm. Quercetin of 0.01–0.1 mg/mL was also measured by the methods described above for the standard curve.

The methods of crude triterpenoid content were described by Wei et al. [12]. Briefly, a 3 g sample was added to 100 mL of 50% ethanol and shaken (100 rpm) for 8 h at room temperature. The solution was centrifuged at 3000 rpm and the supernatant was removed to a new tube. Next, 100 mL 95% ethanol was added in the precipitate and shaken (100 rpm) for 12 h at room temperature. The new solution was centrifuged at 3000 rpm and the two supernatants were mixed. The new supernatant was filtrated by Advantec No.1 filter papers and stoved at 50 °C for three days. The 100 mL saturated $NaHCO_3$ (pH 3.5) and ethyl acetate were used to wash the remains to remove the impurity. The ethyl acetate solution was collected and dried to constant weight. The dry remains weight was measured and calculated as the formula described as follows.

$$\text{Crude triterpenoid content (\%)} = \text{Dry remains weight/ Sample weight} \times 100\% \quad (1)$$

2.2. *Animal Experiment Design*

PWMC was used to test its effect on broiler fat metabolism, nutrient absorption, antioxidant function, and inflammatory regulation. The experiment was carried out at National Chung Hsing University, Taiwan and all protocol followed that of the Animal Care and Use Committee in NCHU (IACUC: 108-049). Every broiler could drink and eat freely and had enough active space (about 20 broilers/9 m^2). If the chickens had become infected during rearing, they would have been quarantined and given antibiotics or medication at the discretion of the veterinarian, however every broiler remained healthy throughout the whole rearing period. Euthanasia was used to reduce broiler suffering. The methods for animal design, testing intestinal morphology, blood and serum characteristics, and mRNA expression were slightly modified from those of Chuang et al. [13]. Briefly, 240 male 1-day-old Ross 308 broilers were used for 35-day experiments and each was placed in one of four groups: basal diet (control), 0.5%, 1%, or 2% PWMC supplementation. Every group had 60 broilers with three replicates (20 per pen). The initial body weight of each chick was similar (48.0 ± 0.7 g/bird) among all groups. The temperature was kept at 33 °C for 1-day-old chicks then slowly decreased to 22 °C after 30 days. To achieve or exceed the nutritional requirements of the broilers, according to NRC 1994, all diets were recalculated and the proximate composition was analyzed according to the AOAC (2012) [14] to determine the gross energy and crude protein levels (Table 1). The starter diet was provided for the first 21 days, and the finisher diet provided for days 22–35. Body weight gain, feed intake, and feed conversion rate (FCR) were measured on 21- and 35-day-old broilers. On day 35, the ileum, abdominal fat, spleen, and liver were collected for qPCR analysis.

2.3. *Serum Characteristics*

At the end of the experiment, 5 mL whole blood was collected from the broilers (three from each pen, nine per treatment) and left to stand for 4 to 5 h at 4 °C before measurement. The samples were centrifuged at 3000 rpm for 10 min to separate the blood cells and serum, before the serum was frozen at −20 °C until use. The concentrations of oxytocin (OXT), corticosterone), tumor necrosis factor alpha (TNF-α), glutathione peroxidase (Gpx), superoxidase dismutase (SOD) A), and malondialdehyde

(MDA) in the broiler serum were measured. All analysis methods followed the manufacturer's protocol. Briefly, the protein analytic was bound by the antibody and the enzyme-linked immunosorbent assay reader was used to measure absorbance at wavelengths provided by the manufacturer. Other blood cells and serum characteristics were measured using the automatic biochemical analyzer (Hitachi, 7150 auto-analyzer, Tokyo, Japan).

Table 1. Composition and calculated analysis (g/kg as fed) of the basal diet for broilers (1–35 days) [1].

Ingredients	Starter Diet (1–21 Days)	Finisher Diet (22–35 Days)
	g/kg	
Corn, yellow	520	572
Soybean meal (CP 44%)	369	247
Full-fat soybean meal	0.00	50.0
Soybean oil	40.6	41.4
Fish meal (CP 65%)	30.0	50.0
Calcium carbonate	16.5	16.5
Monocalcium phosphate	11.2	11.2
DL-Methionine	2.00	2.00
L-Lysine-HCl	3.70	3.70
NaCl	3.90	3.90
Choline-Cl	0.80	0.80
Vitamin premix [2]	1.00	1.00
Mineral premix [3]	1.00	1.00
Total	1000	1000
Calculated nutrient value		
Dry matter, %	88.1	88.2
Crude protein, %DM	23.0	21.0
Crude fiber, %DM [4]	4.71	3.55
Calcium, %DM	1.11	1.21
Total Phosphorus, %DM	0.68	0.72
Available Phosphorus, %DM	0.46	0.52
Methionine + Cysteine, %DM	0.92	0.89
ME, kcal/kg DM	3050	3175
Chemical analysis value		
Dry matter, %	88.7	89.3
Crude protein, %DM	23.1	21.0
Crude fat, DM%	6.85	7.96
Chemical analysis value	PWMC	
Neutral detergent fiber, %DM	60.1 ± 0.70	
Crude protein, %DM	6.50 ± 0.40	
Ash, %DM	10.2 ± 0.20	

[1] Con: basal diet for control group; PWMC: *Pennisetum purpureum* Schum No. 2 waste mushroom compost. PWMC was added directly to the basal diet at different percentages. [2] Vitamins (premix content per kg diet): Vit. A, 15,000 IU; Vit. D3, 3000 IU; Vit. E, 30 mg; Vit. K3, 4 mg; thiamine, 3 mg; riboflavin, 8 mg; pyridoxine, 5 mg; Vit. B12, 25 μg; Ca-pantothenate, 19 mg; niacin, 50 mg; folic acid, 1.5 mg; and biotin, 60 μg. [3] Minerals (premix content per kg diet): Co ($CoCO_3$), 0.255 mg; Cu ($CuSO_4{\cdot}5H_2O$), 10.8 mg; Fe ($FeSO_4{\cdot}H_2O$), 90 mg; Mn ($MnSO_4{\cdot}H_2O$), 90 mg; Zn (ZnO), 68.4 mg; Se (Na_2SeO_3), 0.18 mg. [4] The analysis of crude fiber content in the basal diet, 0.05, 1 and 2% PWMC addition treatments were 4.8, 5.1, 5.3, and 5.7% dry matter (DM) in the starter stage and 3.6, 3.8, 4.1, and 4.5% DM in the finisher stage, respectively.

2.4. Total RNA Isolation and qPCR

Total RNA isolation and qPCR were used to analyze mRNA expression in the liver, spleen, ileum, and abdominal fat. All mRNA was collected from 35-day-old broilers in each treatment group. The methods of total mRNA isolation followed the manufacturer's protocol (SuperScript™ FirstStrand Synthesis System reagent, Thermo Fisher, Waltham, MA, USA). The mRNA purity was determined by the absorbance ratio of 260/280 nm, and the methods for cDNA synthesis and qPCR analysis were as per Chuang et al. [13] Briefly, cDNA was mixed with 2× SYBR GREEN PCR Master Mix-ROX (Gunster Biotech, Co., Ltd., Taipei, Taiwan), deionized water, and each primer at a ratio of 5:1.2:1.8:1. A StepOnePlus™ Real-Time PCR System (Thermo Fisher, Waltham, MA, USA) was used to detect qRT-PCR performance. The $2^{-\Delta\Delta Ct}$ method was used to calculate the relative mRNA expression level, and ß-actin was used as the housekeeping gene for normalization. All primer sequences matched the genes of *Gallus gallus* (chicken) from Genbank (Table 2).

Table 2. The primer sequence of each gene according to Genbank or other research.

Gene Name [1]	Primer Sequence	Genbank No.
ß-actin	F: 5′-CTGGCACCTAGCACAATGAA-3′ R: 5′-ACATCTGCTGGAAGGTGGAC-3′	X00182.1
TNF-α	F: 5′-TGTGTATGTGCAGCAACCCGTAGT-3′ R: 5′-GGCATTGCAATTTGGACAGAAGT-3′	NM 204267
TLR4	F: 5′-TGCACAGGACAGAACATCTCTGGA-3′ R: 5′-AGCTCCTGCAGGGTATTCAAGTGT-3′	NM_001030693
NQO-1	F: 5′-AAGAAGATTGAAGCGGCTGA-3′ R: 5′-GCATGGCTTTCTTCTTCTGG-3′	NM_001277619.1
NFκB	F: 5′- CCAGGTTGCCATCGTGTTCC- 3′ R: 5′- GCGTGCGTTTGCGCTTCT -3′	D13719.1
iNOS	F: 5′-TACTGCGTGTCCTTTCAACG -3′ R: 5′-CCCATTCTTCTTCCAACCTC-3′	U46504
IFN-γ	F: 5′-CTCCCGATGAACGACTTGAG-3′ R: 5′-CTGAGACTGGCTCCTTTTCC-3′	Y07922
IL-1ß	F: 5′-GCTCTACATGTCGTGTGTGATGAG-3′ R: 5′-TGTCGATGTCCCGCATGA-3′	NM_204524
HO-1	F: 5′-GGTCCCGAATGAATGCCCTTG-3′ R: 5′-ACCGTTCTCCTGGCTCTTGG-3′	NM_205344.1
Nrf-2	F: 5′-GGAAGAAGGTGCGTTTCGGAGC-3′ R: 5′-GGGCAAGGCAGATCTCTTCCAA-3′	NM_205117.1
GCLC	F: 5′-CAGCACCCAGACTACAAGCA-3′ R: 5′-CTACCCCCAACAGTTCTGGA-3′	XM_419910.3
GPx	F: 5′- CAGCAAGAACCAGACACCAA-3′ R: 5′- CCAGGTTGGTTCTTCTCCAG-3′	NM_001163245.1
SOD-1	F: 5′- ATTACCGGCTTGTCTGATGG-3′ R: 5′- CCTCCCTTTGCAGTCACATT-3′	NM_205064.155
Claudin-1	F: 5′-GGAGGATGACCAGGTGAAGA-3′ R: 5′-TCTGGTGTTAACGGGTGTGA-3′	NM_001013611.2
MUC-2	F: 5′-GCTACAGGATCTGCCTTTGC-3′ R: 5′-AATGGGCCCTCTGAGTTTTT-3′	JX284122.1
Occludin	F: 5′-GTCTGTGGGTTCCTCATCGT-3′ R: 5′-GTTCTTCACCCACTCCTCCA-3′	NM_205128.1
ZO-1	F: 5′-AGGTGAAGTGTTTCGGGTTG-3′ R: 5′-CCTCCTGCTGTCTTTGGAAG-3′	XM_015278975.1
EAAT3	F: 5′- ACCCCCTTCTGATCACCTCT-3′ R: 5′- TGAGCATGCTGATTCCAAAG-3′	XM_424930.6
FFAR2	F: 5′-GCCCCATAGCAAACTTCT-3′ R: 5′-GGGCAGCCATAAAGAGAG-3′	[15]
GLUT2	F: 5′- CCGCAGAAGGTGATAGAAGC-3′ R: 5′- ACACAGTGGGGTCCTCAAAG-3′	NM_207178.1
SGLT	F: 5′- CATCTTCCGAGATGCTGTCA-3′ R: 5′- CAGGTATCCGCACATCACAC-3′	NM_001293240.1
PEPT-1	F: 5′- CAGGGATCGAGATGGACACT-3′ R: 5′- CACTTGCAAAAGAGCAGCAG-3′	NM_204365.1
KCTD-15	F: 5′-TTAAAAACACCCCGTTCTGC-3′ R: 5′-AAAACAAACCAAGCGACCAC-3′	XM_004944237
Adiponectin	F: 5′-ACTTTCATGGGCTTCCTCCT-3′ R: 5′-GTCCCACGGAAGTCACTTGT-3′	NM_206991
ATGL	F: 5′-CAGCAGGACGTTTGGGTATT-3′ R: 5′-CCACGCAAAGTTGGAGGTAT-3′	EU240627.2
AMPK-α2	F: 5′-GGCATTGAGGAAATCAGGAA-3′ R: 5′-CCTGAACCAATGTGTGTTGC-3′	DQ340396
FAS	F: 5′-GCTGAGAGCTCCCTAGCAGA-3′ R: 5′-TCCTCTGCTGTCCCAGTCTT-3′	NM_205155
FABP4	F: 5′-CAGCATCAATGGTGATGTGA-3′ R: 5′-TCTCTTTGCCATCCCACTTC-3′	NM_204290
CEBPα	F: 5′-GGAGCAAGCCAACTTCTACG-3′ R: 5′-GTCGATGGAGTGCTCGTTCT-3′	NM_001031459
CPT-1	F: 5′-ATCCCAGCTGCAGTGAGTCT-3′ R: 5′-ATTCGCAAGTCAATCCCATC-3′	NM_001012898
IL-6	F: 5′-AGGACGAGATGTGCAAGAAGTTC-3′ R: 5′-TTGGGCAGGTTGAGGTTGTT-3′	NM_204628
PPAR-γ	F: 5′-GATCGCCCAGGTTTGTTAAA-3′ R: 5′-TGCACGTGTTCCGTTACAAT-3′	NM_001001460
PPAR-α	F: 5′-AGGCCAAGTTGAAAGCAGAA-3′ R: 5′-GTCTTCTCTGCCATGCACAA-3′	NM_001001464.1

[1] TNF-α: Tumor necrosis factor alpha; TLR4: Toll-like receptor 4; NQO-1: NADPH dehydrogenase 1; NFκB: Nuclear factor kappa B p 65; iNOS: Inducible nitric oxide synthases; IFN-γ: Interferon-γ; IL-1ß: Interleukin-1ß; HO-1: Heme oxygenase-1; Nrf-2: Nuclear factor erythroid 2–related factor 2; GCLC: Glutamate-cysteine ligase catalytic; Gpx: glutathione peroxidase; SOD-1: Superoxide dismutase-1; MUC-2: Mucin2; ZO-1: Zonula occludens 1; EAAT3: Excitatory amino acid transporter 3; FFAR2: Free fatty acid receptor 2; GLUT2: glucose transporter 2; SGLT: sodium-dependent glucose cotransporters 1; PEPT-1: Peptide transporter 1; KCTD-15: Potassium channel tetramerization domain-containing 15; ATGL: Adipose triglyceride lipase; AMPK-α2: 5′-AMP-activated protein kinase catalytic subunit alpha-2; FAS: Fatty acid synthase; FABP4: Fatty acid binding protein 4; CEBPα: CCAAT-enhancer-binding proteins-alpha; CPT-1: Carnitine palmitoyltransferase I; IL-6: Interleukin-6; PPAR-γ: Peroxisome proliferator-activated receptor gamma; PPAR-α: Peroxisome proliferator-activated receptor alpha.

2.5. Statistical Analysis

All data analysis was calculated using SAS software (SAS® 9.4, 2018, SAS Institute Inc., Cary, NC, USA). The difference between each experimental group was determined by the Tukey range test with a P value less than 0.05. The chemical analysis value present in the unit of "%DM" means analysis value/the percentage of dry matter (DM).

3. Results

3.1. The Characteristics of PWMC

Table 3 reports the levels of the different functional compounds in PWMC, including crude triterpenes, polyphenols, and flavonoids. The total phenol and flavonoids content were 1.84 GAE mg/g DM and 1.20 QE mg/g DM in PWMC. Among the phenol-like compounds, the concentrations of GA, GC, EGC, CC, caffeic acid, EC, EGCG, ECG, and CG in PWMC were 114, 1035, 1493, 7.9, 113, 230, 210, 78, and 401 μg/g DM.

Table 3. The functional chemical composition analysis in PWMC.

Items	PWMC
Functional component analysis	
Crude triterpenes (mg/ g DM [1])	6.25 ± 0.37
Total phenol contents (GAE [2] mg/ g DM)	1.84 ± 0.05
Total flavonoids (QE [3] mg/ g DM)	1.20 ± 0.26
Phenol-like chemical analysis (μg/ g DM)	
Gallic acid	114.0 ± 2.7
Gallocatechin	1035.0 ± 8.0
Epigallocatechin	1493.0 ± 14.0
Catechin	7.9 ± 0.6
Caffeic acid	113.0 ± 0.5
Epicatechin	229.8 ± 2.0
Epigallocatechin gallate	210.2 ± 0.1
Epicatechin gallate	78.1 ± 2.8
Catechin gallate	401.2 ± 5.1

[1] DM: Dry matter; [2] GAE: Gallic acid equivalent; [3] QE: Quercetin. (n = 5).

3.2. Broiler Growth Performance with PWMC Supplementation

Broiler growth performance for days 1–21 and 22–35 are shown in Table 4. Within these two evaluation periods, there was no significant difference in body weight, weight gain, and feed consumption. The best feed conversion rate (FCR) was seen in the 0.5% PWMC group during the starter stage, compared to the control (1.36 vs. 1.28, $p = 0.0374$). However, there were no significant changes in weight gain and FCR for days 22–35 and 1–35.

Table 4. The growth performance of 35-day-old broilers following dietary PWMC [1] supplementation.

Items	Treatments				SEM [2]	p Value
	Con	0.5% PWMC	1% PWMC	2% PWMC		
1–21 d						
Body weight, g/bird	925	956	939	919	30.2	0.848
Weight gain, g/bird	880	907	890	871	30.1	0.853
Feed consumption, g/bird	1197	1161	1184	1219	38.1	0.691
FCR	1.36 [a]	1.28 [b]	1.33 [ab]	1.40[a]	0.03	0.037
22–35d						
Body weight, g/bird	2019	2103	2021	2009	30.7	0.184
Weight gain, g/bird	1091	1148	1081	1089	56.7	0.335
Feed consumption, g/bird	2007	2204	2142	2114	107	0.641
FCR	1.84	1.92	1.98	1.94	0.09	0.742
1–35 d						
Feed consumption, g/bird	3213	3123	3135	3176	30.9	0.190
Weight gain, g/bird	1971	2055	1972	1960	122	0.965
FCR	1.63	1.52	1.59	1.62	0.058	0.588

[1] Con: basal diet for control group; PWMC: *Pennisetum purpureum* Schum No. 2 waste mushroom compost. PWMC was added directly to the basal diet at different percentages. [2] SEM: Standard error of mean. [a,b] Means within the same rows without the same superscript letter are significantly different ($p < 0.05$).

3.3. Intestinal Barrier and Nutrient Absorption-Related mRNA Expression in 35-Day-Old Broilers

There was no significant difference in occludin (*OCCN*) content within each group ($p > 0.05$). However, mucin 2 (*MUC2*), the major protein that maintains the function and structure of mucus, increased about 2.5, 2.5, and 3.5 times in the 0.5%, 1%, and 2% PWMC groups, respectively (Figure 1A, $p < 0.0001$). Both claudin-1 (*CLDN-1*) and zonula occludens 1 (*ZO-1*), which are tight junction-related proteins, increased in the PWMC-supplemented groups, especially 2% PWMC (Figure 1A, $p < 0.0001$ and $=0.0245$, respectively). Among all the groups, only 2% PWMC supplementation enhanced *ZO-1* mRNA expression. The amount of PWMC supplementation was positively correlated to *CLDN1* mRNA expression but there was no significant difference between the 0.5% and 1% PWMC groups.

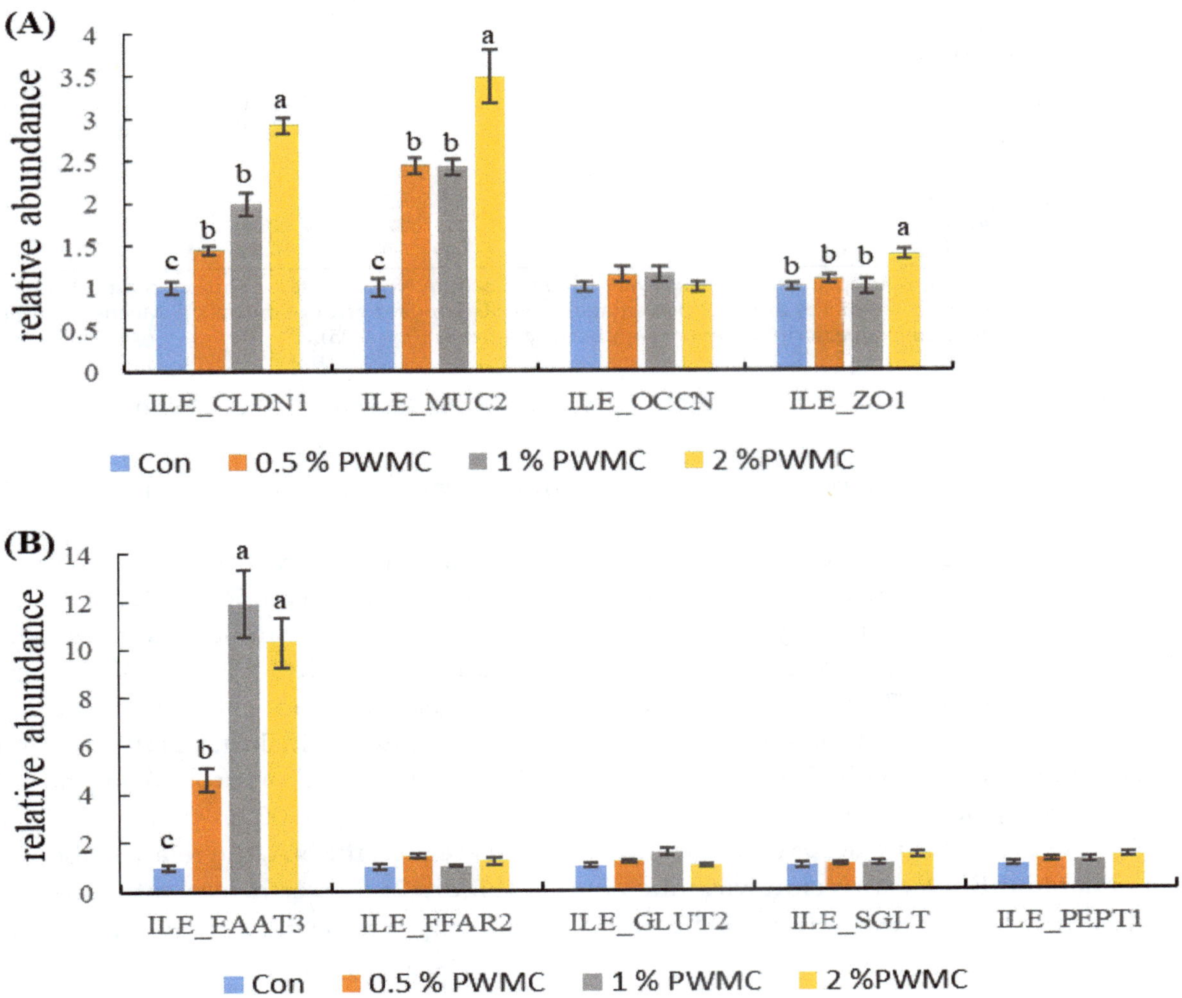

Figure 1. Intestinal barrier (**A**) and nutrient absorption-related (**B**) mRNA expression in the ileum (ILE) for 35-day-old broilers. CLDN: Claudin-1; MUC-2: Mucin2; OCCN: Occludin; ZO-1: Zonula occludens 1; EAAT3: Excitatory amino acid transporter 3; FFAR2: Free fatty acid receptor 2; GLUT2: Glucose transporter 2; SGLT: Sodium-dependent glucose cotransporter 1; PEPT-1: Peptide transporter 1. [a-c] Means within the same rows without the same superscript letter are significantly different ($p < 0.05$).

Most nutrient absorption is through transporters on the intestinal epithelial cells. By adding PWMC to broiler feed, excitatory amino acid transporter 3 (*EAAT3*) levels significantly increased by about 5, 12, and 11 times for the 0.5%, 1%, and 2% PWMC groups, respectively (Figure 1B, $p < 0.0001$). However, the expression of other nutrient absorption-related mRNA, like free fatty acid receptor 2 (*FFAR2*), glucose transporter 2 (*GLUT2*), sodium-dependent glucose cotransporter 1 (*SGLT*), and peptide transporter 1 (*PEPT1*), was not significantly different from the control (Figure 1B).

3.4. *Intestinal Morphology in 35-Day-Old Broilers*

The addition of PWMC did not significantly change the villus height or crypt depth in the jejunum or ileum of 35-day-old broilers, however the *Tunica muscularis* in the jejunum increased significantly by 1.47, 1.34, and 1.06 times in the 0.5%, 1%, and 2% PWMC groups ($p = 0.0104$) (Table 5). Only the 0.5% PWMC group was significantly different from the control in jejunum *Tunica muscularis* ($p = 0.010$), while the 2% PWMC group had a villus:crypt ratio about 1.22 times higher than the control.

Table 5. Jejunum and ileum morphology of 35-day-old broilers following PWMC [1] supplementation.

Items	Treatments				SEM [2]	*p* Value
	Con	0.5% PWMC	1% PWMC	2% PWMC		
Jejunum						
Villus height (μm)	1212	1301	1367	1467	69.9	0.109
Crypt depth (μm)	192	208	203	190	9.37	0.496
Tunica muscularis (μm)	152 ^{b}	224 ^{a}	204 ^{ab}	162^{ab}	22.1	0.010
Villus:crypt	6.32 ^{b}	6.29 ^{b}	6.73 ^{b}	7.75^{a}	0.24	0.002
Ileum						
Villus height (μm)	1166	1110	1102	1038	38.3	0.175
Crypt depth (μm)	171	159	168	158	10.8	0.798
Tunica muscularis (μm)	166	137	144	173	25.6	0.712
Villus:crypt	6.84	6.99	6.58	6.99	0.52	0.932

[1] Con: basal diet for control group; PWMC: *Pennisetum purpureum* Schum No. 2 waste mushroom compost. PWMC was added directly to the basal diet at different percentages. [2] SEM: Standard error of means. [a,b] Means within the same rows without the same superscript letter are significantly different ($p < 0.05$).

3.5. *Adipogenesis- and Adipolysis-Related mRNA Expression in the Liver and Adipocytes of 35-Day-Old Broilers*

The mRNA expression of potassium channel tetramerization domain-containing 15 (*KCTD-15*), adiponectin, adipose triglyceride lipase (*ATGL*), and 5′-AMP-activated protein kinase catalytic subunit alpha-2 (AMPK-α2) were significantly increased in the livers (L) of the treatment groups ($p < 0.0001$, < 0.0001, $=0.0024$ and $= 0.0015$, respectively), thereby increasing the rate of adipolysis (Figure 2A). There was no significant difference among the PWMC-supplemented groups. Similar results are shown in Figure 2C, where the adipolysis-related mRNA, including adiponectin, KCTD15, ATGL, and carnitine palmitoyltransferase 1 (*CPT-1*) (all of the $p < 0.0001$), increased in adipocytes (F) by at least four times. However, the adipogenesis-related mRNA shown in Figure 2B, including fatty acid synthase (*FAS*), fatty acid binding protein 4 (*FABP4*), and CCAAT-enhancer-binding proteins-alpha (*CEBPα*), had a compensatory one to three times increase in the adipocytes ($p < 0.0001$, <0.0001 and $= 0.0052$) (Figure 2D). When comparing adipogenesis- and adipolysis-related mRNA, the relative abundance of adipolysis-related mRNA expression was much higher (Figure 2). Nevertheless, interleukin-6 (IL-6) mRNA expression in adipocytes decreased ($p < 0.0001$), thereby decreasing inflammation.

3.6. *Serum Characteristics in 35-Day-Old Broilers*

As the data shows in Table 6, serum glucose content decreased in the 0.5% and 2% PWMC-supplemented groups ($p < 0.0001$), but there was no significant difference in MDA, SGOT, SGPT, or Alk-P levels. Gpx activity was not significantly different among the different groups ($p = 0.6412$), however SOD, the antioxidant-related enzyme, increased significantly from 523.9 mU/mL to 865.2 mU/mL, 842.6 mU/mL, and 879.3 mU/mL ($p < 0.0001$), respectively, and showed no dose effect. Serum TNF-α levels decreased 1.2 to two times ($p = 0.001$), and adiponectin increased the most in the 1% PWMC group (from 5948 ng/mL to 11820 ng/mL, compared to the control, $p < 0.0001$). OXT content also increased with a dose effect ($p = 0.0016$). The data showed that cholestenone (CHOL) levels, including high density lipoprotein-cholestenone (HDL-C) and low density lipoprotein- cholestenone (LDL-C), were not affected by the addition of PWMC, but triglyceride content decreased significantly ($p < 0.0001$).

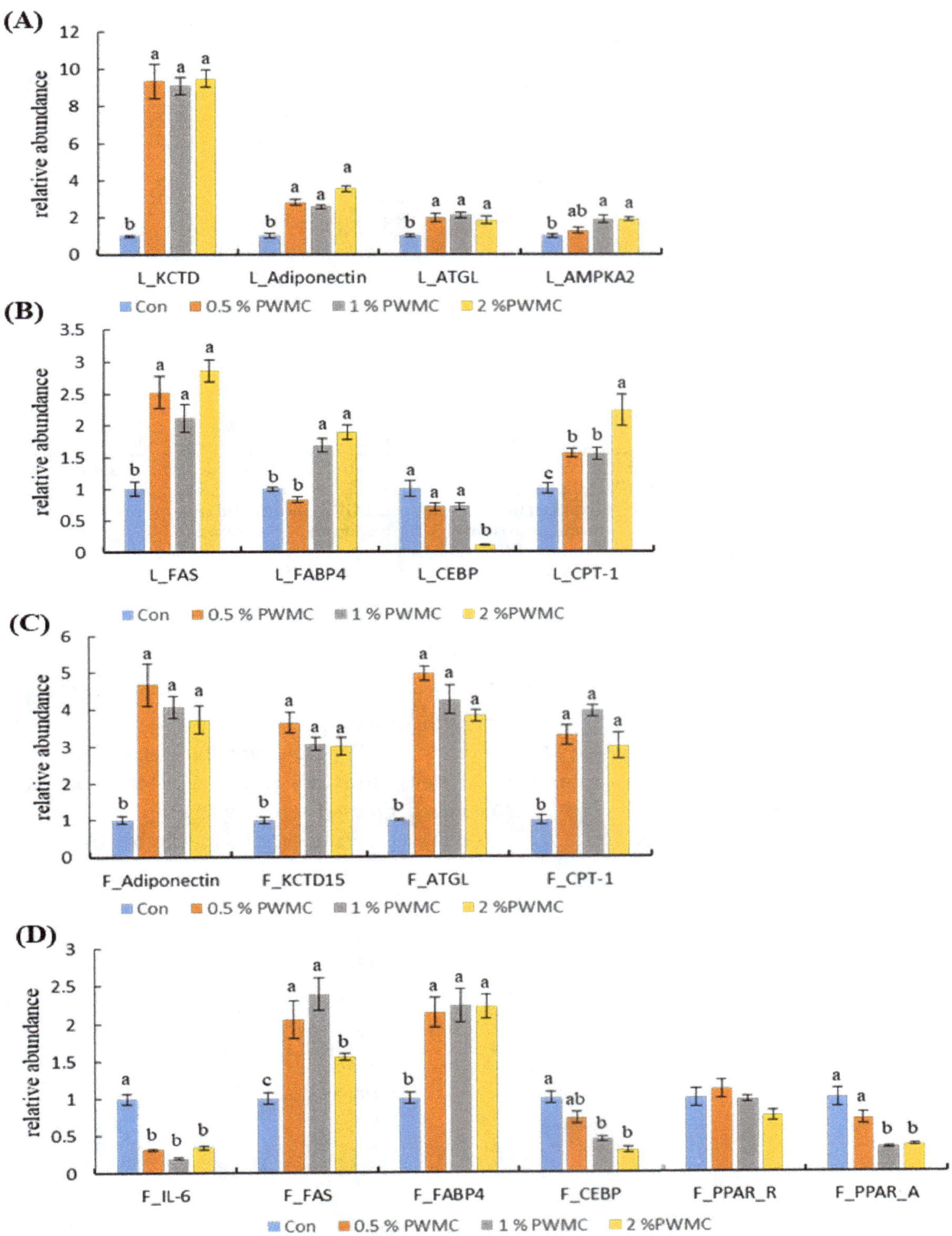

Figure 2. Adipolysis- and adipogenesis-related mRNA expression in the liver (L) and adipocytes (F) of 35-day-old broilers. (**A**) Potassium channel tetramerization domain-containing 15 (KCTD-15), adiponectin, adipose triglyceride lipase (ATGL) and 5′-AMP-activated protein kinase catalytic subunit alpha-2 (AMPK-α2) mRNA expression in L; (**B**) fatty acid synthase (FAS), fatty acid binding protein 4 (FABP4), CCAAT-enhancer-binding proteins-alpha (CEBPα), and carnitine palmitoyltransferase 1 (CPT-1) mRNA expression in L; (**C**) adiponectin, KCTD-15, ATGL, and CPT-1 mRNA expression in F; (**D**) interleukin-6 (IL-6), FAS, FABP4, CEBPα, peroxisome proliferator-activated receptor gamma (PPAR-γ), and peroxisome proliferator-activated receptor alpha (PPAR-α). ^{a,b} Means within the same rows without the same superscript letter are significantly different ($p < 0.05$).

Table 6. Serum characteristics of 35-day-old broilers following PWMC [1] supplementation.

Items [3]	Treatments				SEM [2]	*p* Value
	Con	0.5% PWMC	1% PWMC	2% PWMC		
GLU (mg/dL)	276 [a]	243 [b]	260 [ab]	224 [c]	6.10	<0.001
MDA (uM)	10.1	15.2	13.3	11.4	1.70	0.234
Protein						
SGOT (U/ L)	332	286	273	342	41.0	0.568
SGPT (U/ L)	4.20	4.00	4.50	6.00	0.690	0.184
Alk-P (IU/ L)	1086	937	1366	1282	277	0.694
Gpx (nmol/min/mL)	115	126	129	115	9.59	0.641
SOD (mU/ mL)	524 [b]	865 [a]	843 [a]	879 [a]	20.7	<0.001
TNF-α (pg/mL)	285 [a]	162 [c]	176 [c]	232 [b]	16.6	0.001
Oxytocin (pg/mL)	259 [b]	447 [b]	874 [a]	963 [a]	93.9	0.002
Adiponectin (ng/mL)	5948 [c]	5709 [c]	11820 [a]	7938 [b]	303	<0.001
Corticosterone (pg/mL)	133	121	126	118.2	3.45	0.064
Lipid						
CHOL (mg/dL)	125	124	137	122	7.00	0.379
TG (mg/dL)	54.8 [a]	38.3 [b]	43.7 [ab]	40.2 [b]	2.30	<0.001
HDL-C (mg/dL)	78.0	76.2	83.7	73.2	3.70	0.263
LDL-C (mg/dL)	42.2	42.5	48.0	42.5	3.40	0.577

[1] Con: basal diet for control group; PWMC: *Pennisetum purpureum* Schum No. 2 waste mushroom compost. PWMC was added directly to the basal diet at different percentages. [2] SEM: Standard error of means. [3] GLU: Glucose; MDA: malondialdehyde; SGOT: Serum glutamic-oxaloacetic transaminase; SGPT: Serum glutamic-pyruvic transaminase; Alk-P: Alkaline phosphatase; Gpx: Glutathione peroxidase; SOD: Superoxide dismutase; TNF-α: Tumor necrosis factor alpha; CHOL: Cholestenone; TG: Triglyceride; HDL-C: High density lipoprotein-cholestenone; LDL-C: Low density lipoprotein-cholestenone. [a,b,c] Means within the same rows without the same superscript letter are significantly different ($p < 0.05$).

3.7. Antioxidation and Inflammation-Related mRNA Expression in 35-Day-Old Broilers

When testing for inflammation-related mRNA expression in the spleens of 35-day-old broilers, there were no significant differences in TNF-α, NADPH dehydrogenase 1 (*NQO-1*), nuclear factor kappa B p 65 (*NFκB*), and inducible nitric oxide synthase (*iNOS*) levels ($p > 0.05$). However, the levels of toll-like receptor 4 (*TLR4*), a major lipopolysaccharide receptor, were higher in the 0.5% and 1% PWMC groups ($p = 0.015$). Both interferon-γ (*IFN-γ*) and interleukin-1ß (*IL-1ß*) levels decreased significantly ($p < 0.0001$ and $= 0.0002$, respectively), by about three to five times for *IFN-γ*and 1.1 to 1.6 times for *IL-1ß* (Figure 3A).

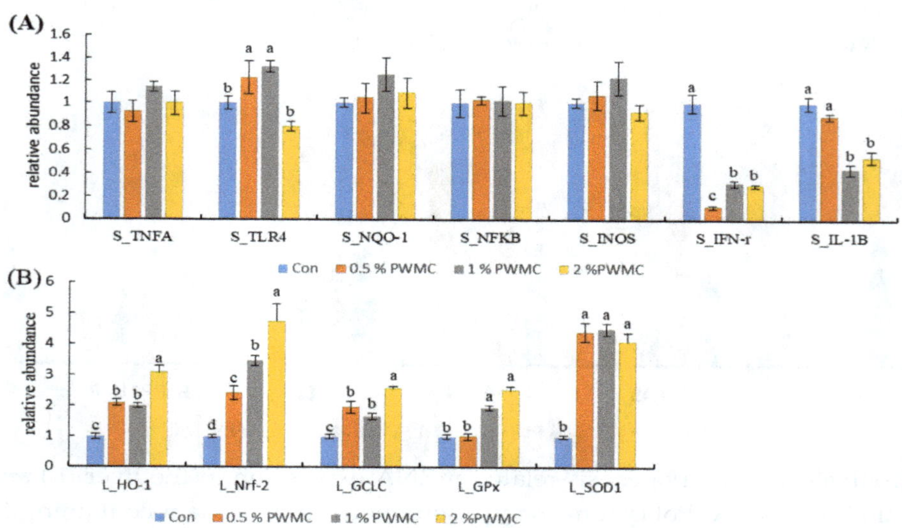

Figure 3. Inflammation- (**A**) and antioxidant-related (**B**) mRNA expression in the spleen (S) and liver (L) of 35-day-old broilers. TNF-α: Tumor necrosis factor alpha; TLR4: Toll-like receptor 4; NQO-1: NADPH dehydrogenase 1; NFκB: Nuclear factor kappa B p 65; iNOS: Inducible nitric oxide synthases; IFN-γ: Interferon-γ; IL-1ß: Interleukin-1ß; HO-1: Heme oxygenase-1; Nrf-2: Nuclear factor erythroid 2–related factor 2; GCLC: Glutamate-cysteine ligase catalytic; Gpx: glutathione peroxidase; SOD-1: Superoxide dismutase-1. [a-d] Means within the same rows without the same superscript letter are significantly different ($p < 0.05$).

The expression of all antioxidant-related mRNA, including heme oxygenase-1 (HO-1), nuclear factor erythroid 2–related factor 2 (Nrf-2), glutamate-cysteine ligase catalytic (GCLC), Gpx, and *SOD-1*, increased significantly following PWMC supplementation (Figure 3B), but while *HO-1*, *Nrf-2*, *GCLC*, and *Gpx* increased with PWMC dosage, *SOD-1* increased equally in each group (Figure 3B).

4. Discussion

Food intake is the only way for animals to gain energy, maintain growth, and survive. Traditionally, due to its low energy content, fiber has not been used in feed [3]. However, research proves that adding fiber to animal feed enhances gut barrier expression, increases antioxidant capacity, and decreases the inflammatory response [16]. There are many polyprotein structures on intestinal epithelial tissue. These structures can block exogenous toxins and bacteria in the intestinal lumen [17]. The immune response is triggered and macrophages are activated when these toxins invade [18]. Due to the complexity of gut microbiota, animals have developed a set of modes that can coexist with microorganisms. Within a single day, animals can adjust the composition of gut microbiota through their diet, further affecting their health and the state of the intestinal barrier [16,17]. Fiber supplementation in animal diets could significantly increase gut barrier function and alter metabolism [16,19]. After PWMC supplementation, the gut barrier in a broiler's ileum increases, by increments that are positively related to the amount of supplement provided. It was also determined that PWMC supplementation increases the expression of *EAAT3*, one of the major transporters of glutamate and cysteine. Furthermore, under the oxidant stress the *EAAT3* expression would decrease in broilers [20]. In this study, high-fiber PWMC supplementation had no negative impact on broiler growth, and even slightly increased FCR performance in the 0.5% PWMC group during the starter stage; there was no significant difference in the other groups.

In animals, the main tissues for adipose synthesis are the liver and adipocytes. CCAAT-enhancer-binding proteins (*CEBP*) and peroxisome proliferator-activated receptor (*PPAR*) are involved in the early differentiation of adipocytes [21]; however, both genes were downregulated in the PWMC-supplemented groups. *FAS* and *FABP4* are mainly involved in the late differentiation of fat cells and the accumulation of fatty oil droplets [22,23]. The former is an upstream gene for fat synthesis and the latter is one of the major proteins that transport fatty acids [22,23]. Therefore, PWMC could suppress adipocyte differentiation by blocking *CEBP* and *PPAR* but would not affect adipose synthesis. Both KCTD-15 and adiponectin promote adipose metabolism, but adiponectin also promotes leptin production [24,25]. It is well-known that leptin stimulates fat metabolism in animals and enhances muscle-cell development [26]. In this study, both mRNA and protein types adiponectin significantly improved with 1% and 2% PWMC supplementation. While both *ATGL* and *CPT-1* enhance adipose metabolism, the former separates glycerol from fatty acids, while the latter enhances fatty acid metabolism [27,28]. The increase of *ATGL* mRNA expression leads to a decrease in the triglyceride (TG) content in the serum of broilers.

PWMC has a high amount of phenolic compounds. Phenolic compounds, which exist in plant-based ingredients have been regarded as a functional antioxidant component for decades [9,29]. Among them, catechins were regarded as one the most functional compounds on both antioxidant and anti-adipogenesis grounds [8,29]. Past research has pointed out that catechins such as EGCG (210 μg/g PWMC) can assist in adipose degradation and improve the antioxidant capacity of animals [29]. Traditionally, OXT has been considered a contractile stimulator of smooth muscle that also helps regulate animal mood [30]. However, recent studies have shown that OXT can also help muscle cells take up glucose, enhance fat metabolism, and promote muscle hyperplasia [31]. In addition, OXT can help solve the problem of fat infiltration caused by obesity [31].

PWMC can enhance adipolysis in both the liver and adipocytes, thus compensating for the gene for adipogenesis. As the increase in adipogenesis-related genes is much lower than adipolysis-related gene expression, the overall adipose content in broilers is reduced. Although adipose is rapidly metabolized, broiler weight was still not significantly different from that of the control. This may be due to the transfer of energy from the breakdown of fat to the production of muscle. Previous research also indicates that dietary fiber supplementation could increase broiler muscle content and the slaughter rate of poultry [32]. However, to our best knowledge, there are only a few studies into the cause of increased muscle weight. Although we only measured adipose metabolism-related mRNA and protein expression in this study, we still confirmed that adding extra fiber to the diet can improve the pattern of adipose metabolism, which may be the reason why high fiber diets can promote muscle mass gains in poultry [19].

Animals may suffer from oxidative stress caused by exogenous reactive oxygen species (ROS) which causes incomplete replication of DNA, and induce mutations or cell lesions [33]. After the mature *Pleurotus eryngii* is removed, the remaining culture medium, also known as PWMC, contains many mycelia. Mushrooms have superb antioxidant capacities and can increase antioxidant capacity in animals [5]. Wang et al. [5,34] indicated that mushroom compost can improve the antioxidant capacity of poultry whether in vivo or *in vitro*. Animal antioxidant systems can be classified as enzymatic or non-enzymatic systems. The former contains enzymes that reduce hydrogen peroxide, such as catalase, and the latter contains small molecules that can directly neutralize ROS, such as glutathione (GSH) [35]. Nrf2 is an upstream antioxidant gene that normally binds to Kelch-like ECH-associated protein 1 in the cytoplasm. When *Nrf2* is activated, Nrf2 enters the nucleus and forms a dimer with the Maf protein. It then reacts with the antioxidant responsive element (ARE) and promotes the expression of downstream antioxidant genes, including *GCLC* and *HO-1* [36,37]. Both *GCLC* and *HO-1* are involved in the antioxidant enzymatic system. *GCLC* is a rate-limiting enzyme in the GSH synthesis pathway [38]. GSH can receive electrons from ROS and become reduced glutathione disulfide (GSSG), which can then be reoxidized by Gpx into GSH. As a catalyst, *HO-1* can indirectly reduce the harm caused by H_2O_2 by cleaving hemoglobin into biliverdin [39]. Previous research determined that phenols have antioxidant capacities and will activate *Nrf2* expression [40], therefore mushroom additives can improve the enzymatic antioxidant system and enhance animals' adaptability to oxidative stress [34]. In addition, as a phenolic compound, epigallocatechin (1.493 mg/g PWMC) can enhance the antioxidant capacity of animals by chelating metal ions or increasing antioxidant enzymes [29]. In addition to ROS, Lee et al. [18] pointed out that inflammation is a double-edged sword. Although inflammation can help animals fight pathogens, it may also increase oxidative stress and cause cell damage [18]. Therefore, it is important to appropriately suppress the inflammatory response under non-disease conditions. *IL-1ß*, one of the most upstream inflammatory factors, is involved in upregulating the expression of inflammatory genes like *iNOS* and *NF-κB*, and may lead to the production of reactive nitrogen species [13]. However, suppressed *IL-1ß* performance would not alter the inflammatory response under normal conditions. *IFN-γ* levels are an indicator of infection. When an animal is infected with a virus or bacteria, *IFN-γ* levels increase and activate *NF-κB*, which further increases inflammation. The glucan content in PWMC might decrease *IFN-γ* expression [13]. Overall, PWMC could increase gut barrier function, antioxidant capacity, and alter the adipose metabolism pattern.

5. Conclusions

Dietary supplementation with PWMC can significantly change the adipose metabolism pattern in broilers and accelerate adipose degradation. Although higher amounts of PWMC could increase intestinal barrier function and antioxidant capacity, supplementation with 0.5% and 2% PWMC resulted in similar levels of adipose metabolism. Furthermore, the 0.5% PWMC group had better FCR. Therefore, to improve health and growth efficiency in broilers, our results suggest that supplementation with 0.5% PWMC is the most effective way to improve fat metabolism and antioxidant capacity. However, further research is recommended to detect the actual mechanism of fat metabolism and antioxidant capacity of PWMC in broilers.

Author Contributions: Conceptualization, W.Y.C. and T.-T.L.; materials analysis, W.Y.C., C.L.L. and H.D.S.; methodology, W.Y.C. and Y.M.S.; supervision, T.-T.L. and C.F.T.; investigation, W.Y.C. and T.-T.L.; writing—original draft preparation, W.Y.C., W.C.L. and T.-T.L.; writing—review and editing, W.Y.C., W.C.L., S.C.C. and T.-T.L. All authors have read and agreed to the published version of the manuscript.

Acknowledgments: The authors thank the Ministry of Science and Technology (MOST 107-2313-B-005-037-MY2) and the iEGG and Animal Biotechnology Center from The Feature Areas Research Center Program within the framework of the Higher Education Sprout Project by the Ministry of Education (MOE) in Taiwan for supporting this study.

Abbreviations

PWMC	*Pennisetum purpureum* schum No.2 waste mushroom compost
PP	*Pennisetum purpureum* schum No.2
OXT	Oxytocin
TNF-α	Tumor necrosis factor alpha
Gpx	Glutathione peroxidase
SOD	Superoxidase dismutase
MDA	Malondialdehyde
TLR4	Toll-like receptor 4
NQO-1	NADPH dehydrogenase 1
NFκB	Nuclear factor kappa B p 65
iNOS	Inducible nitric oxide synthases
IFN-γ	Interferon-γ
IL-1ß	Interleukin-1ß
HO-1	Heme oxygenase-1
Nrf-2	Nuclear factor erythroid 2–related factor 2
GCLC	Glutamate-cysteine ligase catalytic
MUC2	Mucin2
ZO-1	Zonula occludens 1
EAAT3	Excitatory amino acid transporter 3
FFAR2	Free fatty acid receptor 2
GLUT2	Glucose transporter 2
SGLT	Sodium-dependent glucose cotransporters 1
PEPT-1	Peptide transporter 1
KCTD-15	Potassium channel tetramerization domain-containing 15
ATGL	Adipose triglyceride lipase
AMPK-α2	5′-AMP-activated protein kinase catalytic subunit alpha-2
FAS	Fatty acid synthase
FABP4	Fatty acid binding protein 4
CEBPα	CCAAT-enhancer-binding proteins-alpha
CPT-1	Carnitine palmitoyltransferase I
IL-6	Interleukin-6
PPAR-γ	Peroxisome proliferator-activated receptor gamma
PPAR-α	Peroxisome proliferator-activated receptor alpha
FCR	Feed conversion rate
SEM	Standard error of mean
GLU	Glucose
SGOT	Serum glutamic-oxaloacetic transaminase
SGPT	Serum glutamic-pyruvic transaminase
Alk-P	Alkaline phosphatase
CHOL	Cholestenone
TG	Triglyceride
HDL-C	High density lipoprotein-cholestenone
LDL-C	Low density lipoprotein-cholestenone
ROS	Reactive oxygen species
GSH	Glutathione
ARE	Anti-oxidant responsive element
GSSG	Glutathione disulfide

References

1. Al-Harthi, M.A.; Attia, Y.A.; Al-sagan, A.; Elgandy, M.F. Nutrients profile, protein quality and energy value of whole prosopis pods meal as a feedstuff for poultry feeding. *Ital. J. Anim. Sci.* **2018**, *18*, 30–38. [CrossRef]
2. Hemery, Y.; Rouau, X.; Lullien, V.; Barron, C.; Abécassis, J. Dry processes to develop wheat fractions and products with enhanced nutritional quality. *J. Cereal Sci.* **2007**, *46*, 327–347. [CrossRef]
3. Buxton, D.R.; Redfearn, D.D. Plant limitations to fiber digestion and utilization. *J. Nutr.* **1997**, *127*, 8145–8185. [CrossRef] [PubMed]
4. O'Brien, B.J.; Milligan, E.; Carver, J.; Roy, E.D. Integrating anaerobic co-digestion of dairy manure and food waste with cultivation of edible mushrooms for nutrient recovery. *Bioresour. Technol.* **2019**, *285*, 121312. [CrossRef] [PubMed]
5. Wang, C.C.; Lin, L.J.; Chao, Y.P.; Chiang, C.J.; Lee, M.T.; Chang, S.C.; Yu, B.; Lee, T.T. Antioxidant molecular targets of wheat bran fermented by white rot fungi and its potential modulation of antioxidative status in broiler chickens. *Br. Poult. Sci.* **2017**, *58*, 262–271. [CrossRef] [PubMed]
6. Maleko, D.; Mwilawa, A.; Msalya, G.; Pasape, L.; Mtei, K. Forage growth, yield and nutritional characteristics of four varieties of napier grass (Pennisetum purpureum Schumach) in the west Usambara highlands. *Afr. Crop Sci. J.* **2019**, *6*, e00214. [CrossRef]
7. Mahfuz, S.; He, T.; Liu, S.; Wu, D.; Long, S.; Piao, X. Dietary inclusion of mushroom (flammulina velutipes) stem waste on growth performance, antibody response, immune status, and serum cholesterol in broiler chickens. *Animals* **2019**, *9*, 692. [CrossRef]
8. Huang, J.B.; Zhang, Y.; Zhou, Y.B.; Wan, X.C.; Zhang, J.S. Effects of epigallocatechin gallate on lipid metabolism and its underlying molecular mechanism in broiler chickens. *J. Anim. Physiol. Anim. Nutr.* **2015**, *99*, 719–727. [CrossRef]
9. Lee, T.T.; Ciou, J.Y.; Chiang, C.J.; Chao, Y.P.; Yu, B. Effect of pleurotus eryngii stalk residue on the oxidative status and meat quality of broiler chickens. *J. Agric. Food Chem.* **2012**, *60*, 11157–11163. [CrossRef]
10. Tabart, J.; Kevers, C.; Pincemail, J.; Defraigne, J.O.; Dommes, J. Comparative antioxidant capacities of phenolic compounds measured by various tests. *Food Chem.* **2009**, *113*, 1226–1233. [CrossRef]
11. Pourmorad, F.; Hosseinimehr, S.J.; Shahabimajd, N. Antioxidant activity, phenol and flavonoid contents of some selected Iranian medicinal plants. *Afr. J. Biotechnol.* **2006**, *5*, 1142–1145.
12. Wei, L.; Zhang, W.; Yin, L.; Yan, F.; Xu, Y.; Chen, F. Extraction optimization of total triterpenoids from Jatropha curcas leaves using response surface methodology and evaluations of their antimicrobial and antioxidant capacities. *Electron. J. Biotechnol.* **2015**, *18*, 88–95. [CrossRef]
13. Chuang, W.Y.; Lin, W.C.; Hsieh, Y.C.; Huang, C.M.; Chang, S.C.; Lee, T.T. Evaluation of the combined use of Saccharomyces Cerevisiae and Aspergillus Oryzae with phytase fermentation products on growth, inflammatory, and intestinal morphology in broilers. *Animals* **2019**, *9*, E1051. [CrossRef] [PubMed]
14. AOAC. *Official Methods of Analysis of AOAC International*, 19th ed.; AOAC International: Gaithersburg, MD, USA, 2012.
15. Meslin, C.; Desert, C.; Callebaut, I.; Djari, A.; Klopp, C.; Pitel, F.; Leroux, S.; Martin, P.; Froment, P.; Guilbert, E. Expanding duplication of free fatty acid receptor-2 (GPR43) genes in the chicken genome. *Genome Biol. Evol.* **2015**, *7*, 1332–1348. [CrossRef] [PubMed]
16. Dos Santos, T.T.; Baal, S.C.S.; Lee, S.A.; E Silva, F.R.O.; Scheraiber, M.; da Silva, A.V.F. Influence of dietary fibre and betaine on mucus production and digesta and plasma osmolality of broiler chicks from hatch to 14 days of age. *Livest. Sci.* **2019**, *220*, 67–73. [CrossRef]
17. Barekatain, R.; Chrystal, P.V.; Howarth, G.S.; McLaughlan, C.J.; Gilani, S.; Nattrass, G.S. Performance, intestinal permeability, and gene expression of selected tight junction proteins in broiler chickens fed reduced protein diets supplemented with arginine, glutamine, and glycine subjected to a leaky gut model. *Poult. Sci.* **2019**, *98*, 6761–6771. [CrossRef] [PubMed]
18. Lee, M.T.; Lin, W.C.; Lee, T.T. Potential crosstalk of oxidative stress and immune response in poultry through phytochemicals—A review. *Asian-Australas. J. Anim. Sci.* **2019**, *32*, 309–319. [CrossRef]
19. Al-Harthi, M.A.; Attia, Y.A.; Al-sagan, A.; Elgandy, M.F. The effects of autoclaving or/and multi-Enzymes complex supplementation on performance, egg quality and profitability of laying hens fed whole Prosopis juliflora pods meal diet. *Europ. Poult. Sci.* **2018**, *82*. [CrossRef]

20. Ebrahimi, R.; Jahromi, M.F.; Liang, J.B.; Farjam, A.S.; Shokryazdan, P.; Idrus, Z. Effect of Dietary Lead on Intestinal Nutrient Transporters mRNA Expression in Broiler Chickens. *Biomed Res. Int.* **2015**, *1*, 149745. [CrossRef]
21. Barbosa-da-Silva, S.; Souza-Mello, V.; Magliano, D.C.; Marinho, T.S.; Aguila, M.B.; Mandarim-de-Lacerda, C.A. Singular effects of PPAR agonists on nonalcoholic fatty liver disease of diet-Induced obese mice. *Life Sci.* **2015**, *127*, 73–81. [CrossRef]
22. Wang, Q.; Guan, T.; Li, H.; Bernlohr, D.A. A novel polymorphism in the chicken adipocyte fatty acid-Binding protein gene (FABP4) that alters ligand-Binding and correlates with fatness. *Comp. Biochem. Physiol. B* **2009**, *154*, 298–302. [CrossRef] [PubMed]
23. De Silva, G.S.; Desai, K.; Darwech, M.; Naim, U.; Jin, X.; Adak, S.; Harroun, N.; Sanchez, L.A.; Semenkovich, C.F.; Zayed, M.A. Circulating serum fatty acid synthase is elevated in patients with diabetes and carotid artery stenosis and is LDL-Associated. *Atherosclerosis* **2019**, *287*, 38–45. [CrossRef] [PubMed]
24. Liang, S.S.; Ouyang, H.J.; Liu, J.; Chen, B.; Nie, Q.H.; Zhang, X.Q. Expression of variant transcripts of the potassium channel tetramerization domain-Containing 15 (KCTD15) gene and their association with fatness traits in chickens. *Domest. Anim.* **2015**, *50*, 65–71. [CrossRef] [PubMed]
25. Tahmoorespur, M.; Ghazanfari, S.; Nobari, K. Evaluation of adiponectin gene expression in the abdominal adipose tissue of broiler chickens: Feed restriction, dietary energy, and protein influences adiponectin messenger ribonucleic acid expression. *Poult. Sci.* **2010**, *89*, 2092–2100. [CrossRef] [PubMed]
26. Murase, D.; Namekawa, S.; Ohkubo, T. Leptin activates chicken growth hormone promoter without chicken STAT3 in vitro. *Comp. Biochem. Physiol. B* **2016**, *191*, 46–52. [CrossRef] [PubMed]
27. Habets, D.D.J.; Coumans, W.A.; Hasnaoui, M.E.; Zarrinpashneh, E.; Bertrand, L.; Viollet, B.; Kiens, B.; Jensen, T.E.; Richter, E.A.; Bonen, A.J.; et al. Crucial role for LKB1 to AMPKα2 axis in the regulation of CD36-Mediated long-Chain fatty acid uptake into cardiomyocytes. *BBA-Mol. Cell Biol. L.* **2009**, *1791*, 212–219. [CrossRef]
28. Andler, S.M.; Wang, L.S.; Goddard, J.M.; Rotello, V.M. Chapter One-Preparation of Biocatalytic Microparticles by Interfacial Self-Assembly of Enzyme–Nanoparticle Conjugates Around a Cross-Linkable Core. *Methods Enzymol.* **2016**, *516*, 1–17.
29. Yan, J.; Zhao, Y.; Suo, S.; Liu, Y.; Zhao, B. Green tea catechins ameliorate adipose insulin resistance by improving oxidative stress. *Free Radic. Biol. Med.* **2012**, *52*, 1648–1657. [CrossRef]
30. Arrowsmith, S. Oxytocin and vasopressin signaling and myometrial contraction. *Curr. Opin. Physiol.* **2020**, *13*, 62–70. [CrossRef]
31. Iwasa, T.; Matsuzaki, T.; Mayila, Y.; Yanagihara, R.; Yamamoto, Y.; Kawakita, T.; Kuwahara, A.; Irahara, M. Oxytocin treatment reduced food intake and body fat and ameliorated obesity in ovariectomized female rats. *Neuropeptides* **2019**, *75*, 49–57. [CrossRef]
32. Nassara, M.K.; Lyua, S.; Zentekc, J.; Brockmann, G.A. Dietary fiber content affects growth, body composition, and feed intake and their associations with a major growth locus in growing male chickens of an advanced intercross population. *Livest. Sci.* **2019**, *227*, 135–142. [CrossRef]
33. Poulsen, H.E. Oxidative DNA modifications. *Exp. Toxicol. Pathol.* **2005**, *57*, 161–169. [CrossRef] [PubMed]
34. Wang, C.C.; Chang, C.H.; Chang, S.C.; Fan, G.J.; Lin, M.J.; Yu, B.; Lee, T.T. In vitro free radicals scavenging activity and antioxidant capacity of solid-state fermented wheat bran and its potential modulation of antioxidative molecular targets in chicken PBMC. *R. Bras. Zootec.* **2016**, *45*, 451–457. [CrossRef]
35. Jung, K.A.; Kwak, M.K. The Nrf2 System as a potential target for the development of indirect antioxidants. *Molecules* **2010**, *15*, 7266–7291. [CrossRef] [PubMed]
36. Dickinson, D.; Levonen, A.L.; Moellering, D.R.; Arnold, E.K.; Zhang, H.; Darley-Usmer, V.M.; Forman, H.J. Human glutamate cysteine ligase gene regulation through the electrophile response element. *Arch. Biochem. Biophys.* **2004**, *423*, 116–125. [CrossRef]
37. Sahin, K.; Orhan, C.; Smith, M.O.; Sahin, N. Molecular targets of dietary phytochemicals for the alleviation of heat stress in poultry. *Poult. Sci.* **2013**, *69*, 113–124.
38. Teskey, G.; Abrahem, R.; Cao, R.; Gyurjian, K.; Islamoglu, H.; Lucero, M.; Martinez, A.; Paredes, E.; Salaiz, O.; Robinson, B.; et al. Glutathione as a marker for human disease. *Adv. Clin. Chem.* **2018**, *87*, 141–159.

39. Araujo, J.A.; Zhang, M.; Yin, F. Heme oxygenase-1, oxidation, inflammation, and atherosclerosis. *Front. Pharmacol.* **2012**, *3*, 1–17. [CrossRef]

40. Arredondo, F.; Echeverry, C.; Abin-Carriquiry, J.A.; Blasina, F.; Antunez, K.; Jones, D.P.; Go, Y.M.; Liang, Y.L. After cellular internalization, quercetin causes Nrf2 nuclear translocation, increases glutathione levels, and prevents neuronal death against an oxidative insult. *Free Radic. Biol. Med.* **2010**, *49*, 738–747. [CrossRef]

5

Effects of the Diet Inclusion of Common Vetch Hay Versus Alfalfa Hay on the Body Weight Gain, Nitrogen Utilization Efficiency, Energy Balance, and Enteric Methane Emissions of Crossbred Simmental Cattle

Wuchen Du [1], Fujiang Hou [2,*], Atsushi Tsunekawa [3,*], Nobuyuki Kobayashi [3], Toshiyoshi Ichinohe [4] and Fei Peng [5,6]

1 The United Graduate School of Agricultural Sciences, Tottori University, Tottori 680-8550, Japan; duwuchen1990@126.com
2 State Key Laboratory of Grassland Agro-Ecosystems, Key Laboratory of Grassland Livestock Industry Innovation, Ministry of Agriculture, College of Pastoral Agriculture Science and Technology, Lanzhou University, Lanzhou 730020, China
3 Arid Land Research Center, Tottori University, Tottori 680-0001, Japan; kobayashi.nobuyuki@alrc.tottori-u.ac.jp
4 Faculty of Life and Environmental Science, Shimane University, Matsue 690-8504, Japan; toshi@life.shimane-u.ac.jp
5 International Platform for Dryland Research and Education, Tottori University, Tottori 680-0001, Japan; pengfei@tottori-u.ac.jp
6 Key Laboratory of Desert and Desertification, Northwest Institute of Eco-Environment and Resources, Chinese Academy of Sciences, Lanzhou 730000, China
* Correspondence: cyhoufj@lzu.edu.cn (F.H.); tsunekawa@tottori-u.ac.jp (A.T.).

Simple Summary: Nitrogen utilization efficiency and enteric methane emission from ruminants remain the primary concerns when developing ruminant feed globally. Nitrogen utilization efficiency is the ratio of retained nitrogen in body tissue to the total nitrogen intake, which is the main factor in the body weight gain of ruminants, and usually range from 15% to 40%. The methane emissions of ruminants are an inevitable by-product when feeds have been fermented in the rumen and represents a 2% to 12% loss of diet energy. The low nitrogen utilization of ruminants can damage air quality and lead to soil nitrification and acidification, whereas high methane emissions from ruminants can increase global warming. Our study investigated the effects of two kinds of legumes (alfalfa and common vetch) with different levels (20% vs. 40%) of total dry matter allowance on body weight gain, nutrient digestibility, nitrogen utilization efficiency, and enteric methane emissions for crossbred Simmental cattle. Our results suggested that nitrogen utilization efficiency and methane emissions are significantly affected by the legume species and proportions. These results could be beneficial for the development of regional or national ruminant feeding systems, thereby improving nitrogen utilization efficiency and reducing methane emissions.

Abstract: A low nitrogen utilization efficiency (NUE, the ratio of retained N to N intake) and high methane (CH_4) emissions of ruminants can lead to potentially high diet protein wastage and directly contribute to global warming. Diet manipulation is the most effective way to improve NUE or reduce CH_4 emissions. This study investigated how replacing oat hay with alfalfa hay (AH) or common vetch hay (CVH) with different proportions (20% (20) and 40% (40) of the total dry matter (DM) allowance) affects the body weight gain (BWG), NUE, and CH_4 emissions of crossbred Simmental cattle. The forage dry matter intake (DMI) and the total DMI of cattle fed on a CVH40 diet were significantly higher than the values for those fed on AH20 or AH40 diets ($p < 0.05$). There were no differences in the BWG for the four treatments observed, however, nutrient digestibility significantly decreased in the AH40 diet as compared with the AH20 diet ($p < 0.05$). The NUE was significantly lower in AH40 than in CVH20. The CH_4 emissions were significantly lower for the CVH40 diet than with the AH20 diet ($p < 0.05$). Our findings suggest that a 20% AH and 40% CVH substitution for oat hay are the optimal proportions to maintain the BWG, NUE, nutrient digestibility, and reduce the CH_4 emissions of crossbred Simmental cattle. Overall, CVH has a greater potential to reduce CH_4 emissions than AH.

Keywords: leguminous forage; digestibility; energy utilization efficiency; nitrogen metabolism; dryland

---

1. Introduction

The impacts of the low nitrogen utilization efficiency (NUE) and high enteric methane (CH_4) emissions of beef cattle remain the primary concerns in the development of ruminant feeding systems [1]. A low NUE could contribute more ammonia emissions to the air and more manure N outputs to the soil [2], which could damage air quality [3] and lead to soil nitrification and acidification [2]. The enteric CH_4 emissions from ruminants not only represent a loss of diet energy [4] but could also contribute to global warming [5]. The development of a diet that can improve the NUE and reduce enteric CH_4 emissions is in demand and would be beneficial to both animal husbandry and in facing global environmental challenges [6,7].

Grass occupies an important role in the ruminant feeding system as it represents a low-cost and abundant source of dry matter (DM). However, grass only is not capable of sustaining the required levels of animal production due to its low feeding value [8]. Hence, the interest in supplementing legumes into a grass-based diet because they are rich in protein and energy [9]. Previous studies in sheep have shown that the intake of organic matter (OM) and crude protein (CP), as well as ruminal ammonia nitrogen (N) concentrations increased with a 3:1 grass/legume mixture diet as compared to a diet of only grass [10]; the total tract digestibility of CP and digestible CP was significantly higher in a 1:1 alfalfa/oat mixture diet than oat only hay diet [11]. Moreover, alfalfa (78%) and grass (22%) pastures could reduce energy loss through CH_4 emissions of cows as compared to grass-only pastures [12]. However, these studies focused on diets where supplementation of legumes was the only factor considered. Few studies explained the effects of diets with different levels of legumes on feed intake, digestibility, and CH_4 emissions. In another study, the inclusion of 30% common vetch hay (CVH) was more optimal in reducing CH_4 emissions than 0%, 10%, and 20% CVH diets but significantly depressed digestibility as compared to a 20% CVH diet [13]. Recently, Kobayashi et al. [14,15] concluded that 8% to 14% alfalfa hay (AH) in the warm season and 8% to 21% in the cool season were optimal when considering body weight gain (BWG), metabolizable energy (ME) intake, and increased economic benefits of growing beef cattle on a corn- and straw-based diet.

Alfalfa (*Medicago sativa* L.) is the mostly widely planted perennial legume crop in the world and has been studied for many years [12,14]. Common vetch (*Vicia sativa* L.), a multipurpose annual cereal legume for livestock feed [16], not only plays an important role in dryland mixed farming systems [16] for grazing [17] or cutting for hay [18] but also meets the structural forage deficit in winter, which is linked to the seasonality of other feed sources [9]. Previous studies have shown that CP, digestible OM intake, and in vitro OM digestibility are significantly higher with the oat and common vetch mixture diet than with the oat-only diet for cattle [18], and the growth performance of animals is significantly higher with common vetch supplementation than without [19]. However, until now, there has been no available information on whether common vetch could substitute alfalfa in the ruminant feeding system or whether optimal proportions of common vetch could replace alfalfa. Therefore, the objective of this study was to investigate how CVH versus AH affects BWG, N metabolism (i.e., N digestibility, ruminal ammonia-N, and blood urea N (BUN) concentrations), and CH_4 emissions associated with ruminal fermentation parameters using two different proportions (20% (20) and 40% (40) of the total DM allowance) for growing crossbred Simmental cattle in dryland environments at similar CP and predicted ME levels.

2. Materials and Methods

The Animal Ethics Committee of Gansu Province, China, approved the experimental protocols (file No. 2010-1 and 2010-2). This study was conducted in Linze Grassland Agriculture Trial Station, Lanzhou University, Zhangye City, Gansu Province, China (latitude 39.24°N, longnitude 00.06°E, 1390 m a.s.l.), which is characterized as a typical temperature continental climate because its average annual precipitation is 121.5 mm and annual average temperature is 7.7 °C. In this study, the AH was second cut, and common vetch (*Vicia sativa* L.cv. Lanjian No. 3) was harvested at the flowering stage and restored as common vetch hay (CVH). Oat hay (OH; *Avena sativa* L.) was purchased from a forage company (Sanbao Agricultural Company, Zhangye, Gansu, China). The ingredients for the concentrate (maize, soybean meal, and wheat bran) were acquired from a local source. The chemical composition of the forage and ingredients of the concentrate are shown in Table 1.

Table 1. Chemical composition of alfalfa hay, oat hay, common vetch hay, and ingredients of the concentrate used in the experimental diets.

Item †	Alfalfa Hay	Oat Hay	Common Vetch Hay	Soybean Meal	Wheat Bran	Maize
OM, g/kg DM	905	942	918	935	931	983
CP, g/kg DM	168	60	177	465	182	83
NDF, g/kg DM	458	559	413	166	454	100
ADF, g/kg DM	347	407	302	102	186	20
Ether extract, g/kg DM	22	18	23	26	55	44
GE, MJ/kg DM	17.9	16.8	17.7	19.6	19.4	18.5
MEC §, MJ/kg DM	8.7	9.0	9.5	13.0	10.9	13.4
MPC ¶, g/kg DM	62	68	71	87	73	90

† OM, organic matter; CP, crude protein; NDF, neutral detergent fiber; ADF, acid detergent fiber; GE, gross energy; MEC, metabolizable energy concentration; MPC, metabolizable protein concentration; §, ¶ They were calculated by the Agricultural and Food Research Council (1993) and the Chinese Feeding Standard for Beef Cattle (2004), see details in Methods and Materials.

2.1. Animals, Treatments, and Diets

The Animal Ethics Committee of Gansu Province, China, approved the experimental protocols. This experiment involved 16 crossbred male Simmental cattle (Simmental × local cattle) with initial body weights (BWs) of 216 ± 24.4 kg (mean ± standard deviation, 10 months of age) at the start of the experimental period. The experiment used a randomized block experimental design with a 2 × 2 factorial arrangement of diets. All 16 cattle were allocated to one of the 4 treatments. The forage to concentrate ratio was fixed (60:40, DM basis) for all diets. Diet treatments used two kinds of legumes (AH and CVH) and two different OH-to-AH/CVH ratios in the diet (40:20 or 20:40, DM basis), indicated as follows: 20% CVH and 40% OH (CVH20), 40% CVH and 20% OH (CVH40), 20% AH and 40% OH (AH20), and 40% AH and 20% OH (AH40). This experiment consisted of 2 feeding periods. Each period consisted of a 14 day diet adaptation in the cowshed and 32 day data collection period in the chambers.

The target BWG for each cattle was set at 1.5 kg/d. All experimental diets were formulated to provide sufficient ME and metabolizable protein (MP) to meet the target BWG for cattle according to the published estimation equations and values of the Agricultural and Food Research Council [20] and BW of cattle (measured every 8 days). The diet composition required to fulfill the ME and MP requirements was calculated based on the tabulated values of digestible energy and ruminal CP degradation parameters for OH, AH, and the concentrate ingredients established by the Chinese

Feeding Standard for Beef Cattle [21]. The digestibility of ruminal CP and energy and ruminal degradation parameters for CVH were taken from Larbi et al. [16]. The CP, ME, and MP levels of all diets are shown in Table 2. Throughout this experimental period of 8 weeks, all cattle were given free access to water and 10 g/day of mineral mixture containing (minimum values in mg) manganese, 720; copper, 30; biotin, 0.05; folic acid, 0.4; vitamin B_1, 50; vitamin B_2, 2.5; vitamin B_6, 0.5; and vitamin B_{12}, 0.1. The daily mixed forage was divided into two equal parts and offered as separate meals twice a day (08:00 and 19:00). The mixed concentrate was fed once a day (14:00).

Table 2. Composition of the feed ingredients and the target metabolizable energy concentration and metabolizable protein concentration of all diets.

Feed Formula	Experimental Diet †			
	CVH20	CVH40	AH20	AH40
Forage				
Leguminous forage (g/kg DM)	200	400	200	400
Oat hay (g/kg DM)	400	200	400	200
Concentrate				
Maize (g/kg DM)	30	80	48	120
Soybean meal (g/kg DM)	92	25	107	56
Wheat bran (g/kg DM)	278	295	245	224
Nutrient value ‡				
CP (g/kg DM)	156.3	156.4	156.4	156.4
MEC § (MJ/kg DM)	10.05	10.05	10.05	10.05
MPC ¶ (g/kg DM)	102.9	94.6	106.1	101.4

† CVH20, 20% common vetch + 40% oat hay; CVH40, 40% common vetch + 20% oat hay; AH20, 20% alfalfa + 40% oat hay; AH40, 40% alfalfa + 20% oat hay. ‡ CP, crude protein, MEC, metabolizable energy concentration, MPC, metabolizable protein concentration. §,¶ These values were calculated by the Agricultural and Food Research Council (1993) and the Chinese Feeding Standard for Beef Cattle (2004); see details in Methods and Materials.

2.2. Chamber Description

The four indirect open-respiration calorimeter chambers used in the present study were equipped with a computer-controlled air-handling system with air conditioning units set to a temperature of 18 ± 1 °C and relative humidity of 60% ± 10%. The calorimeter chambers were built with double Perspex walls fitted in aluminum frames [22], with a total volume of approximately 18 m^3 (4.2 m long, 1.95 m wide, and 2.2 m high). Each chamber was equipped with a gas flow meter (GFM57, Aalborg, Orangeburg, New York, NY, USA) at the outflow site for recording the total airflow and an engine to ensure a slight negative pressure within the chamber. All chambers were ventilated by suction pumps with a flow rate of 45 to 50 m^3/h. The exhaust air was removed for volume, temperature, and humidity measurement and analysis in the bottom, middle, and upper areas, inside each chamber. The concentrations of CO_2, CH_4, and O_2 in the air moving into and out of each chamber were measured every 16 min (the interval for each chamber) using a multigas analyzer (VA-3000, Horiba Ltd., Tokyo, Japan) in a general control room. The analyzer was calibrated using standard gases (O_2-free N_2 and a known quantity of CH_4, CO_2 and O_2, Dalian Special Gases Co., Ltd., Liaoning, China) at the beginning of the gas exchange collection period in each experiment. The determined concentrations were in an absolute range of 0–500 μL/L for CH_4, 0–2000 μL/L for CO_2, and 0%–25% (*v*/*v*) for O_2 (with linearity within this range). The recovery rate of CH_4 was determined by comparing the CH_4 release into the chamber with a given concentration as well as the CH_4 concentration at the outlet. The gas recovery rate was approximately 100% ± 2% for all chambers, as highlighted recently by Gerrits et al. [23]. Each chamber was designed with a dedicated door, which was located next to the animal trough. The staff only opened the door to feed the animal immediately after the completion of each data collection in the chamber during the 3 day gas exchange data collection period. This minimized the effects of feeding activity (less than 1 min) on the gas concentrations inside. The CH_4 emissions were expressed as the average CH_4 emissions (g/day) from 3-day measurements divided by BWG, which was calculated from the BW change between moving in and moving out the chamber.

2.3. *Energy Balance*

ME intake (MEI) was calculated as the difference between GEI, excreted fecal energy (FE), and the sum of UE and CH_4 energy (CH_4-E) output. Retained energy (RE) was calculated using the equation MEI − heat production (HP). CH_4-E was calculated from CH_4 emissions (L/day) and the conversion coefficient (39.54 kJ/L; [24]). The CH_4 emissions was converted to grams from the CH_4 emissions (L/day) using the conversion coefficient (0.716 g/L, [24]). HP (kJ/day) was calculated with the following equation [24]:

$$\begin{aligned}\text{HP (kJ/day)} =\ & 16.18 \times \text{O}_2 \text{ consumption (L/day)} + 5.02 \times \text{CO}_2 \text{ production (L/day)} \\ & - 2.17 \times \text{CH}_4 \text{ production (L/day)} - 5.99 \times \text{N excretion (urinary N, g/day)}\end{aligned}$$

2.4. *Sample Collection and Procedures*

The amount of offered forage and concentrate and all leftovers was weighed daily throughout the experimental period to calculate the daily DM intake (DMI) for individual cattle. On day 15 of the experimental period, after the 14 day acclimation period for target feeds, one cattle was randomly selected from each diet group and moved to one of the four chambers for 8 days. On day 22, these cattle were moved to the individual pens in the cowshed, and another 4 cattle, randomly selected from the remaining cattle of the four diet groups, entered the chambers and left on day 30. This process continued until day 46 for the first feeding period, when all 16 cattle had completed 8 days of measurement. These acclimation and chamber measurements for metabolism and gas exchange were repeated for another 46 days with the 16 cattle randomly allocated to the four diets. The BWs of all cattle were measured in the morning with an empty stomach to calculate CH_4 emissions, energy, and N balance based on the metabolic BW when exchanging cattle between the chambers and the cowshed. The BWG (kg/day) was calculated by the difference of the BW at the start and end time of each feeding period. During the 8 days of measurements in the chamber, the cattle were kept for acclimation for the first 2 days. We collected the digestibility data over the following 3 days and gas exchange data (O_2 consumption, CH_4, and CO_2 emission) over the remaining 3 days. During the digestibility data collection period, the total weight of the daily excreted feces and urine was recorded. Feces, which were excreted onto a plastic mat placed under the cattle, were collected right after excretion with a shovel and placed into a plastic container (around 15 times a day but varied according to individuals) and weighed, mixed, and sampled once per day. A total of 10% of each feces sample was stored at −20 °C for later chemical analysis. All urine was collected once a day through a handmade urine bag into a bucket containing 200 mL 10% (*v*/*v*) H_2SO_4 to reduce ammonia loss. Acidified urine was checked for pH with a portable pH instrument (PHBJ-260, Shanghai INESA Scientific Instrument Co., Ltd., Shanghai, China). A total of 20% of the daily urine was removed with a 500 mL cylinder (deviation ± 5 mL) and stored at −20 °C for chemical analysis.

Rumen fluid samples were taken from each cattle 4 h post forage supply every morning using stomach tubing on the last 2 days of each feeding period after these cattle were moved to the cowshed. The collected samples were immediately measured for pH using a portable pH meter (PHBJ-260, Shanghai INESA Scientific Instrument Co., Ltd., Shanghai, China), strained through two layers of muslin (mesh size 1 mm^2) and stored at −20 °C for volatile fatty acid (VFA) analysis. An additional 1 mL of strained rumen fluid was deproteinated by adding 0.2 mL metaphosphoric acid (215 g/L) and 0.1 mL internal standard (Crotonic acid), and the VFA concentrations were determined by a gas chromatograph (Trace1300, Thermo Ltd., Rodano Milan, Italy) fitted with a polar capillary column. The plasma urea N concentration was assumed to be equivalent to the BUN concentration in the serum, since urea diffuses freely into and out of blood cells [25].

2.5. Chemical Analysis

After the chamber measurement, the stored feces samples were thawed at room temperature for 12 h, and the feces samples from each cattle over the three days were then mixed. A portion of the thawed feces sample was used for the N measurement according to the Association of Official Analytical Chemists methods, method 976.05 [26]. The CP concentration was calculated by multiplying the N concentration by 6.25. The remaining samples were oven dried at 65 °C for 48 h and then ground to pass through a 1 mm screen. A portion of each dried sample, mixed forage, and concentrate samples were used to measure ash by combustion using a muffle furnace at 550 °C for 10 h (method 942.05 [26]). The organic matter (OM) content (g/kg DM) was calculated by 1000 ash content (g/kg DM). Another portion of each dried sample was finely ground to measure gross energy (GE), neutral detergent fibers (NDFs), and acid detergent fibers (ADFs). The GE of the samples was determined with an automatic isoperibol calorimeter (6400, PARR Instrument Company, Moline, IL, USA). The NDF and ADF concentrations were analyzed sequentially in an ANKOM 2000 fiber analyzer (ANKOM Technology, Fairport, NY, USA) following the protocol described by Van Soest [27]. The ash was included to facilitate the NDF and ADF analysis of all the forage, concentrate, and feces samples. The α-amylase for NDF analysis was used only for the concentrate samples. The urine samples from each cattle over the three days were also thawed at room temperature for 12 h and then mixed before determining their urinary energy (UE) by using an automatic isoperibol calorimeter (see above), and N was determined using Kjeldahl procedure described previously by the Association of Official Analytical Chemists [26]. For the UE measurement, 4 mL fully mixed urine was taken and absorbed by a filter paper of a known weight, and then the total energy of the filter paper with a urine sample was measured by an automatic isoperibol calorimeter after it became dry at room temperature. There were another 5 samples using the same filter paper (known weight) to be measured for energy content (MJ/kg), which was used to calculate the UE. The measurements of CP, NDF, and GE of the forage and concentrate of the diets also followed the above methods and instruments. The ether extract of the forage and concentrate was analyzed using an ANKOM XT15 Extractor (ANKOM Technology, Fairport, NY, USA).

2.6. Statistical Analysis

A one-way analysis of variance (ANOVA) and generalized linear model analysis were used to investigate the effects of legume species (LS), legume proportion (LP), and their interactions (LS × LP) on DMI, BWG, nutrient digestibility, energy balance, N metabolism, and energy/N utilization efficiency. Differences among the means were considered to be significant at a $p \leq 0.05$ on the basis of Tukey's test, unless otherwise stated. The statistical program used in the current study was IBM SPSS Statistics for Windows, version 19.0 (IBM Corp., Armonk, NY, USA).

3. Results

3.1. Feed Intake, Apparent Nutrient Digestibility, and BWG

LS significantly influenced the forage DMI and total DMI ($p < 0.05$, Table 3). In detail, the forage DMI and total DMI of cattle were significantly higher when fed on a CVH40 diet than on AH20 and AH40 diets ($p < 0.05$, Figure 1a). However, no significant differences were found in the concentrate DMI under LS ($p > 0.05$, Table 3). In addition, there were no significant differences in the forage DMI, concentrate DMI, and total DMI of cattle under LP ($p > 0.05$, Table 3, Figure 1a).

Table 3. A general linear model analysis of legume species (LS), legume proportion (LP), and their interaction effect on feed intake, digestibility, growth performance, and CH_4 emissions (n = 8).

Item †	LS ‡	LP ‡	LS × LP ‡
Dry matter intake (DMI)			
Forage DMI (g/kg BW$^{0.75}$/day)	5.783 *	0.932	0.498
Concentrate DMI (g/kg BW$^{0.75}$/day)	1.108	1.189	0.001
Total DMI (g/kg BW$^{0.75}$/day)	5.207 *	0.109	0.598
Digestibility			
DM digestibility (%)	0.215	5.671 *	1.303
OM digestibility (%)	0.306	6.744 *	1.582
NDF digestibility (%)	1.177	18.476 ***	0.001
Apparent N digestibility (%)	5.515 *	5.949 *	0.265
Growth performance			
BWG (kg/day)	0.205	0.403	1.389
FCR (kg BWG/kg DMI)	0.077	2.515	5.796 *
CH_4 emissions			
CH_4 emissions (g/kg BW$^{0.75}$/24 h)	5.907 *	7.056 *	0.815
CH_4 emissions (g/kg DMI/24 h)	1.698	5.604 *	0.000

† DMI, dry matter intake; DM, dry matter; OM, organic matter; NDF, neutral detergent fiber; BWG, body weight gain; FCR, feed conversion ratio (ratio of BWG divided by the total DMI). ‡ values are the F value, * $p < 0.05$, and *** $p < 0.001$.

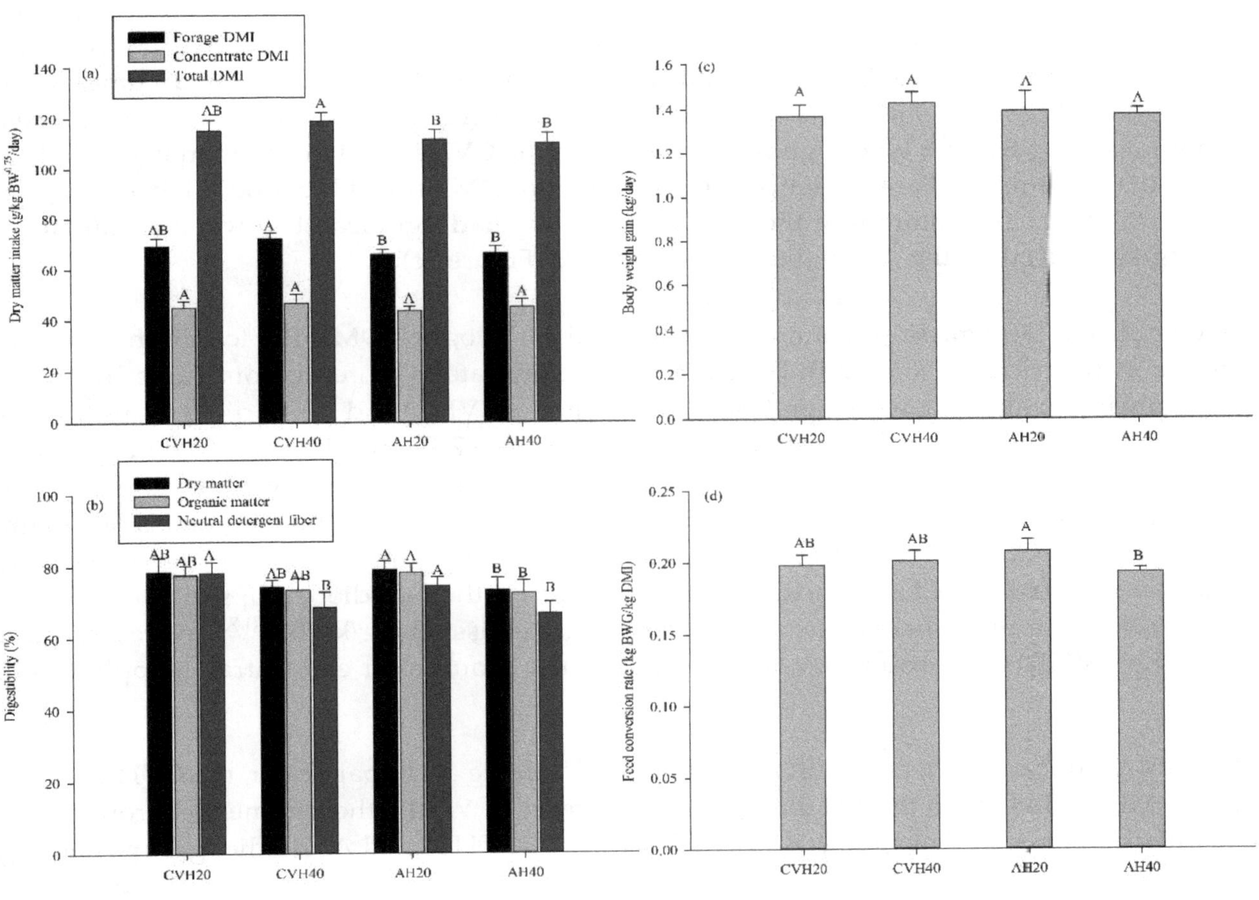

Figure 1. The dry matter intake (DMI) (**a**), digestibility (**b**), body weight gain (BWG) (**c**) and feed conversion rate (**d**) of cattle among the four diet groups. Values are presented as the mean ± standard deviation (SD). Uppercase letters only represent the difference among the four diet groups.

LP significantly affected the nutrient digestibility of cattle, including the digestibility of DM, OM, NDF, and apparent N ($p < 0.05$, Table 3). Specifically, the digestibility of the DM, OM, and NDF of cattle when fed on an AH40 diet were significantly lower than those on an AH20 diet ($p < 0.05$, Figure 1b). In the CVH diet groups, only NDF digestibility was significantly lower in the CVH40 diet group than in the CVH20 diet group ($p < 0.05$, Figure 1b).

Both LS and LP did not significantly influence BWG and the feed conversion rate (FCR) of cattle ($p > 0.05$, Table 3), but the interaction between LS and LP had a significant effect on the FCR of cattle ($p < 0.05$, Table 3). In detail, the AH40 diet group had a significantly lower FCR than that in the AH20 diet group ($p < 0.05$, Figure 1d), whereas there was no difference between the CVH20 and CVH40 diet groups ($p > 0.05$, Figure 1d).

3.2. Enteric CH_4 Emission, Energy Balance, and Energy Utilization Efficiency

CH_4 emissions, expressed on a milligram scale every 15 min per kilogram of metabolic BW and on a gram scale per kilogram DMI over 24 h post feeding, are shown in Figure 2a,b, respectively. There were intermittent peaks throughout the day and it was apparent that the peaks occurred a short time after feed supply. Moreover, the peaks of CH_4 emissions (mg/kg $BW^{0.75}$ or g/kg DMI) were relatively higher after the concentrate supply than after the forage supply (Figure 2a,b).

Both LS and LP could significantly affect CH_4 emissions (g/kg $BW^{0.75}$) over 24 h ($p < 0.05$, Table 3). Individually, the CVH diet groups had lower accumulated CH_4 emissions (g/kg $BW^{0.75}$) than the AH diet groups (Figure 2c), and the CVH40 and AH40 diet groups had relatively lower accumulated CH_4 emissions (g/kg $BW^{0.75}$) than the CVH20 and AH20 diet groups (Figure 2c). In addition, the accumulated CH_4 emissions (g/kg $BW^{0.75}$) were significantly lower in the CVH40 diet group than in the AH20 diet group ($p < 0.05$, Figure 2c). For CH_4 emissions per kilogram DMI in a 24 h period, LP had a significant effect ($p < 0.05$, Table 3). In summary, the CVH40 diet group had significantly lower accumulated CH_4 emission (g/kg DMI) than the AH20 diet group ($p < 0.05$, Figure 2c).

Accumulated CH_4 emissions per metabolic BW and per kilogram DMI after forage and concentrate supply in a 24 h period are shown in Figure 2d–f. Accumulated CH_4 emissions (g/kg $BW^{0.75}$) after forage supply in the morning were significantly lower in the CVH40 and AH40 diet groups than in the CVH20 and AH20 diet groups, respectively ($p < 0.05$, Figure 2d), and the CVH40 diet group had a significantly lower value than the AH20 diet group, regardless of forage supply in the morning or night ($p < 0.05$, Figure 2d,f). There was a similar trend between the accumulated CH_4 emissions in grams per kilogram of metabolic BW and grams per kilogram DMI (Figure 2d). There were no significant differences in accumulated CH_4 emissions (g/kg DMI) among the four diet groups ($p > 0.05$, Figure 2e) after concentrate supply, whereas the accumulated CH_4 emissions (g/kg $BW^{0.75}$) were significantly higher in the CVH20 diet group than in the CVH40 diet group after concentrate supply ($p < 0.05$, Figure 2e).

LS only significantly affected MEI and HP ($p < 0.05$, Table 4). In particular, the CVH diet groups had higher MEI and HP than the AH diet groups (Figure 3a,e). Within the legume diet groups, LP only significantly influenced FE output ($p < 0.05$, Table 4). The CVH40 and AH40 diet groups had higher FE output than the CVH20 and AH20 diet groups (Figure 3d), whereas FE only significantly differed between the AH20 and AH40 diet groups ($p < 0.05$, Figure 3d). For energy utilization efficiency, LP only significantly influenced the ratio of FE to GEI ($p < 0.05$, Table 4). In detail, this ratio was significantly higher in the AH40 diet group than in the CVH20 and AH20 diet groups ($p < 0.05$, Figure 3d).

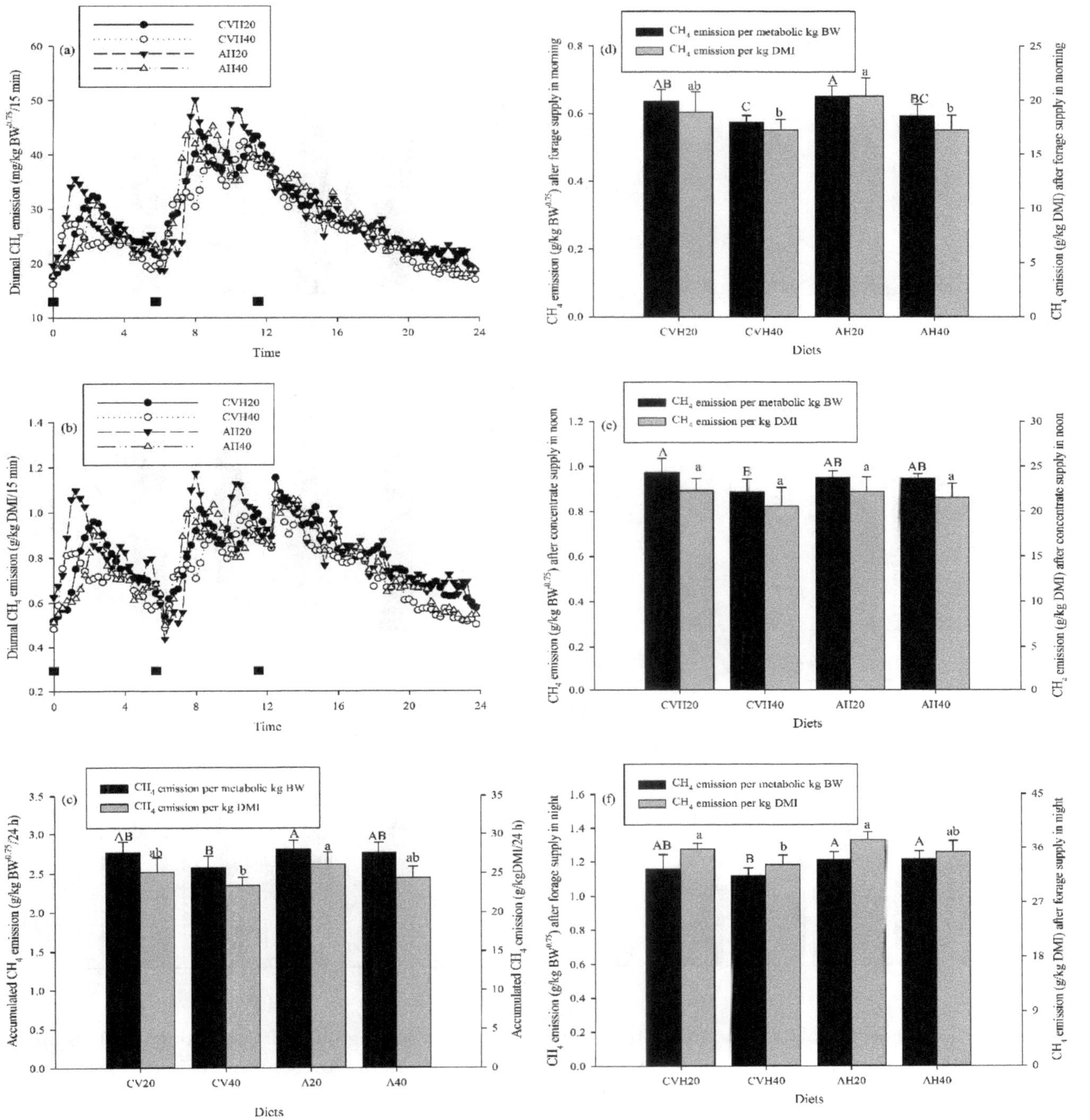

Figure 2. Diurnal CH_4 emissions (**a**) and (**b**), accumulated CH_4 emissions (**c**) after forage supply in the morning (**d**), after the concentrate supply in noon (**e**) and after forage supply in the night (**f**) for cattle among the four diet groups. Values are presented as the mean ± standard deviation (SD). Uppercase letters represent the differences among the four diet groups per kilogram of metabolic body weight, and lowercase letters represent the difference among the four diet groups per kilogram dry matter intake (DMI).

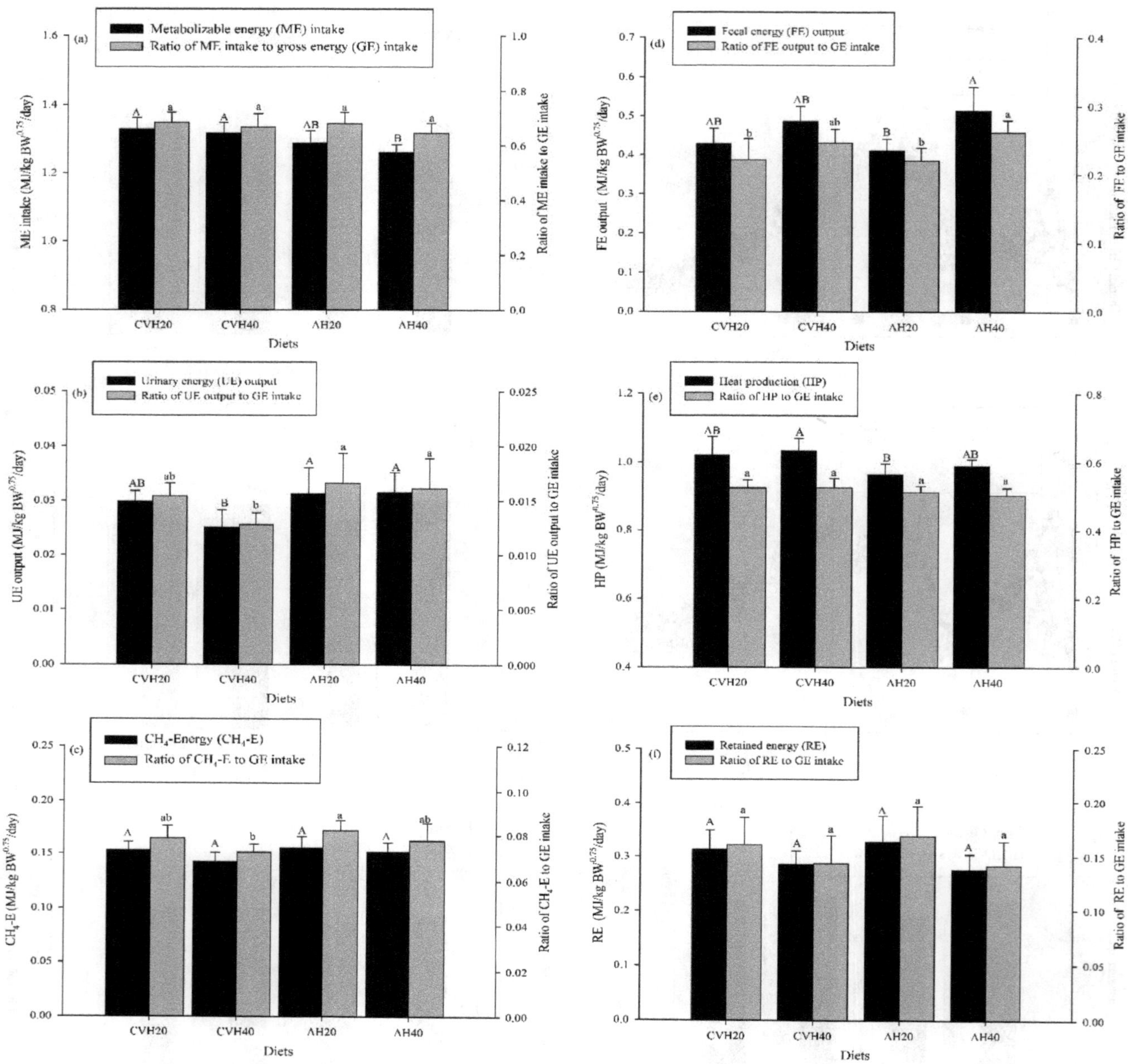

Figure 3. The energy balance and utilization efficiency of cattle among the four diet groups. Values are presented as the mean ± standard deviation (SD). Uppercase letters represent the differences among the four diet groups in energy balance, and lowercase letters represent the differences among the four diet groups in energy utilization. (**a**–**f**) represent ME intake, UE output, CH_4-E, FE output, HP, and RE, respectively, as well as their proportion of GE intake.

Table 4. A general linear model analysis of legume species (LS), legume proportion (LP) and their interaction effects on energy balance/nitrogen balance and energy/nitrogen utilization efficiency (n = 8).

Item †	LS ‡	LP ‡	LS × LP ‡
Energy balance			
GE intake (MJ/kg $BW^{0.75}$/day)	1.302	2.783	0.126
ME intake (MJ/kg $BW^{0.75}$/day)	6.749 *	1.132	0.127
FE output (MJ/kg $BW^{0.75}$/day)	0.042	13.739 **	1.054
UE output (MJ/kg $BW^{0.75}$/day)	4.675	1.584	1.992
CH_4-E (MJ/kg $BW^{0.75}$/day)	1.604	2.225	0.684
HP (MJ/kg $BW^{0.75}$/day)	6.208 **	1.198	0.170
RE (MJ/kg $BW^{0.75}$/day)	0.012	4.758	0.469

Table 4. *Cont.*

Item †	LS ‡	LP ‡	LS × LP ‡
Energy utilization efficiency			
Ratio of ME intake to GE intake (MJ/MJ)	0.436	1.589	0.224
Ratio of FE output to GE intake (MJ/MJ)	0.392	8.630 *	0.504
Ratio of UE output to GE intake (MJ/MJ)	4.647	2.254	1.025
Ratio of HP to GE intake (MJ/MJ)	2.189	0.148	0.171
Ratio of CH_4-E to GE intake (MJ/MJ)	2.332	3.644	0.066
Ratio of RE to GE intake (MJ/MJ)	0.051	2.993	0.178
Nitrogen balance			
N intake (g/kg $BW^{0.75}$/day)	2.956	1.317	0.168
FN output (g/kg $BW^{0.75}$/day)	8.792 *	21.653 ***	0.207
UN output (g/kg $BW^{0.75}$/day)	9.602 **	0.046	0.176
RN (g/kg $BW^{0.75}$/day)	21.681 ***	3.876	3.038
N metabolism			
Ruminal ammonia N (mmol/L)	2.044	12.989 **	1.685
Blood urea N (mmol/L)	14.243 **	6.884 *	0.970
Urinary ammonia N (mmol/L)	0.241	1.420	0.140
Nitrogen utilization efficiency			
Ratio of FN output to N intake (g/g)	3.464	12.862 **	0.459
Ratio of UN output to N intake (g/g)	16.116 **	0.311	0.398
Ratio of RN to N intake (g/g)	5.992 *	4.759	1.252

† GE, gross energy; ME, metabolizable energy; FE, fecal energy; UE, urinary energy; CH_4-E, methane energy; HP, heat production; RE, retained energy; N intake, nitrogen intake; FN, fecal N; UN, urinary N; RN, retained N. ‡ Values are the F value, * $p < 0.05$, ** $p < 0.01$, and *** $p < 0.001$.

3.3. N Balance, N Metabolism, and NUE

LS did not affect the NI of cattle, but it significantly affected the FN, UN and RN outputs in N balance ($p < 0.05$, Table 4). Although the UN output of the CVH20 and CVH40 diet groups was significantly lower than that in the AH40 diet group ($p < 0.05$, Figure 4b), the CVH diet groups had a relatively higher FN output, especially between the CVH40 and AH20 diet groups ($p < 0.05$, Figure 4c). As a consequence, the RN of cattle in the CVH20 and CVH40 diet groups was significantly higher than that in the AH40 diet group ($p < 0.05$, Figure 4d). For the effect of LP on N balance, the CVH40 and AH40 diet groups had relatively higher FE outputs than the CVH20 and AH20 diet groups, but FE output only significantly differed between the AH20 and AH40 diet groups ($p < 0.05$, Table 4, Figure 4c).

LP could significantly influence ruminal ammonia-N concentration ($p < 0.05$, Table 4), which was significantly lower in the AH40 diet group than in the AH20 and CVH20 diet groups ($p < 0.05$, Figure 4e). Both LS and LP significantly affected BUN concentration ($p < 0.05$, Table 4), and AH20 had a significantly lower BUN than the AH20, CVH20, and CVH40 diet groups ($p < 0.05$, Figure 4f). No differences were found for the urinary ammonia-N concentration among the four diet groups ($p > 0.05$, Table 4).

LP significantly affected the ratio of FN to NI ($p < 0.05$, Table 4). The AH20 diet group had a significantly lower FN to NI than that in the AH40 and CVH40 diet groups ($p < 0.05$, Figure 4c). Moreover, LS significantly influenced the ratio of UN to NI ($p < 0.05$, Table 4) and the ratio of RN to NI ($p < 0.05$, Table 4). The CVH40 diet group had a significantly lower UN:NI than the AH20 and AH40 diet groups ($p < 0.05$, Figure 4b), and the CVH20 diet group had a significantly higher RN:NI than the AH40 diet group ($p < 0.05$, Figure 4d).

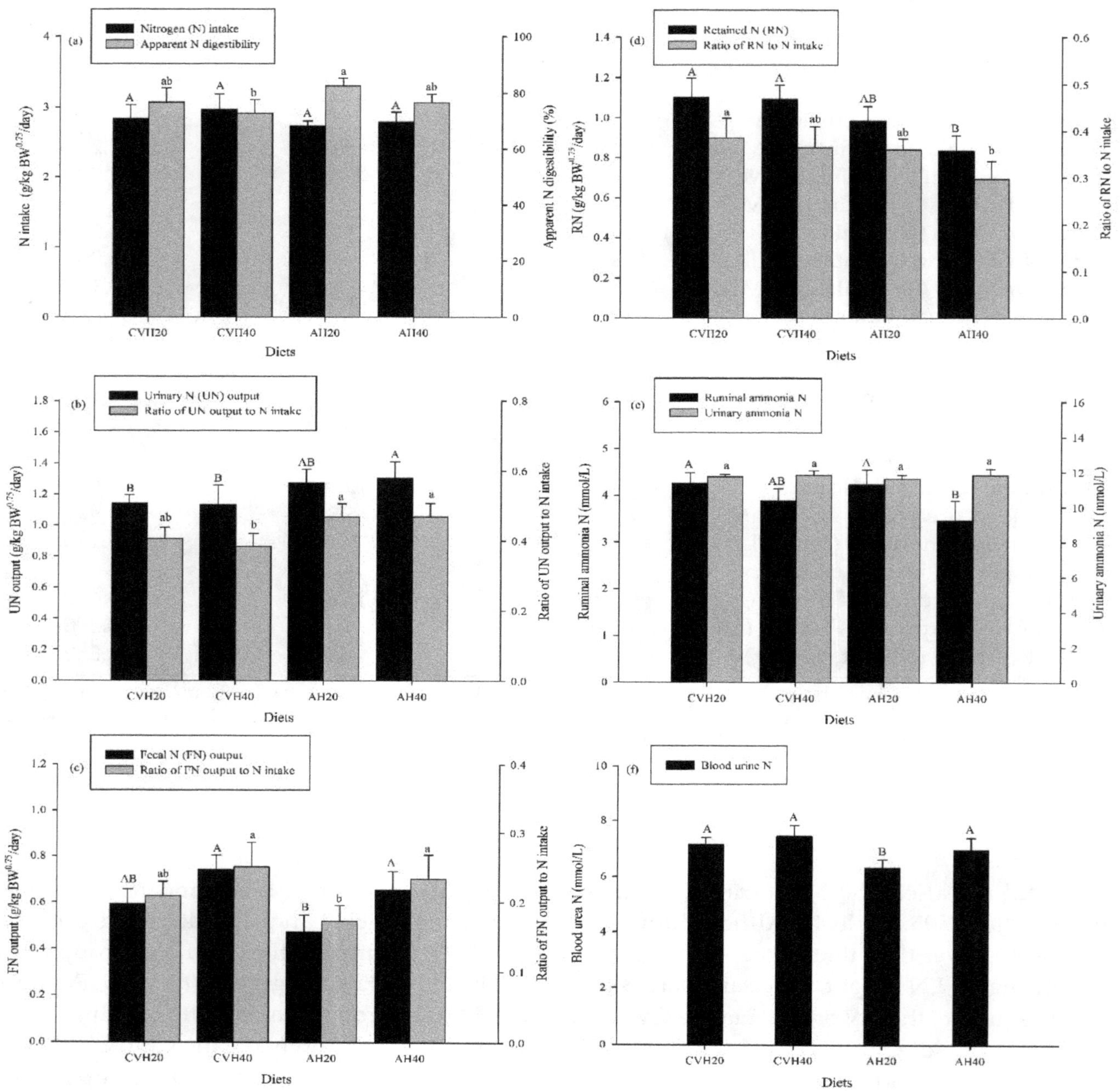

Figure 4. Nitrogen balance and utilization efficiency of cattle among the four diet groups. Values are presented as the mean ± standard deviation (SD). Uppercase letters represent the differences among the four diet groups in nitrogen balance and lowercase letters represent the differences among the four diet groups in nitrogen utilization. (**a**) represents N intake and apparent N digestibility; (**b–d**) represent UN output, FN output, and RN, respectively, as well as their proportion of N intake; (**e**) represents ruminal ammonia N and urinary ammonia N concentrations; (**f**) represents blood urea N.

3.4. Ruminal Fermentation Parameters

The total VFA and pH of the ruminal fluid did not significantly differ among the four diet treatments ($p > 0.05$, Table 5). However, the molar proportion of acetate was significantly lower in the CVH40 and AH40 diet groups than in the CVH20 and AH20 diet groups, respectively ($p < 0.05$, Table 5). Additionally, the molar proportion of acetate was also significantly lower in the CVH diet groups than in the AH diet groups ($p < 0.05$, Table 5). The molar proportions of propionate in the CVH40 and AH40 diet groups were significantly higher than those in the CVH20 and AH20 diet groups, respectively

($p < 0.05$, Table 5). As a consequence, the ratio of acetate to propionate was significantly lower in the CVH40 and AH40 diet groups than in the CVH20 and AH20 diet groups ($p < 0.05$, Table 5).

Table 5. Effects of different diets on the ruminal fermentation parameters in Simmental crossbred cattle.

Item	Experimental Diet †				Variance Analysis ‡		
	CVH20	CVH40	AH20	AH40	LS	LP	LS × LS
Total VFA, mmol/L	75.4 ± 6.73	72.5 ± 7.22	77.8 ± 3.32	75.7 ± 9.98	0.536	0.423	0.011
pH	6.07 ± 0.16	6.12 ± 0.25	6.05 ± 0.08	6.01 ± 0.06	0.686	0.009	0.293
Molar proportions (mol/100 mol)							
Acetate	72.3 ± 1.24	70.8 ± 0.56	73.8 ± 0.64	72.7 ± 1.13	11.967 **	6.503 *	0.122
Propionate	14.4 ± 0.24	15.7 ± 1.08	13.9 ± 0.76	15.2 ± 0.75	1.382	10.576 **	0.007
Butyrate	10.2 ± 1.11	10.2 ± 1.44	9.2 ± 0.72	8.9 ± 0.46	4.747	0.072	0.042
Iso-butyrate	1.1 ± 0.06	1.2 ± 0.19	1.1 ± 0.20	1.1 ± 0.15	0.173	0.640	0.539
Valerate	0.7 ± 0.12	0.6 ± 0.08	0.6 ± 0.18	0.6 ± 0.05	1.444	1.950	0.544
Iso-valerate	1.23 ± 0.09	1.44 ± 0.24	1.32 ± 0.23	1.37 ± 0.18	0.003	1.780	0.679
Acetate/propionate ratio	5.01 ± 0.11	4.53 ± 0.30	5.32 ± 0.30	4.78 ± 0.31	3.987	11.522 **	0.036

† CVH20, 20% common vetch + 40% oat hay; CVH40, 40% common vetch + 20% oat hay; AH20, 20% alfalfa + 40% oat hay; AH40, 40% alfalfa + 20% oat hay. ‡ Values are presented as the mean ± standard deviation (SD); LS, legume species; LP, legume proportion; LS × LP, interaction between LS and LP. Values are the F value; * $p < 0.05$; ** $p < 0.01$.

4. Discussion

4.1. Feed Intake, Nutrient Digestibility, and BWG

In general, feed intake is restricted by the capacity of the rumen [22] and NDF content, which is a measure of cell wall content, and the digestibility of forage [28,29]. In this study, the higher forage DMI in the CVH diet groups than in the AH diet groups ($p < 0.05$, Table 3, Figure 1a) could be attributed to the lower NDF content in CVH (Table 1). This indicates that feeds equal in digestibility but different in NDF content would result in different intakes [22]. The similar DM digestibility (Figure 1b) in CVH20 and AH20, as well as CVH40 and AH40 (Figure 1b), along with the higher DMI in the CVH diet groups, confirm the above deduction.

The digestibility of mixed feed is affected by the feed's chemical composition [22]. For example, forage intake with increasing legume proportions could promote the passage rate of feedstuff in the rumen [12] because legumes have lower fiber content than grass, which reduces the retention time of forage in the rumen [22,30]. In this study, the lower NDF digestibility in diets with higher legume proportions than those in the lower legume proportion diets (Table 3, Figure 1b) confirms the above finding. Compared to grasses, highly lignified cell walls could decrease cell wall digestion in legumes and then decrease OM digestion in the rumen [31]. The lower OM digestibility in higher legume proportion diets as compared to those with lower proportions supports the previous finding (Figure 1b).

4.2. Enteric CH_4 Emission and Ruminal Fermentation

There is a clear relationship between forage type, concentrate feed or starch intake, OM digestibility, and patterns of ruminal fermentation [4]. In this study, the lower CH_4 emissions in the diets with higher proportions of legumes than in those with lower ones, regardless of the per kilogram metabolic BW or per kilogram DMI ($p < 0.05$, Table 3, Figure 2c), indicate that a diet with a higher proportion of legumes could decrease CH_4 emissions. This is consistent with the findings of Lee et al. [32], who reported that increasing the percentage of white clover feed with perennial ryegrass could decrease CH_4 emissions. This could be attributed to the polyphenolic compounds in legumes, such as condensed tannins, which have been previously shown to be negatively correlated with CH_4 emissions [33]. For rumen fermentation, there is a negative relationship between CH_4 emissions and propionate formation in the rumen, which could depress the activity of methanogens [4,5]. In this study, the lower ratios of acetate to propionate (Table 5) correspond to lower CH_4 emissions after forage supply (Figure 2d) in diet groups with a higher proportion of legumes and this is consistent with the above finding. In addition, it has been reported that lipid supplementation could reduce CH_4 emissions [34,35]. In this study, legumes have higher concentrations of crude fat (ether extract) than grasses (Table 2), which led to a higher crude fat concentration per unit of DM in the diets with higher proportions of legume than in those with lower proportions. The lower CH_4 emissions in the CVH40 and AH40 diet groups than in the CVH20 and AH20 groups could also be explained in the same way. More importantly, feed intake is the single most important determinant of CH_4 emissions [36]. In this study, there was no difference in the DMI between the diets with higher or lower proportions of legume (Figure 1a), but the lower OM digestibility (1b) of the diet with a higher proportion of legumes and a higher passage rate [12] leaves less time for microorganisms to ferment the feedstuff in the rumen [31]. Therefore, lower CH_4 emissions were observed in CVH40 and AH40 than in CVH20 and AH20 (Figure 2c).

In addition to the effects of LP on CH_4 emissions, LS also affected CH_4 emissions, especially on the basis of per kilogram metabolic BW (Table 3). The relatively lower CH_4 emissions (g/kg $BW^{0.75}$) of the CVH diet groups as compared to the AH diet groups at the same LP (Figure 2c) indicate that CVH has better potential to inhibit CH_4 emissions than AH, especially after forage supply (Figure 2d,f). This might be due to the lower content of NDF and ADF in CVH than in AH, which is in agreement with Beauchemin [35], who reported that lower CH_4 emissions for animals fed legumes could often be explained by the lower fiber content of their diets. Moreover, the production of propionate over acetate in the rumen could also reduce CH_4 emissions in the rumen [35]. In this study, these changes in propionate and acetate also confirmed that the acetate molar proportion was lower in the CVH diet groups than in the AH diet groups ($p < 0.05$, Table 5), although there were no differences in the propionate molar proportion (Table 5). The ratio of acetate to propionate was around 4.77 in the CVH diet groups (Table 5), which was higher than the result of the study by Calabrò et al. [36], who reported a value of 2.28 for an OH and CVH mixture diet using an in vitro gas production technique. This could be attributed to differences that may exist in vivo and in vitro. For example, increases of in vivo rumen propionate concentrations were lower than those observed in vitro [37].

Increasing the inclusion of concentrates in the diet, especially starch content, was regarded as another way to reduce CH_4 emissions [34,38]. In this study, the CVH40 and AH40 diet groups have a relatively higher proportion of maize than the CVH20 and AH20 diet groups (Table 2). As a consequence, lower CH_4 emissions were observed in the CVH40 and AH40 diet groups, even though this value only significantly differed between the CVH20 and CVH40 diet groups per kilogram metabolic BW (Figure 2e). Moreover, CH_4 emissions still tended to be lower in the higher proportion of maize diet groups per kilogram DMI, although this was not considered significant (Figure 2e). These results suggest that starch intake could suppress CH_4 emissions, in accordance with previous studies [4,31].

4.3. Energy Balance

In ruminants, energy is lost in the form of feces, urine, and methane emissions [39]. In this study, FE output and the ratio of FE output to GE intake were greater in the CVH40 and AH40 diet groups than in the CVH20 and AH20 diet groups (Figure 3d). This could be explained by the higher passage rate of the diets with a higher proportion of legumes in the rumen [12], as well as decreased DM digestibility (Figure 1b) because the greater the DM excretion, the greater the FE loss. The ratio of UE output to GE intake, which in previous studies was found to range from 0.9% to 4.8% [39,40], is an indispensable element of energy loss, and high UE loss is more common when animals are fed a silage diet [41]. In this study, the mean 1.4% for the ratio of UE output to GE intake fell within the lower range of the quoted studies, but LP did not significantly influence UE output or the ratio of UE output to GE intake (Figure 3b). The relatively lower values of the ratio of CH_4-E to GE intake in the CVH40 and AH40 diet groups could be explained by the lower OM digestibility (Figure 1b), which reduced the retention time of the feedstuff in the rumen.

ME intake, expressed as per kilogram metabolic BW, was higher in the CVH diet groups than in the AH diet groups (Table 3, Figure 3a), which could be attributed to the higher forage DMI in the

CVH diet groups (Figure 1a) because CVH had a higher ME concentration (MEC) than AH (Table 1). However, no differences were found for the ratio of ME intake to GE intake among the four diet groups (Figure 3a). Nevertheless, the higher ratio of FE output to GE intake in the diet with a higher proportion of legumes (Table 4 and Figure 3d), which accounted for the largest part of the feed energy that could not be utilized by the animals [19], still tended to be lower in the CVH40 and AH40 than in the CVH20 and AH20 diet groups (Figure 3a). Additionally, the ratio of ME intake to GE intake for the crossbred Simmental cattle, in this study, was around 0.67, which was higher than the previously reported 0.47 for mature Simmental cows [42]. This could be attributed to a higher OM digestibility (averaged 75.4%) in this study ascompared to that (62.4%) in the previous study [42]. The higher ME intake (Table 3 and Figure 3a) alongside a lack of differences in RE (Figure 3f) in the CVH diet groups as compared to the AH diet groups could be attributed to an increased HP for the CVH diet groups than the AH diet groups (Table 4 and Figure 3e). This is consistent with the finding of Ferrell and Jenkins [43] that HP increased alongside increasing ME intake for crossbred beef cattle.

4.4. N Balance, N Metabolism, and N Utilization Efficiency (NUE)

N excretion in feces and urine represents a considerable N loss from ruminant husbandry [7,22]. In this study, N losses were affected by LS and LP, although LS and LP did not influence total N intake (Table 3). For example, the significantly higher FN output and the ratio of FN output to N intake corresponded with a higher proportion of legume (CVH40 vs. CVH20 and AH40 vs. AH20, Figure 4c). These were likely caused by the decreased nutrient digestibility (Figure 1b), as well as decreased apparent N digestibility (Figure 4a), which usually lead to more N being excreted in feces. As a result, the higher FN output (Figure 4c) (but no different UN output, Figure 4b) in the diet with a higher proportion of legumes (Figure 4b) led to a reduced RN in the lower proportion of the legume diets ($p = 0.073$, Table 3, and Figure 4d). The UN, FN, and RN outputs were influenced by LS (Table 3). The UN output in the CVH diet groups was lower than that in the AH diet groups (Figure 4b), whereas the FN output presented an opposite result (Figure 4c). The greater shift of N excretion from urine to feces in the CVH diet groups than in the AH diet groups was regarded as a way to reduce the impact of volatile N excretion on the environment [6] because urinary urea is rapidly hydrolyzed to ammonium and then converted to ammonia which is readily volatilized and lost from the farm system to the environment [44]. By contrast, fecal ammonia production is generally low due to the slow mineralization rates of organic nitrogenous compounds [3,7]. As a consequence, the RN in the CVH diet groups was higher than that in the AH diet groups (Figure 4d). Therefore, the CVH diet has a greater potential to reduce the effects of volatile N excretions on the environment than the AH diet.

Generally, high ruminal ammonia-N concentrations for optimal OM degradation will result in an increase in the loss of N through urine [45]. In this study, ammonia-N concentrations in the rumen tended to be lower with a higher proportion of legumes, especially in AH diets (Figure 4e). This difference could possibly be due to the relatively higher passage rate of feedstuff in the rumen with increasing legume proportions [12], thereby, yielding a lower OM digestibility (Figure 1b) and ammonia-N concentrations in the diets with higher proportions of legumes than in the diets with lower proportions of legumes.

In addition, BUN levels reflected the protein status of cattle and positively corresponded with changes in the ammonia-N concentration in rumen fluid [46]. In this study, BUN tended to be higher in the diets with a higher proportion of legumes (Figure 4f), which was inconsistent with ruminal ammonia-N concentrations (Figure 4e). This might be attributable to the lowest pH in AH40 (Table 5), which depressed the transport of ammonia across the rumen wall. Studies have shown that the permeability of the rumen wall for ammonia is pH dependent, and has a positive correlation with pH [47]. Additionally, although the ruminal ammonia-N concentration tended to be lower in the diets with higher proportions of legumes than in the diets with lower proportions of legumes, there was no reduction in BWG (Figure 1c). This suggests that adequate ruminal available N was provided from the diet to maximize microbial fermentation in the rumen under a ruminal ammonia-N concentration of around 4.0 mmol/L.

5. Conclusions

The results of this study suggest the following: (1) a higher proportion of legumes in the diet could reduce CH_4 emissions and minimize the impact of volatile N excretion to the environment; (2) increasing legume proportions in the diet could reduce nutrient digestibility, whereas the degree of reduction differs between common vetch hay and alfalfa hay; and (3) common vetch hay has great potential to minimize the negative effects of CH_4 emissions and N excretion into the environment. Therefore, an opportunity for strategic feeding exists by using alfalfa hay (20%) and common vetch hay (40%) to reduce the direct impact of volatile N excretion and CH_4 emissions on the environment while maintaining BWG, as well as nutrient digestibility for crossbred Simmental cattle, in dryland environments.

Author Contributions: Conceptualization, A.T., F.H., and T.I.; methodology, A.T., F.H., N.K., T.I., and W.D.; data curation, W.D.; writing—original draft preparation, W.D.; writing—review and editing, F.P., and T.I.; visualization, W.D.; supervision, A.T. and F.P.; project administration, A.T. and F.H.; funding acquisition, A.T. and F.H.

Acknowledgments: The authors thank Chang Shenghua, Zhang Cheng, and the students of the College of Pastoral Agriculture Science and Technology (Lanzhou University, China) for supporting the operation of the respiration chambers and for analyzing the feed, fecal, and urinary samples. The authors also wish to thank the Academician Nan Zhibiao for supporting common vetch seeds (*Vicia sativa* L.cv. Lanjian No. 3) for this study.

References

1. National Research Council (NRC). *Air Emissions from Animal Feeding Operations: Current Knowledge, Future Needs*; National Academies Press: Washington, DC, USA, 2003.
2. Montes, F.; Meinen, R.; Dell, C.; Rotz, A.; Hristov, A.N.; Oh, J.; Waghorn, G.; Gerber, P.J.; Henderson, B.; Makkar, H.P.S.; et al. Special topics—Mitigation of methane and nitrous oxide emissions from animal operations: II. A review of manure management mitigation options. *J. Anim. Sci.* **2013**, *91*, 5070–5094. [CrossRef]
3. Kebreab, E.; Dijkstra, J.; Bannink, A.; France, J. Recent advances in modeling nutrient utilization in ruminants. *J. Anim. Sci.* **2009**, *87*, E111–E122. [CrossRef] [PubMed]
4. Hristov, A.N.; Oh, J.; Firkins, J.L.; Dijkstra, J.; Kebreab, E.; Waghorn, G.; Makkar, H.P.S.; Adesogan, A.T.; Yang, W.; Lee, C.; et al. Special topics—Mitigation of methane and nitrous oxide emissions from animal operations: I. A review of enteric methane mitigation options. *J. Anim. Sci.* **2013**, *91*, 5045–5069. [CrossRef] [PubMed]
5. Dong, L.; Li, B.; Diao, Q. Effects of Dietary Forage Proportion on Feed Intake, Growth Performance, Nutrient Digestibility, and Enteric Methane Emissions of Holstein Heifers at Various Growth Stages. *Animals* **2019**, *9*, 725. [CrossRef] [PubMed]
6. Yan, T.; Frost, J.P.; Keady, T.W.J.; Agnew, R.E.; Mayne, C.S. Prediction of nitrogen excretion in feces and urine of beef cattle offered diets containing grass silage. *J. Anim. Sci.* **2007**, *85*, 1982–1989. [CrossRef]
7. Waldrip, H.M.; Todd, R.W.; Cole, N.A.H. Prediction of nitrogen excretion by beef cattle: A meta-analysis. *J. Anim. Sci.* **2013**, *91*, 4290–4302. [CrossRef]
8. Givens, D.I.; Owen, E.; Omed, H.M.; Axford, R.F.E. *Forage Evaluation in Ruminant Nutrition*; CABI Publishing: Wallingford, UK, 2000; pp. 9–10.
9. Graham, P.H.; Vance, C.P. Legumes: Importance and constraints to greater use. *Plant Physiol.* **2003**, *131*, 872–877. [CrossRef]
10. Abreu, A.; Carulla, J.E.; Lascano, C.E.; Diaz, T.E.; Kreuzer, M.; Hess, H.D. Effects of Sapindus saponaria fruits on ruminal fermentation and duodenal nitrogen flow of sheep fed a tropical grass diet with and without legume. *J. Anim. Sci.* **2004**, *82*, 1392–1400. [CrossRef]
11. Doran, M.P.; Laca, E.A.; Sainz, R.D. Total tract and rumen digestibility of mulberry foliage (*Morus alba*), alfalfa hay and oat hay in sheep. *Anim. Feed. Sci. Technol.* **2007**, *138*, 239–253. [CrossRef]
12. McCaughey, W.P.; Wittenberg, K.; Corrigan, D. Impact of pasture type on methane production by lactating beef cows. *Can. J. Anim. Sci.* **1999**, *79*, 221–226. [CrossRef]

13. Wuchen, D.; Fujiang, H.; Atsushi, T.; Kobayashi, N.; Fei, P.; Ichinohe, T. Effects of oat hay and leguminous forage mixture feeding on body weight gain, enteric methane emission and energy utilization in crossbred Simmental male beef cattle. *Anim. Sci. J.* **2019**. under review.
14. Kobayashi, N.; Hou, F.J.; Tsunekawa, A.; Chen, X.J.; Yan, T.; Ichinohe, T. Effects of substituting alfalfa hay for concentrate on energy utilization and feeding cost of crossbred Simmental male calves in Gansu Province, China. *Grassl. Sci.* **2017**, *63*, 245–254. [CrossRef]
15. Kobayashi, N.; Hou, F.J.; Tsunekawa, A.; Chen, X.J.; Yan, T.; Ichinohe, T. Appropriate level of alfalfa hay in diets for rearing Simmental crossbred cattle in dryland China. *Asian Australas. J. Anim. Sci.* **2018**, *31*, 1881–1889. [CrossRef] [PubMed]
16. Larbi, A.; El-Moneim, A.A.; Nakkoul, H.; Jammal, B.; Hassan, S. Intra-species variations in yield and quality determinants in Vicia species: 3. Common vetch (*Vicia sativa ssp. sativa* L.). *Anim. Feed Sci. Technol.* **2011**, *164*, 241–251. [CrossRef]
17. Rihawi, S.; Iniguez, L.; Knaus, W.F.; Zaklouta, M.; Wurzinger, M.; Soelkner, J.; Larbi, A.; Bomfim, M.A.D. Fattening performance of lambs of different Awassi genotypes, fed under cost-reducing diets and contrasting housing conditions. *Small Rumin. Res.* **2010**, *94*, 38–44. [CrossRef]
18. Assefa, G.; Ledin, I. Effect of variety, soil type and fertiliser on the establishment, growth, forage yield, quality and voluntary intake by cattle of oats and vetches cultivated in pure stands and mixtures. *Anim. Feed Sci.Technol.* **2001**, *92*, 95–111. [CrossRef]
19. Hernández-Ortega, M.; Heredia-Nava, D.; Espinoza-Ortega, A.; Sánchez-Vera, E.; Arriaga-Jordán, C.M. Effect of silage from ryegrass intercropped with winter or common vetch for grazing dairy cows in small-scale dairy systems in Mexico. *Trop. Anim. Health Prod.* **2011**, *43*, 947–954. [CrossRef]
20. Agricultural and Food Research Council. *Energy and Protein Requirements of Ruminants. An Advisory Manual Prepared by the AFRC Technical Committee on Responses to Nutrients*; Centre for Agriculture and Bioscience International: Wallingford, UK, 1993; pp. 1–38.
21. Chinese Feeding Standard for Beef Cattle (CFSBC). Ministry of Agriculture of the People's Republic of China. 2004. Available online: https://wenku.baidu.com/view/154d59bdf121dd36a32d8269.html (accessed on 4 October 2019).
22. Zhao, Y.G.; Aubry, A.; O'Connell, N.E.; Annett, R.; Yan, T. Effects of breed, sex, and concentrate supplementation on digestibility, enteric methane emissions, and nitrogen utilization efficiency in growing lambs offered fresh grass. *J. Anim. Sci.* **2015**, *93*, 5764–5773. [CrossRef]
23. Gerrits, W.; Labussiere, E.; Dijkstra, J.; Reynolds, C.; Metges, C.; Kuhla, B.; Lund, P.; Weisbjerg, M.R. Letter to the Editors: Recovery test results as a prerequisite for publication of gaseous exchange measurements. *Animal* **2018**, *12*, 4. [CrossRef]
24. Brouwer, E. Report of sub-committee on constants and factors. In Proceedings of the 3rd Symposium on Energy Metabolism of Farm Animals of European Association for Animal Production; Blaxter, K.L., Ed.; Academic Press: Scotland, UK, 1965.
25. Kohn, R.A.; Dinneen, M.M.; Russek-Cohen, E. Using blood urea nitrogen to predict nitrogen excretion and efficiency of nitrogen utilization in cattle, sheep, goats, horses, pigs, and rats. *J. Anim. Sci.* **2005**, *83*, 879–889. [CrossRef]
26. Association of Official Analytical Chemists (AOAC). *Official Methods of Analysis of the Association of Official Analytical Chemists*, 15th ed; Association of Official Analytical Chemists: Arlington, VA, USA, 1990.
27. Van Soest, P.J.; Robertson, J.B.; Lewis, B.A. Methods for dietary fiber, neutral detergent fiber, and nonstarch polysaccharides in relation to animal nutrition. *J. Dairy Sci.* **1991**, *74*, 3583–3597. [CrossRef]
28. Buxton, D.R. Quality-related characteristics of forages as influenced by plant environment and agronomic factors. *Anim. Feed Sci. Technol.* **1996**, *59*, 37–49. [CrossRef]
29. Karabulut, A.; Canbolat, O.; Kalkan, H.; Gurbuzol, F.; Sucu, E.; Filya, I. Comparison of in vitro gas production, metabolizable energy, organic matter digestibility and microbial protein production of some legume hays. *Asian Australas. J. Anim. Sci.* **2007**, *20*, 517–522. [CrossRef]
30. Patra, A.K. Effects of supplementing low–quality roughages with tree foliages on digestibility, nitrogen utilization and rumen characteristics in sheep: A meta–analysis. *J. Anim. Physiol. Anim. Nutr.* **2010**, *94*, 338–353. [CrossRef] [PubMed]
31. Archimède, H.; Eugène, M.; Marie Magdeleine, C.; Boval, M.; Martin, C.; Morgavi, D.P.; Lecomte, P.; Doreau, M. Comparison of methane production between C3 and C4 grasses and legumes. *Anim. Feed Sci. Technol.* **2011**, *166*, 59–64. [CrossRef]

32. Lee, J.M.; Woodward, S.L.; Waghorn, G.C.; Clark, D.A. Methane emissions by dairy cows fed increasing proportions of white clover (*Trifolium repens*) in pasture. *Proc. N. Z. Grassl. Assoc.* **2004**, *66*, 151–155.
33. Guglielmelli, A.; Calabrò, S.; Primi, R.; Carone, F.; Cutrignelli, M.I.; Tudisco, R.; Piccolo, G.; Ronchi, B.; Danieli, P.P. In vitro fermentation patterns and methane production of sainfoin (*Onobrychis viciifolia* Scop.) hay with different condensed tannin contents. *Grass Forage Sci.* **2011**, *66*, 488–500. [CrossRef]
34. Grainger, C.; Beauchemin, K.A. Can enteric methane emissions from ruminants be lowered without lowering their production? *Anim. Feed Sci. Technol.* **2011**, *166*, 308–320. [CrossRef]
35. Beauchemin, K.A.; Kreuzer, M.; O'Mara, F.; McAllister, T.A. Nutritional management for enteric methane abatement: A review. *Aust. J. Exp. Agric.* **2008**, *48*, 21–27. [CrossRef]
36. Calabrò, S.; Carone, F.; Cutrignelli, M.; D'Urso, S.; Piccolo, G.; Tudisco, R.; Angelino, G.; Infascelli, F. The effect of haymaking on the neutral detergent soluble fraction of two intercropped forages cut at different growth stages. *Ital. J. Anim. Sci.* **2006**, *5*, 327–339. [CrossRef]
37. Fievez, V.; Dohme, F.; Danneels, M.; Raes, K.; Demeyer, D. Fish oils as potent rumen methane inhibitors and associated effects on rumen fermentation in vitro and in vivo. *Anim. Feed Sci. Technol.* **2003**, *104*, 41–58. [CrossRef]
38. Yan, T.; Agnew, R.E.; Gordon, F.J.; Porter, M.G. Prediction of methane energy output in dairy and beef cattle offered grass silage-based diets. *Livest. Prod. Sci.* **2000**, *64*, 253–263. [CrossRef]
39. Chaokaur, A.; Nishida, T.; Phaowphaisal, I.; Sommart, K. Effects of feeding level on methane emissions and energy utilization of Brahman cattle in the tropics. *Agric. Ecosyst. Environ.* **2015**, *199*, 225–230. [CrossRef]
40. Zou, C.X.; Lively, F.O.; Wylie, A.R.G.; Yan, T. Estimation of the maintenance energy requirements, methane emissions and nitrogen utilization efficiency of two suckler cow genotypes. *Animal* **2016**, *10*, 616–622. [CrossRef]
41. Kirkpatrick, D.E.; Steen, R.W.J.; Unsworth, E.F. The effect of differing forage: Concentrate ratio and restricting feed intake on the energy and nitrogen utilization by beef cattle. *Livest. Prod. Sci.* **1997**, *51*, 151–164. [CrossRef]
42. Estermann, B.L.; Sutter, F.; Schlegel, P.O.; Erdin, D.; Wettstein, H.R.; Kreuzer, M. Effect of calf age and dam breed on intake, energy expenditure, and excretion of nitrogen, phosphorus, and methane of beef cows with cattle. *J. Anim. Sci.* **2002**, *80*, 1124–1134. [CrossRef]
43. Ferrell, C.L.; Jenkins, T.G. Body composition and energy utilization by steers of diverse genotypes fed a high-concentrate diet during the finishing period: I.; Angus, Belgian Blue, Hereford, and Piedmontese sires. *J. Anim. Sci.* **1998**, *76*, 637. [CrossRef]
44. Koenig, K.M.; Beauchemin, K.A. Effect of feeding condensed tannins in high protein finishing diets containing corn distillers grains on ruminal fermentation, nutrient digestibility, and route of nitrogen excretion in beef cattle. *J. Anim. Sci.* **2018**, *96*, 4398–4413. [CrossRef]
45. Ipharraguerre, I.R.; Clark, J.H. Varying protein and starch in the diet of dairy cows. II. Effects on performance and nitrogen utilization for milk production. *J. Dairy Sci.* **2005**, *88*, 2556–2570. [CrossRef]
46. Dong, R.L.; Zhao, G.Y.; Chai, L.L.; Beauchemin, K.A. Prediction of urinary and fecal nitrogen excretion by beef cattle. *J. Anim. Sci.* **2014**, *92*, 4669–4681. [CrossRef]
47. Abdoun, K.; Stumpff, F.; Martens, H. Ammonia and urea transport across the rumen epithelium: A review. *Anim. Health Res. Rev.* **2006**, *7*, 43–59. [CrossRef] [PubMed]

6

Pre-Exposure of Early-Weaned Lambs to a Herb-Clover Mix does not Improve their Subsequent Growth

Lukshman Jay. Ekanayake [1,*], Rene Anne Corner-Thomas [2], Lydia Margaret Cranston [2], Paul Richard Kenyon [2], Stephen Todd Morris [2] and Sarah Jean Pain [2]

1 Department of Animal Science, Faculty of Agriculture, University of Peradeniya, Peradeniya 20000, Sri Lanka
2 School of Agriculture and Environment, Private Bag 11-222, Massey University, Palmerston North 4442, New Zealand; R.Corner@massey.ac.nz (R.A.C.-T.); L.Cranston@massey.ac.nz (L.M.C.); P.R.Kenyon@massey.ac.nz (P.R.K.); S.T.Morris@massey.ac.nz (S.T.M.); S.J.Pain@massey.ac.nz (S.J.P.)
* Correspondence: jayampathiekn@gmail.com.

Simple Summary: Exposure of lambs to herbage-based diets prior to weaning may facilitate the development of the rumen which may subsequently increase animal performance after early weaning. The aim of this study was to estimate the effects of varying durations of exposure of lambs to a herb–clover mix containing chicory, plantain, red clover, and white clover prior to early weaning (at ~45 days of age) on their subsequent growth and rumen development at conventional weaning age. Prolonged exposure of lambs to the herb–clover mix prior to early weaning had no impact on lamb growth or rumen development, suggesting that using this management option will not improve performance of lambs after early weaning.

Abstract: Twin sets of lambs were randomly allocated to one of six treatments: (1) lambs born and managed on ryegrass–clover-based pasture until conventional weaning approximately at 99 days of age (Grass–Grass$_{CW}$); (2) lambs born on ryegrass–clover-based pasture and early weaned onto a herb–clover mix at ~45 days of age (Grass–Herb$_{EW}$); (3) lambs born on ryegrass–clover-based pasture, transferred with their dam onto a herb–clover mix at ~45 days of age until conventional weaning (Grass–Herb$_{CW}$); (4) lambs born on ryegrass–clover-based pasture, transferred with their dam onto a herb–clover mix at ~15 days of age and early weaned onto a herb–clover mix at ~45 days of age (Grass–Herb$_{D15EW}$); (5) lambs born and managed on herb–clover mix until conventional weaning (Herb–Herb$_{CW}$); (6) lambs born on herb–clover mix and weaned early onto a herb–clover mix at ~45 days of age (Herb–Herb$_{EW}$). In both years, Herb–Herb$_{CW}$ lambs had greater ($p < 0.05$) growth rates than lambs in other treatments. The liveweight gains and rumen papillae development of Herb–Herb$_{EW}$, Grass–Herb$_{D15EW}$ and Grass–Herb$_{EW}$ lambs did not differ ($p > 0.05$). The weight of the empty digestive tract components at either early weaning or conventional weaning did not differ ($p > 0.05$) between treatments. Exposing early-weaned lambs to the herb mix for a prolonged period, prior to early weaning, does not improve their subsequent growth.

Keywords: early weaning; lamb; rumen development; herb–clover mix

---

1. Introduction

Weaning lambs early can be a useful management option that farmers can use when herbage production is inadequate or of poor quality, to reduce the overall ewe feed demand and to support the growth of early-weaned lambs to achieve liveweight targets [1,2]. Lambs weaned early at 4–8 weeks of age onto a ryegrass–clover-based pasture have consistently displayed lower live weight gains and poorer survival compared to unweaned lambs on ryegrass–clover-based pastures, to approximately 14 weeks of age [3,4]. This is likely explained by the inadequate nutritional quality of ryegrass–clover-based pastures. In contrast, a herb–clover mix containing plantain (*Plantago lanceolata*), chicory (*Cichorium intybus*), red clover (*Trifolium pretanse*) and white clover (*Trifolium repens*) has been shown to have a greater nutritional quality compared to a ryegrass–clover-based pasture [5–7]. This suggests that a herb–clove mix can be used to support lamb growth, post-early weaning. However, the liveweight gain of lambs weaned early, at a minimum live weight of 14 to 16 kg, onto a herb–clover mix, has been inconsistent across studies, compared to unweaned lambs on ryegrass–clover-based pastures [8–11]. In contrast, the live weight of ewes whose lambs were weaned early has consistently been greater than those left unweaned on ryegrass–clover [10,11]. The inconsistency in lamb liveweight gain has not yet been explained. The reasons for the variation in lamb growth post-early weaning on a herb–clover mix appear not to be driven by variation of metabolisable energy [9,11] or crude protein content [9–12] of the herbage. If farmers are to utilize early weaning, a consistent method for achieving suitable lamb growth post weaning is required. The impact of prior exposure to the herb–clover mix on lamb growth, post early weaning, has not been previously examined. If prior exposure to the herb–clover mix was found to result in a consistent positive effect on lamb growth, management guidelines for farmers could be developed.

Early development of the reticulo-rumen and its microbial population is important for efficient utilization of herbage-based diets after weaning [13,14]. Early exposure to herbage-based diets can help develop a strong host rumen microbiota [15,16] and familiarity with a herbage type can influence a lamb's preference for that herbage later in life [17]. Sheep avoid grazing novel herbages [18,19] for several reasons including taste aversion [20], as a strategy to maintain effective rumen micro-flora [21] and to meet their ideal protein intake [22,23]. Combined, these studies suggest that early exposure of lambs to herbage may facilitate rumen development and increase subsequent animal performance. To date the rumen development of early-weaned lambs offered herb–clover mix has not been studied. It was hypothesized that prolonged exposure of lambs to a herb–clover mix, prior to early weaning, would have a positive effect on their subsequent growth and rumen development. The aims of the present study were to test the effects of (i) weaning lambs early at approximately 45 days of age onto a herb–clover mix and (ii) early prolonged exposure of lambs to a herb–clover mix prior to early weaning at approximately 45 days of age on their subsequent growth and rumen development until conventional weaning age (~99 days of age).

2. Materials and Methods

2.1. Herbage Treatments

Eight paddocks (12.85 ha in total land area) of herb–clover mix and seven paddocks (14.85 ha in total land area) of ryegrass–clover-based pasture containing perennial ryegrass (*Lolium perenne* L.) and white clover (*Trifolium repens*) were used over two spring periods in 2017 and 2018. Herb–clover mix paddocks were sown during autumn in 2012 and 2013 with a seed mixture of chicory (*Cichorium intybus*; 6 kg/ha), plantain (*Plantago lanceolata*; 6 kg/ha), white clover (4 kg/ha) and red clover (*Trifolium pratense*; 6 kg/ha). During the experimental period (August to December in both years), a rotational grazing system was used whereby the post-grazing sward surface heights were maintained at a minimum of 5 to 7 cm for both ryegrass–clover-based pasture and herb–clover mix, respectively, in order to provide ad-libitum dry matter intakes for lambs and ewes [24,25]. The management of the herbages in this study was similar to the management applied in previous studies [8–12].

2.2. Experimental Design

The studies were conducted at Massey University's Keeble farm, 5 km southeast of Palmerston North, New Zealand (40°24′ S and 175°36′ E). All animal manipulations were approved by the

Massey University Animal Ethics Committee (MUAEC 17-40). Each year, Romney ewes, from a large commercial flock, were naturally bred during a 17-day breeding period. A random subset of ewes diagnosed as twin-bearing using transabdominal ultrasound at approximately 90 days of pregnancy, were included in each year. Throughout the gestation period, within each year, ewes were managed as a single flock (group) under commercial pastoral farming conditions on a ryegrass–clover-based pasture.

Each year, twin-bearing ewes were allocated to treatments three days prior to lambing using a stratified random sampling procedure, in order to balance the groups for ewe live weight. A summary of the experimental design can be found in Table 1. Briefly, in 2017, the study included six treatments (Grass–Grass$_{CW}$, Grass–Herb$_{EW}$, Grass–Herb$_{CW}$, Grass–Herb$_{D15EW}$, Herb–Herb$_{CW}$ and Herb–Herb$_{EW}$, Table 1), while in 2018, Grass–Grass$_{CW}$ was omitted resulting in five treatments (Grass–Herb$_{EW}$, Grass–Herb$_{CW}$, Grass–Herb$_{D15EW}$, Herb–Herb$_{CW}$ and Herb–Herb$_{EW}$). The Grass–Grass$_{CW}$ treatment had previously been examined over three years [10,11] and was determined to be unnecessary as the overall aim was to evaluate the effects of early exposure of lambs to a herb–clover mix prior to early weaning on their subsequent growth and rumen development until a conventional weaning age. Three days prior to lambing, ewes (n = 117 in 2017 and n = 161 in 2018) were allocated to Grass–Herb$_{EW}$, Grass–Herb$_{CW}$, and Grass–Herb$_{D15EW}$ and Grass–Grass$_{CW}$ (only in 2017) treatments and grazed on a ryegrass–clover-based pasture. Similarly, on the same date, ewes (n = 80 in 2017 and n = 82 in 2018) were allocated to Herb–Herb$_{CW}$ and Herb–Herb$_{EW}$ treatments begun grazing the herb–clover mix.

Table 1. Summary of the experimental design with a description of treatments, year, number of lambs in each treatment, treatment herbage description and age of weaning.

Treatment	Year	Number of Lambs	Treatment Description	Lamb Age [1]			
				0	15	47–51	87–99
Grass–Grass$_{CW}$	2017	36	Lambs born and remained on ryegrass–clover-based pasture until conventional weaning	◆	◆	◆	◆,•
Grass–Herb$_{EW}$	2017 2018	36 42	Lambs born on ryegrass–clover-based pasture and early weaned onto herb–clover mix	◆	◆	★,•	★
Grass–Herb$_{CW}$	2017 2018	36 44	Lambs born on ryegrass–clover-based pasture, transferred with dam onto herb–clover mix at early weaning and remained until conventional weaning	◆	◆	★	★,•
Grass–Herb$_{D15EW}$	2017 2018	30 42	Lambs born on ryegrass–clover-based pasture, transferred with dam onto herb–clover mix at 15 days of age and early weaned	◆	★	★,•	★
Herb–Herb$_{CW}$	2017 2018	34 36	Lambs born and remained on herb–clover mix until conventional weaning	★	★	★	★,•
Herb–Herb$_{EW}$	2017 2018	36 40	Lambs born on herb–clover mix and early weaned onto herb–clover mix	★	★	★,•	★

[1] Management of the lambs during each time period. ◆ Lambs on ryegrass–clover-based pasture; ★ Lambs on herb–clover mix; • Weaning age; Early weaned ewes were removed and grazed on ryegrass–clover-based pastures; early weaning occurred at L51 and L44 (average days of age) in 2017 and 2018, respectively; conventional weaning occurred at L99 and L87 days of age in 2017 and 2018, respectively. Grass–GrassCW, Grass–HerbEW and Grass–HerbCW lambs have been used in a previous study [11].

Lambing began on 28 and 27 August in 2017 and 2018 (L0), respectively. Lambs were born over 26- and 28-day period in 2017 and 2018 (mid-point of the lambing period each year was defined as L1, i.e., average day of age at that point was one). All lambs were weighed, ear tagged and identified to their dam within 24 h of birth. In total, n = 136 and n = 138 lambs were born on herb–clover mix in 2017 and 2018, respectively. On the ryegrass–clover-based pasture n = 224 and n = 302 lambs were born in 2017 and 2018, respectively. In accordance with veterinary advice, at L27 (average 27 days of age) in 2017 and L26 in 2018, lambs were drenched with an oral triple combination drench (Matrix, Merial Ancare, Manukau City, New Zealand) and thereafter at 28-day intervals throughout the study at a rate of 1 mL per 5 kg live weight to control internal parasites. This method has been successfully used in previous studies [8–11], therefore, lamb growth in this study was unlikely to be affected by internal parasites.

At L15 (average 15 dayss of age) in both years, ewes and lambs allocated to the Grass–Herb$_{D15EW}$ treatment were transferred from ryegrass–clover-based pasture to a herb–clover mix where they remained until early weaning (at L51 and L44 in 2017 and 2018, respectively, Table 1). Ewes and lambs allocated to the remaining treatments were maintained on their respective herbage until early weaning. At L51 in 2017 and at L44 in 2018 (early weaning), ewes rearing twin lambs (n = 123 and n = 112 in 2017 and 2018, respectively) that were both a minimum of 14 kg live weight (n = 246 and n = 224, in 2017 and 2018, respectively) were included in the study. Any ewes and lambs that did not fulfil these criteria were excluded from the remainder of the study. Lambs in each treatment (n = 42, 44, 38, 42, 46, 34 in Herb–Herb$_{CW}$, Herb–Herb$_{EW}$, Grass–Herb$_{D15EW}$, Grass–Herb$_{CW}$, Grass–Herb$_{EW}$, Grass–Grass$_{CW}$ in 2017 and n = 36, 36, 42, 34, 34 in Herb–Herb$_{CW}$, Herb–Herb$_{EW}$, Grass–Herb$_{D15EW}$, Grass–Herb$_{CW}$, Grass–Herb$_{EW}$ in 2018, respectively) remained on these treatments until conventional weaning at L99 in 2017 and L87 in 2018.

After early-weaning, ewes in the Grass–Herb$_{EW}$, Grass–Herb$_{D15EW}$ and Herb–Herb$_{EW}$ treatments (weaned ewes) were managed with Grass–Grass$_{CW}$ treatment in 2017 until L99 or with a commercial flock under pastoral farming conditions on ryegrass–clover-based pasture until L87 in 2018, receiving similar allowances in terms of pasture offered. Early-weaned lambs (Grass–Herb$_{EW}$, Grass–Herb$_{D15EW}$ and Herb–Herb$_{EW}$) were managed with the Herb–Herb$_{CW}$ ewes and lambs until L99 and L87 in 2017 and 2018, respectively. Within 1 to 2 h of birth, ewes develop an exclusive bond with their lambs and reject any alien lamb that attempts to suck for the remainder of the lactation [26]. Therefore, there was no concern regarding the early-weaned lambs suckling from unweaned ewes.

2.3. *Animal Measurements*

Lambs and ewes were weighed within 1 h of removal from herbage at L15, L51, L65 and L99 in 2017 and L15, L44, L61 and L87 in 2018. Ewe body condition score (BCS, scale 1–5, including half units, [27]) was assessed at each weighing by a single experienced operator.

2.4. *Anatomical and Histological Examination of the Digestive System*

A subset of male lambs from the complete twin sets were selected for anatomical and histological examination of the digestive system at L44 (n = 12) and L87 (n = 24) in 2018 to determine the effect of feed type prior to early weaning, and the combination of feed type and early weaning on the rumen development, respectively. Lambs were selected using stratified random sampling procedure, based on live weights, in order to obtain a representative sample of lambs from each treatment. At L44, lambs in the Grass–Herb$_{EW}$ (n = 6, average live weight 19.0 ± 1.5) and Herb–Herb$_{EW}$ (n = 6, average live weight 21.0 ± 1.5) treatments were euthanised. At L87, 24 male lambs were selected from the Grass–Herb$_{EW}$ (n = 6; average live weight 27.4 ± 2.0), Grass–Herb$_{CW}$ (n = 6; average live weight 31.0 ± 2.0), Herb–Herb$_{CW}$ (n = 6; average live weight 35.0 ± 2.0) and Herb–Herb$_{EW}$ (n = 6; average live weight 32.0 ± 2.0) treatments and euthanised. Lambs were weighed within 1 h of removal from herbage and were euthanised using captive bolt stunning and exsanguination.

Immediately after euthanasia, their skins removed, they were eviscerated, and the head was separated from the carcass. Hot carcass weight was then recorded. The digestive system was removed by severing the esophagus which was tied at the cardia of the stomach. All the organs attached to the digestive system were removed and then weighed. The digestive system was then separated into its components (reticulo-rumen, small intestine and large intestine) which were cleaned using tap water. Empty component weights were recorded to the nearest gram. The cleaned reticulo-rumen was dissected according to the procedure of [28] and the reticulo-rumen was opened and laid flat symmetrically. Two samples (1 cm^2 each) were collected from the right side of the caudal dorsal sac as [29] reported that samples from this region were representative of rumen development.

Rumen tissue samples (two from each animal) were immediately fixed in a 10% formalin saline solution for 24 h. Samples were then processed using a Excelsior ES Tissue Processor (ThermoFisher®, Waltham, MA, USA). Dehydration was carried out by ascending graded series of ethyl alcohol (70, 95% and absolute Alcohol) for 12 h, followed by clearing in Xylene at 60 °C. The samples were then embedded and impregnated in melted Paraffin wax at 55–58 °C using a HistoStar Embedder (ThermoFisher®, MA, USA). Embedded samples were cut using Rotary Microtome (MicroTec®, Wetzlar, Germany) into sections (4 μm). Two sections were produced, 150 μm apart from each tissue sample resulting in 4 sections per animal (replicates). The sections were floated in a Tissue Bath (ThermoFisher®,MA, USA) at 43 °C and were mounted onto adhesive pre-cleaned slides (90° ground edges, 76 mm × 26 mm). The sections were stained using Haemytoxelin and Eosin on an Autostainer XL (Leica®, Wetzlar, Germany). The stained slides, 2 slides per lamb, each contained two sections were covered using a Leica® CV5030 cover slip. All complete papillae in each section were examined microscopically to determine their length and width at the base (Figure 1).

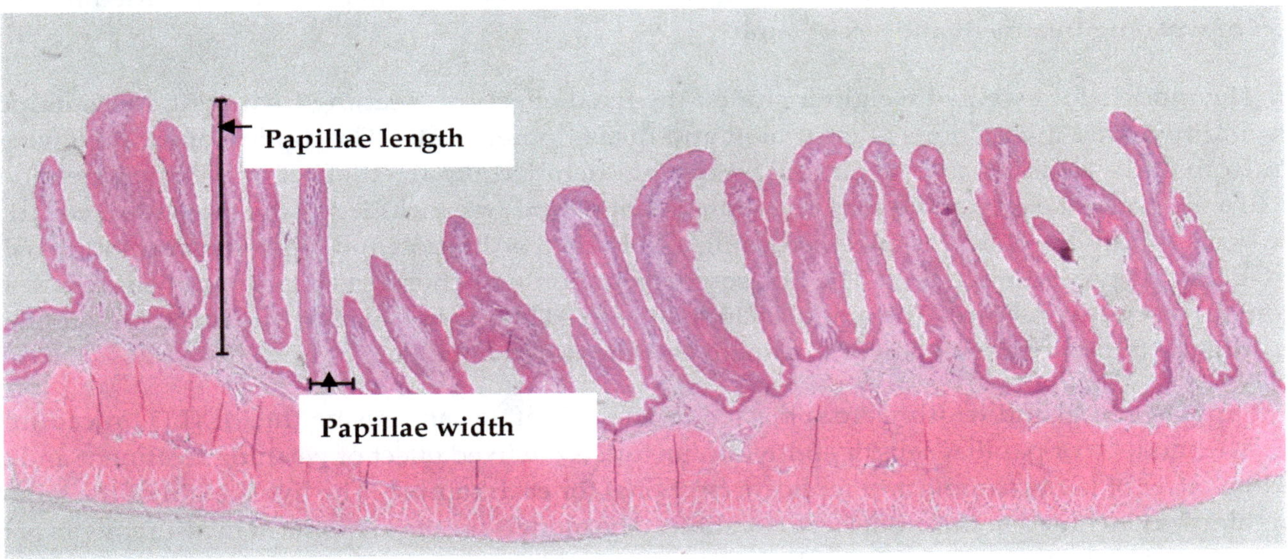

Figure 1. Example of histomorphometric measurements (papillae length and width) of the rumen of lambs measured using a computerised micrometre at 100× magnification.

2.5. Herbage Measurements

Herbage masses were measured at L15, L51 and L99 in 2017 and L15, L44 and L87 in 2018. Four random quadrat cuts (0.1 m^2 each) were taken to ground level from each herbage type at each sampling date using an electric shearing handpiece [30]. Samples were then oven dried to a constant weight to estimate herbage mass. In addition, four composite herbage samples, each containing ten grab samples per herbage type, were also collected, to mimic post grazing height, at each sampling date to determine the botanical and nutritional composition [30]. To determine the botanical composition, a subsample from each composite sample (4 per herbage type) was sorted into each species (herb–clover mix: plantain, chicory, red clover, white clover; ryegrass–clover-based pasture: ryegrass, clover; other grasses (combined), weeds, and dead matter) and then oven dried and weighed to determine the botanical composition. The remaining sample was then frozen, dried, ground, sieved (1 mm) and analysed using in vitro methods to determine the nutritional quality. These measures included dry matter digestibility (DMD, [31]), digestible organic matter digestibility (DOMD, [31]), percentage crude protein (CP; "Dumas" procedure, AOAC method 968.06 using a Leco total combustion method, LECO Corporation, St. Joseph, MI, USA). Percentage acid detergent fibre (ADF) was analysed by a Tecator Fibretec System [32]. Metabolisable energy (ME) content of herbages was calculated from the organic matter digestibility (DOMD × 0.16 MJ/Kg DM, [31]). These herbage measurements have been used in previous studies [8–12] to estimate the dry matter mass, botanical and chemical composition of both ryegrass–clover-based pasture and herb–clover mix.

2.6. Statistical Analysis

The individual animal was considered the experimental unit for these analyses. Live weight of lambs and ewes were subjected to analysis of variance for repeated measures using the MIXED procedure in SAS (Statistical Analysis System, version 9.2; SAS Institute Inc., Cary, NC, USA). Analyses were performed separately for each year due to the differences in the days on which measurements were collected, number of treatments in 2017 and 2018, differences in climate and herbage quality, and different animals used between years. The model for lamb live weight at L51 and L99 and at L44 and L87 in 2017 and 2018, respectively, included the fixed effects of weaning treatment, sex of lamb (male, female), measurement date and the two-way interactions of treatment and measurement date. The live weight of lambs at the start of the treatment period was included in the model as a covariate. The model for lamb liveweight gain from early weaning to conventional weaning included the fixed effect of weaning treatment and sex of lamb.

The models for ewe live weight included the fixed effects of weaning treatment, measurement date and two-way interactions of treatment and measurement date. The model for ewe liveweight gain from early weaning to conventional weaning included the fixed effect of weaning treatment. The live weight of the ewe at the start of the treatment period was included in the model as a covariate. Ewe body condition score was analysed using a Poisson distribution and logit transformation using the GENMOD procedure in SAS. The Poisson distribution was chosen as it is a nonlinear regression model for discrete outcomes. The model included the fixed effects of weaning treatment, measurement date and the two-way interaction of treatment and measurement date.

Papillae length and width were subjected to analysis of variance using the MIXED procedure in SAS. The models for papillae length and width included the fixed effect of weaning treatment and the random effect of lamb, sample and replicate (sections) for each sample. The live weight of the lamb at slaughter was included in the model as a covariate. The models for the anatomical measurements of the rumen included the fixed effect of weaning treatment and lamb live weight at slaughter as a covariate.

The botanical composition of herbages was subjected to an analysis of variance for repeated measures using the MIXED procedure in SAS. The model included fixed effects of plant species and measurement date. Herbage masses were analysed using a model that included herbage type and measurement date as fixed effects. The nutritional quality data were analysed using the MIXED procedure in a model that included the fixed effects of herbage type and measurement date.

3. Results

3.1. Botanical Composition, Herbage Mass and Nutritional Quality of Herbage

The percentage of chicory and clover was 16% greater ($p < 0.05$) and plantain was 23% lower ($p < 0.05$) in the herb–clover mix at L44 in 2018 than at L47 in 2017 (Figure 2). The percentage of plantain was 20% greater ($p < 0.05$) and dead matter content was 35% lower ($p < 0.05$) in the herb–clover mix at L87 in 2018 than at L99 in 2017. Other plant components in the herb–clover mix did not differ ($p > 0.05$) between years and time points. At L15, the percentage of ryegrass in the ryegrass–clover-based pasture was 22% greater ($p < 0.05$) in 2018 than 2017. Other plant components in the ryegrass–clover-based pasture did not differ ($p > 0.05$) between years and time points. In both years, herbage mass of both herbage types was above 1600 kg DM/ha at all sampling dates throughout the study (Table 2).

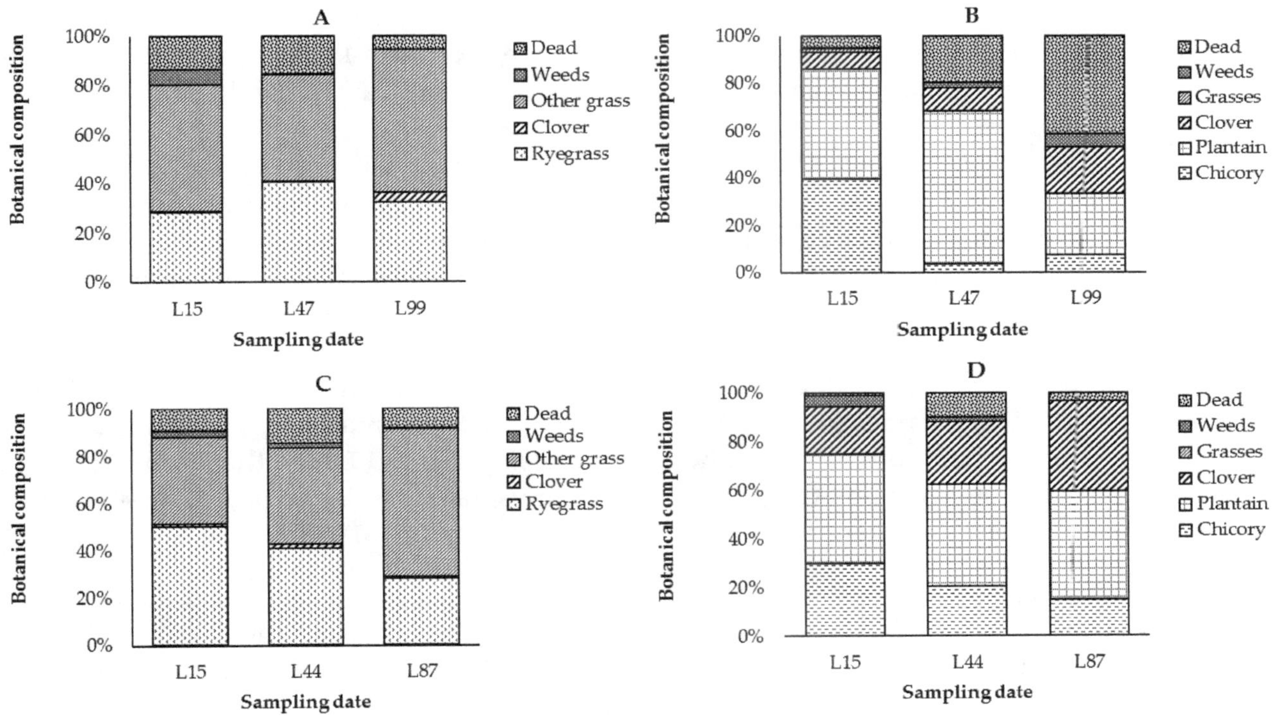

Figure 2. The botanical composition of the components of ryegrass–clover-based pasture (**A**), herb–clover mix (**B**) in 2017 and ryegrass–clover-based pasture (**C**) and herb–clover mix (**D**) in 2018 on 15, 51 and 99 days of age in 2017 and 2018 (L15, L44 and L87).

Table 2. Herbage mass (HM) offered to both lambs and ewes, crude protein (CP), neutral detergent fibre (NDF) acid detergent fibre (ADF), dry matter digestibility (DMD) and metabolisable energy content (ME) of herbages collected L15, L51, L99 in 2017 and 2018 (L15, L44, L87) (least-squares mean ± SEM).

Year	Herbage Type	Study Day	HM (kg DM/ha)	CP (%)	NDF (%)	ADF (%)	DMD (%)	ME (MJ/Kg)
2017	Herb–clover mix	L15	2742 ^{b,c} ± 260	10.9 ^{a} ± 1.1	50.4 ^{c,d} ± 1.4	35.6 ^{b} ± 1.3	65.7 ^{b} ± 0.7	9.2 ^{a,b} ± 0.1
		L51	3221 ^{c} ± 260	17.3 ^{b} ± 1.1	46.7 ^{c} ± 1.4	26.6 ^{a} ± 1.3	65.8 ^{b} ± 0.7	9.5 ^{b} ± 0.1
		L99	3048 ^{b,c} ± 260	15.6 ^{b} ± 1.1	53.7 ^{d} ± 1.4	30.5 ^{b} ± 1.3	63.5 ^{a} ± 0.7	9.1 ^{a} ± 0.1
	Ryegrass–clover-based pasture	L15	2313 ^{a,b} ± 260	16.5 ^{b} ± 1.1	28.7 ^{a} ± 1.4	22.9 ^{a} ± 1.3	75.5 ^{d} ± 0.7	11.0 ^{d} ± 0.1
		L51	1899 ^{a} ± 260	12.0 ^{a} ± 1.1	32.7 ^{b} ± 1.4	26.2 ^{a} ± 1.3	72.0 ^{c} ± 0.7	10.6 ^{c} ± 0.1
		L99	2680 ^{b} ± 260	13.0 ^{a,b} ± 1.1	31.4 ^{a,b} ± 1.4	23.7 ^{a} ± 1.3	73.1 ^{c} ± 0.7	10.6 ^{c} ± 0.1
2018	Herb–clover mix	L15	4298 ^{d} ± 260	14.4 ^{a} ± 1.1	34.0 ^{b} ± 1.4	24.2 ^{b} ± 1.3	71.4 ^{b,c} ± 0.7	10.5 ^{c,d} ± 0.1
		L44	3481 ^{c} ± 260	17.2 ^{b} ± 1.1	27.6 ^{a} ± 1.4	19.5 ^{a} ± 1.3	74.1 ^{d} ± 0.7	11.0 ^{e} ± 0.1
		L87	3857 ^{c} ± 260	12.1 ^{a} ± 1.1	31.7 ^{b} ± 1.4	22.4 ^{a,b} ± 1.3	73.2 ^{c,d} ± 0.7	10.8 ^{d,e} ± 0.1
	Ryegrass–clover-based pasture	L15	1684 ^{a} ± 260	17.3 ^{b} ± 1.1	47.5 ^{d} ± 1.4	26.1 ^{b} ± 1.3	64.7 ^{a} ± 0.7	9.3 ^{a} ± 0.1
		L44	2057 ^{a} ± 260	12.9 ^{a} ± 1.1	48.0 ^{d} ± 1.4	24.6 ^{b} ± 1.3	65.9 ^{a} ± 0.7	9.8 ^{b} ± 0.1
		L87	2068 ^{a} ± 260	18.4 ^{b} ± 1.1	42.4 ^{c} ± 1.4	21.8 ^{a,b} ± 1.3	69.5 ^{b} ± 0.7	10.4 ^{b,c} ± 0.1

L, average days of age. $^{a–e}$ Means with different superscripts within columns are significantly different across years and treatments ($p < 0.05$).

At L15 in 2017, the CP content of ryegrass–clover-based pasture was ~5% greater ($p < 0.05$) than herb–clover mix whereas the reverse was seen at L51. At L99, CP content did not differ ($p > 0.05$) between herbages. At L15 and L99, the ADF content of herb–clover mix was greater (13% and 7%, respectively, $p < 0.05$) than that of ryegrass–clover-based pasture, however, at L51 ADF did not differ ($p > 0.05$). The DMD, neutral detergent fibre (NDF) and ME of herb–clover mix were lower (~6 to 10%, 14 to 22% and 1.1 to 1.8 MJ/Kg, respectively, $p < 0.05$) than ryegrass–clover-based pasture at all sampling times.

In 2018, at L15 and L87, the CP content of herb–clover mix was lower (~3 and 6%, respectively, $p < 0.05$) than ryegrass–clover-based pasture, whereas the CP content of herb–clover mix was ~3% greater ($p < 0.05$) at L44 (Table 2). The ADF content of herb–clover mix did not differ ($p > 0.05$) from that of ryegrass–clover-based pasture at L15 and L87 but was ~5% lower ($p < 0.05$) at L44. The NDF of herb–clover mix was 10 to 20% lower ($p < 0.05$) than that of ryegrass–clover-based pasture at L15, L44 and L87. The DMD and ME of herb–clover mix were greater (~4 to 8% and 0.4 to 1.2 MJ/Kg, respectively, $p < 0.05$) than that of ryegrass–clover-based pasture at L15, L44 and L87.

3.2. Lamb Live Weight and Liveweight Gain

In 2017, at L51 (date of early weaning), the live weights of lambs did not differ ($p > 0.05$) among treatments (Table 3). At L65, the live weight of lambs in the Herb–Herb$_{EW}$, Grass–Herb$_{D15EW}$ and Grass–Herb$_{EW}$ treatments did not differ ($p > 0.05$) but were ~1 kg lighter ($p < 0.05$) than Herb–Herb$_{CW}$, Grass–Herb$_{CW}$ and Grass–Grass$_{CW}$ lambs. At L65, Herb–Herb$_{CW}$ lambs were 0.8 kg heavier ($p < 0.05$) than Grass–Grass$_{CW}$ lambs but neither differed ($p > 0.05$) from Grass–Herb$_{CW}$ lambs. At L99 (at conventional weaning), Herb–Herb$_{CW}$ and Grass–Herb$_{CW}$ lambs did not differ ($p > 0.05$) in live weight but were 4 to 5 kg heavier ($p < 0.05$) than lambs in all other treatments, which did not differ ($p > 0.05$) from one another. Lamb liveweight gains between L51 and L99 in the Herb–Herb$_{CW}$ and Grass–Herb$_{CW}$ treatments were greater ($p < 0.05$) than lambs in the Herb–Herb$_{EW}$, Grass–Herb$_{D15EW}$, Grass–Herb$_{EW}$ and Grass–Grass$_{CW}$ treatments that did not differ ($p > 0.05$).

Table 3. Impact of weaning treatment; Herb–Herb$_{CW}$, Herb–Herb$_{EW}$, Grass–Herb$_{D15EW}$, Grass–Herb$_{CW}$, Grass–Herb$_{EW}$, Grass–Grass$_{CW}$ on live weight of lambs on L51, L65 and L99 in 2017 and L44, L61 and L87 in 2018; and liveweight gain (LWG) during L51–L99 in 2017 and during L44–L87 in 2018 (least-squares mean ± SEM).

Herbage Treatment	Lamb Live Weight (kg)						LWG (g/day)
	n		*n*		*n*		
2017							
		L51		L65		L99	L51–L99
Herb–Herb$_{CW}$	42	19.3 ± 0.2 ^{a}	41	23.5 ± 0.2 ^{d}	42	34.7 ± 0.4 ^{f}	317 ± 7 ^{b}
Herb–Herb$_{EW}$	44	19.4 ± 0.2 ^{a}	43	21.9 ± 0.2 ^{b}	43	31.1 ± 0.4 ^{e}	244 ± 7 ^{a}
Grass–Herb$_{D15EW}$	38	19.1 ± 0.2 ^{a}	38	21.7 ± 0.2 ^{b}	38	30.1 ± 0.4 ^{e}	231 ± 8 ^{a}
Grass–Herb$_{CW}$	42	19.3 ± 0.2 ^{a}	41	23.1 ± 0.2 ^{c,d}	42	34.6 ± 0.4 ^{f}	318 ± 7 ^{b}
Grass–Herb$_{EW}$	46	19.0 ± 0.2 ^{a}	45	21.6 ± 0.2 ^{b}	46	30.6 ± 0.4 ^{e}	239 ± 7 ^{a}
Grass–Grass$_{CW}$	34	19.2 ± 0.2 ^{a}	33	22.7 ± 0.2 ^{c}	34	30.3 ± 0.4 ^{e}	231 ± 8 ^{a}
2018							
		L44		L61		L87	L44–L87
Herb–Herb$_{CW}$	36	19.0 ± 0.1 ^{a}	36	25.0 ± 0.2 ^{c}	35	31.7 ± 0.4 ^{e}	289 ± 8 ^{b}
Herb–Herb$_{EW}$	36	18.9 ± 0.1 ^{a}	34	23.0 ± 0.2 ^{b}	33	28.9 ± 0.4 ^{d}	232 ± 8 ^{a}
Grass–Herb$_{D15EW}$	42	18.9 ± 0.1 ^{a}	41	22.7 ± 0.2 ^{b}	41	28.7 ± 0.3 ^{d}	231 ± 7 ^{a}
Grass–Herb$_{CW}$	34	18.8 ± 0.1 ^{a}	34	24.8 ± 0.2 ^{c}	34	32.1 ± 0.3 ^{e}	321 ± 8 ^{c}
Grass–Herb$_{EW}$	34	18.8 ± 0.1 ^{a}	34	22.6 ± 0.2 ^{b}	34	28.7 ± 0.4 ^{d}	237 ± 8 ^{a}

L, average days of age; Herb–Herb$_{CW}$, lambs born on herb–clover mix and conventional weaning (at ~99 days of age); Herb–Herb$_{EW}$, lambs born on herb–clover mix and early weaned onto herb–clover mix at ~45 days of age; Grass–Herb$_{D15EW}$, lambs born on ryegrass–clover-based pasture and transferred with dam onto herb–clover mix at 15 days of age and early weaned at ~45 days of age; Grass–Herb$_{CW}$, lambs born on ryegrass–clover-based pasture and transferred with dam onto herb–clover mix at ~45 days of age and conventional weaning; Grass–Herb$_{EW}$, lambs born on ryegrass–clover-based pasture and early weaned at ~45 days of age onto herb–clover mix; Grass–Grass$_{CW}$, lambs born on ryegrass–clover-based pasture and conventional weaning. $^{a–f}$ Means with different superscripts are significantly different across years and treatments.

In 2018, at L44 (date of early weaning), the live weights of lambs did not differ ($p > 0.05$) among the treatments (Table 3). At L61 and L87, Herb–Herb$_{CW}$ and Grass–Herb$_{CW}$ lambs were ~2 kg heavier ($p < 0.05$) than Herb–Herb$_{EW}$, Grass–Herb$_{D15EW}$ and Grass–Herb$_{EW}$ lambs that did not differ ($p > 0.05$). Liveweight gain of Grass–Herb$_{CW}$ lambs between L44 and L87 was greater ($p < 0.05$) than Herb–Herb$_{CW}$ lambs, which in turn was greater ($p < 0.05$) than Herb–Herb$_{EW}$, Grass–Herb$_{D15EW}$ and Grass–Herb$_{EW}$ lambs, which did not differ ($p > 0.05$).

3.3. Ewe Live Weight and Liveweight Gain

In 2017, at L51 (date of early weaning), the live weights of ewes did not differ ($p > 0.05$) between treatments (Table 4). At L65, the live weights of Herb–Herb$_{EW}$, Herb–Herb$_{CW}$, Grass–Herb$_{D15EW}$, Grass–Herb$_{CW}$ ewes did not differ ($p > 0.05$). Herb–Herb$_{EW}$ ewes were 3 kg heavier ($p < 0.05$) than Grass–Herb$_{EW}$ and Grass–Grass$_{CW}$ ewes whose live weight did not differ ($p > 0.05$). At L99 (at conventional weaning), Herb–Herb$_{CW}$, Herb–Herb$_{EW}$, Grass–Herb$_{D15EW}$, Grass–Herb$_{CW}$ ewes did not differ ($p > 0.05$) in live weight but were 2 to 5 kg heavier ($p < 0.05$) than Grass–Herb$_{EW}$ and Grass–Grass$_{CW}$ ewes. Between L51 and L99, the liveweight gain of Herb–Herb$_{CW}$, Herb–Herb$_{EW}$, Grass–Herb$_{D15EW}$, Grass–Herb$_{CW}$ and Grass–Herb$_{EW}$ ewes did not differ ($p > 0.05$) but were at least 30 g/day greater ($p < 0.05$) than Grass–Grass$_{CW}$ ewes.

Table 4. Impact of weaning treatment; Herb–Herb$_{CW}$, Herb–Herb$_{EW}$, Grass–Herb$_{D15EW}$, Grass–Herb$_{CW}$, Grass–Herb$_{EW}$, Grass–Grass$_{CW}$ on live weight of ewes on L51, L65 and L99 in 2017 and L44, L61 and L87 in 2018; and liveweight gain (LWG) during L51–L99 in 2017 and during L44–L87 in 2018 (least-squares mean ± SEM).

Weaning Treatment	Ewe Live Weight (kg)						LWG (g/day)
	n		*n*		*n*		
2017							
		L51		L65		L99	L51–L99
Herb–Herb$_{CW}$	21	68.8 ± 0.9 ^{a}	21	71.0 ± 0.9 ^{a,b}	20	76.3 ± 1.0 ^{b}	145 ± 18 ^{b}
Herb–Herb$_{EW}$	22	70.0 ± 0.8 ^{a}	22	73.0 ± 0.8 ^{b}	21	76.7 ± 0.9 ^{b}	129 ± 17 ^{b}
Grass–Herb$_{D15EW}$	18	70.1 ± 0.9 ^{a}	18	71.7 ± 1.0 ^{a,b}	18	76.3 ± 1.0 ^{b}	119 ± 19 ^{b}
Grass–Herb$_{CW}$	21	70.0 ± 0.9 ^{a}	20	71.1 ± 0.9 ^{a,b}	21	77.1 ± 0.9 ^{b}	136 ± 17 ^{b}
Grass–Herb$_{EW}$	23	68.2 ± 0.9 ^{a}	23	70.5 ± 0.9 ^{a}	22	73.9 ± 0.9 ^{a}	112 ± 17 ^{b}
Grass–Grass$_{CW}$	17	68.1 ± 0.9 ^{a}	16	70.4 ± 1.0 ^{a}	17	72.0 ± 1.0 ^{a}	78 ± 19 ^{a}
2018							
		L44		L61		L87	L44–L87
Herb–Herb$_{CW}$	18	69.5 ± 1.0 ^{a}	18	72.2 ± 1.0 ^{a}	18	74.0 ± 1.0 ^{a}	113 ± 19 ^{a}
Herb–Herb$_{EW}$	18	71.2 ± 0.9 ^{a}	17	72.8 ± 1.0 ^{a}	17	74.7 ± 1.0 ^{a}	81 ± 19 ^{a}
Grass–Herb$_{D15EW}$	21	71.0 ± 0.9 ^{a}	21	72.8 ± 0.9 ^{a}	21	74.5 ± 0.9 ^{a}	76 ± 18 ^{a}
Grass–Herb$_{CW}$	17	69.8 ± 0.8 ^{a}	16	73.8 ± 0.8 ^{a}	16	73.9 ± 0.9 ^{a}	89 ± 17 ^{a}
Grass–Herb$_{EW}$	17	70.1 ± 0.9 ^{a}	17	74.3 ± 0.9 ^{a}	14	74.6 ± 1.0 ^{a}	103 ± 19 ^{a}

L, average days of age; Herb–Herb$_{CW}$, lambs born on herb–clover mix and conventional weaning (at ~99 days of age); Herb–Herb$_{EW}$, lambs born on herb–clover mix and early weaned onto herb–clover mix at ~45 days of age; Grass–Herb$_{D15EW}$, lambs born on ryegrass–clover-based pasture and transferred with dam onto herb–clover mix at 15 days of age and early weaned at ~45 days of age; Grass–Herb$_{CW}$, lambs born on ryegrass–clover-based pasture and transferred with dam onto herb–clover mix at ~45 days of age and conventional weaning; Grass–Herb$_{EW}$, lambs born on ryegrass–clover-based pasture and early weaned at ~45 days of age onto herb–clover mix; Grass–Grass$_{CW}$, lambs born on ryegrass–clover-based pasture and conventional weaning. $^{a–b}$ Means with different superscripts are significantly different across years and treatments.

In 2018, at L44, L61 and L87, the live weights of ewes in all five treatments did not differ ($p > 0.05$) (Table 4). Similarly, the liveweight gain of ewes between L44 and L87 in each treatment did not differ ($p > 0.05$).

3.4. Ewe Body Condition Score

In 2017, at L51 (date of early weaning), the BCS of ewes did not differ ($p > 0.05$) between treatments (Table 5). At L65, the BCS of Herb–Herb$_{CW}$, Herb–Herb$_{EW}$, Grass–Herb$_{D15EW}$ and Grass–Herb$_{EW}$ ewes did not differ ($p > 0.05$). The BCS of Grass–Herb$_{CW}$ and Grass–Grass$_{CW}$ ewes did not differ

($p > 0.05$) but were lower ($p < 0.05$) than Herb–Herb$_{EW}$, Grass–Herb$_{D15EW}$ and Grass–Herb$_{EW}$ ewes. At L99 (at conventional weaning), the BCS of Herb–Herb$_{CW}$, Herb–Herb$_{EW}$, Grass–Herb$_{D15EW}$ and Grass–Herb$_{EW}$ and Grass–Herb$_{CW}$ were greater ($p < 0.05$) than Grass–Grass$_{CW}$ ewes.

Table 5. Impact of weaning treatment; Herb–Herb$_{CW}$, Herb–Herb$_{EW}$, Grass–Herb$_{D15EW}$, Grass–Herb$_{CW}$, Grass–Herb$_{EW}$, Grass–Grass$_{CW}$ on the BCS of ewes at L51, L65 and L99 in 2017 and at L44, L61 and L87 in 2018 (results displayed as back transformed logit mean and 95% confidence interval).

Herbage Treatment	Ewe Body Condition Score					
	n		*n*		*n*	
			2017			
		L51		L65		L99
Herb–Herb$_{CW}$	21	2.9 (2.5–2.9) ^{a}	21	3.1 (3.1–3.6) ^{a,b}	20	3.5 (3.1–3.4) ^{b}
Herb–Herb$_{EW}$	22	2.9 (2.8–3.1) ^{a}	22	3.3 (3.1–3.6) ^{b}	21	3.3 (3.1–3.6) ^{b}
Grass–Herb$_{D15EW}$	18	2.9 (2.7–3.1) ^{a}	18	3.4 (3.2–3.7) ^{b}	18	3.5 (3.3–3.7) ^{b}
Grass–Herb$_{CW}$	21	2.7 (2.5–2.9) ^{a}	20	2.8 (2.6–3.1) ^{a}	21	3.3 (3.1–3.4) ^{b}
Grass–Herb$_{EW}$	23	2.9 (2.8–3.1) ^{a}	23	3.4 (3.1–3.6) ^{b}	22	3.4 (3.2–3.5) ^{b}
Grass–Grass$_{CW}$	17	2.8 (2.6–3.0) ^{a}	16	2.9 (2.2–3.0) ^{a}	17	2.9 (2.7–3.1) ^{a}
			2018			
		L44		L61		L87
Herb–Herb$_{CW}$	18	2.5 (2.5–2.9) ^{b}	18	2.8 (2.6–3.0) ^{a,b}	18	3.1 (2.8–3.4) ^{a,b}
Herb–Herb$_{EW}$	18	2.9 (2.7–3.3) ^{b}	17	3.0 (2.8–3.3) ^{b}	17	3.1 (2.8–3.6) ^{a,b}
Grass–Herb$_{D15EW}$	21	2.9 (2.6–3.2) ^{b}	21	3.0 (2.9–3.2) ^{b}	21	3.4 (3.2–3.7) ^{b}
Grass–Herb$_{CW}$	18	2.4 (2.2–2.6) ^{a}	16	2.6 (2.4–2.7) ^{a}	16	2.8 (2.6–3.1) ^{a}
Grass–Herb$_{EW}$	17	2.3 (2.1–2.5) ^{a}	17	2.9 (2.6–3.1) ^{a,b}	14	3.3 (3.3–3.8) ^{b}

L, average days of age; Herb–Herb$_{CW}$, lambs born on herb–clover mix and conventional weaning (at ~99 days of age); Herb–Herb$_{EW}$, lambs born on herb–clover mix and early weaned onto herb–clover mix at ~45 days of age; Grass–Herb$_{D15EW}$, lambs born on ryegrass–clover-based pasture and transferred with dam onto herb–clover mix at 15 days of age and early weaned at ~45 days of age; Grass–Herb$_{CW}$, lambs born on ryegrass–clover-based pasture and transferred with dam onto herb–clover mix at ~45 days of age and conventional weaning; Grass–Herb$_{EW}$, lambs born on ryegrass–clover-based pasture and early weaned at ~45 days of age onto herb–clover mix; Grass–Grass$_{CW}$, lambs born on ryegrass–clover-based pasture and conventional weaning. $^{a-b}$ Means with different superscripts are significantly different across years and treatments.

In 2018, at L44 (date of early weaning), the BCS of Herb–Herb$_{CW}$, Herb–Herb$_{EW}$ and Grass–Herb$_{D15EW}$ ewes were greater ($p < 0.05$) than the Grass–Herb$_{CW}$ and Grass–Herb$_{EW}$ ewes that did not differ ($p > 0.05$). At L61, the BCS of Grass–Herb$_{CW}$ ewes was lower ($p < 0.05$) than Herb–Herb$_{EW}$ and Grass–Herb$_{D15EW}$ ewes but was similar ($p > 0.05$) to Herb–Herb$_{CW}$ and Grass–Herb$_{EW}$ ewes. At L87 (at conventional weaning), the BCS of Herb–Herb$_{CW}$, Herb–Herb$_{EW}$, Grass–Herb$_{D15EW}$ and Grass–Herb$_{EW}$ ewes did not differ ($p > 0.05$). The BCS of Grass–Herb$_{CW}$ ewes was lower ($p < 0.05$) than Grass–Herb$_{D15EW}$ and Grass–Herb$_{EW}$ ewes but was similar ($p > 0.05$) to that of Herb–Herb$_{CW}$ and Herb–Herb$_{EW}$ ewes.

3.5. *Digestive Tract Components and Rumen Papillae Dimensions of Lambs in 2018*

The weight of the empty digestive tract components of lambs between treatments at either L44 (at early weaning) or L87 (at conventional weaning) did not differ ($p > 0.05$; Table 6). At L44, papillae length and width of Grass–Herb$_{EW}$ and Herb–Herb$_{EW}$ lambs did not differ ($p > 0.05$; Table 6). At L87, Grass–Herb$_{EW}$ and Herb–Herb$_{EW}$ lambs had ~200 µm longer ($p < 0.05$) papillae than Grass–Herb$_{CW}$ and Herb–Herb$_{CW}$ lambs, however, the width of ruminal papillae did not differ ($p > 0.05$) between any of the treatments.

Table 6. Mean weight of hot carcass, total digestive tract, reticulo-rumen, empty small intestine and empty large intestine and mean length and width of rumen papillae of Herb–Herb$_{EW}$ and Grass–Herb$_{EW}$ lambs at L44; and Grass–Herb$_{CW}$, Herb–Herb$_{CW}$, Herb–Herb$_{EW}$ and Grass–Herb$_{EW}$ lambs at L87 in 2018 (least-squares mean ± SEM).

Treatment		Hot Carcass Weight (kg)	Total Digestive Tract Weight (kg)	Reticulo-Rumen Weight (g)	Empty Small Intestine Weight (g)	Empty Large Intestine Weight (g)	Papillae Length (μm)	Papillae Width (μm)
Herb–Herb$_{EW}$	L44	8.9 ± 0.3	3.5 ± 0.3	458 ± 40	458 ± 40	296 ± 51	809 ± 136	268 ± 16
Grass–Herb$_{EW}$		9.6 ± 0.3	3.3 ± 0.3	410 ± 40	410 ± 40	234 ± 51	732 ± 148	260 ± 17
Grass–Herb$_{CW}$	L87	14.7 ± 0.3	10.1 ± 0.3	951 ± 50	1353 ± 85	483 ± 80	892 ± 135 ^{a}	221 ± 16
Herb–Herb$_{CW}$		14.6 ± 0.3	10.3 ± 0.3	884 ± 53	1275 ± 90	425 ± 85	1090 ± 152 ^{a}	211 ± 18
Herb–Herb$_{EW}$		14.3 ± 0.3	10.8 ± 0.3	1013 ± 51	1119 ± 86	611 ± 80	1225 ± 137 ^{b}	242 ± 16
Grass–Herb$_{EW}$		14.0 ± 0.3	10.9 ± 0.3	970 ± 55	1280 ± 93	463 ± 87	1333 ± 122 ^{b}	245 ± 14

L, average days of age; Herb–Herb$_{EW}$, lambs born on herb–clover mix and early weaned onto herb–clover mix at ~45 days of age; Grass–Herb$_{EW}$, lambs born on ryegrass–clover-based pasture and early weaned at ~45 days of age onto herb–clover mix; Grass–Herb$_{CW}$, lambs born on ryegrass–clover-based pasture and transferred with dam onto herb–clover mix at ~45 days of age and conventional weaning; Herb–Herb$_{CW}$, lambs born on herb–clover mix and conventional weaning (at ~99 days of age). ^{a,b} Means with different superscripts are significantly different across treatments.

4. Discussion

The duration of the exposure of lambs to a herb–clover mix prior to early weaning, either from birth to early weaning, or from 15 days of age (L15) to early weaning, had no effect on lamb growth rates post-early weaning or their live weights at conventional weaning age, in either year of the current study. This suggests that this management option is not needed to improve the growth of lambs after early weaning onto herb–clover mix. This finding is supported by the limited differences in rumen measurements observed in 2018. In early life, lambs rely primarily on milk to fulfil their nutritional requirements and herbage consumption is low [33]. In the present study, lambs had ad-libitum access to their dams' milk prior to early weaning, potentially limiting their need to consume herbage to fulfil their nutritional requirements thereby mitigating against any potential advantage from prolonged early exposure to the herb–clover mix. Combined, these results, and those from previous studies [9–11], indicate that a four-day adaptation period to the herb–clover mix, before early weaning, is all that is required to ensure adequate growth to a traditional weaning age.

Lambs weaned early onto a herb–clover mix grew at a similar rate, and were as heavy at conventional weaning, as unweaned lambs on a ryegrass–clover-based pasture in 2017. This was the only year that this later treatment was utilised. The growth rates achieved in the current study for early-weaned lambs on a herb–clover mix, were similar to those previously reported for commercially-reared-twin lambs on ryegrass white clover in New Zealand [34]. This suggests that lambs weaned at ~45 days of age and at a minimum live weight of 14 kg onto a herb–clover mix, have the potential to achieve liveweight gains similar to unweaned lambs reared on ryegrass–clover-based pasture. This adequate level of growth of lambs weaned early onto a herb–clover mix was likely driven by two potential factors. Firstly, increased herbage intake, due to herbs and clovers having lower fiber content and faster rumen flow through rates, than ryegrass [35]. Secondly, through greater preferential selection of different plant species [33] as lambs prefer herbs and clovers over ryegrass [36], or a combination of these variables. These results, in combination with previous studies [9–11] indicate early-weaning, onto a herb–clover mix is a management tool that can be used by farmers without risk of poor live weights at ~100 days of age.

In both years of the current study, lambs weaned early onto a herb–clover mix had lower liveweight gains to a conventional weaning age, than unweaned lambs on a herb–clover mix. This was likely due to unweaned lambs having greater total nutrient intake from both milk and herbage, compared to early-weaned lambs that had access only to herbage post-early weaning. Interestingly, the papillae length of early-weaned lambs was 200 μm greater than unweaned lambs on the herb–clover mix at conventional weaning age in 2018. The longer papillae of early-weaned lambs were likely driven by the need for greater dry matter intake to fulfil their nutritional requirements. Dry matter supplementation early in life is known to improve dry matter intake, rumination, the establishment of rumen microflora and alter feeding behaviors of neonatal ruminates [37]. Previously it has been shown that the development of the rumen, either reticulo-rumen capacity or the papillae length, of lambs did not differ at 90 days of age, even under conditions when differences in lamb growth and dry matter intake occurred [38]. That finding, combined with the present finding in regards to papillae length, may suggest that prior to day 90, any potential positive effects of rumen development on lamb growth may not occur and that studies should consider examining lambs for longer term impacts, post a traditional weaning age.

Lambs left unweaned on the herb–clover mix were 4 kg heavier at conventional weaning than those on ryegrass–clover-based pastures in 2017, supporting previous findings [9–11]. In both years of the study, their variations in herbage availability and quality were observed, although not at levels which would have restricted lamb intake [24,25]. Similarly, herbage quality and, in particular, ME and NDF were not inadequate to meet the requirements for lamb growth [39]. These parameters, therefore, do not explain why unweaned lamb growth was greater in the herb–clover mix treatment than unweaned lambs on ryegrass–clover-based pastures. As eluded to earlier, the likely explanation for the greater growth rate of unweaned lambs on herb–clover mix is the greater milk production of their dams, compared to those on ryegrass white clover [40], although the milk and herbage intake of lambs was not measured in this study. The results of this study, and previous studies [9–11] indicate grazing unweaned lambs on herb–clover mix is a management tool that farmers can utilize to increase lamb live weight at the conventional weaning age.

This study also allowed for the examination of the impact of moving ewes and lambs from a ryegrass pasture onto herb–clover mix at approximately 45 days of age without early weaning. The results in 2017 indicated that lambs were heavier and grew faster to conventional weaning than unweaned lambs that remained on ryegrass–clover-based pastures. Farmers, therefore, could consider moving ewes and lambs onto the herb–clover mix as an alternative to lambing ewes on a herb–clover mix. Further, unweaned lambs moved to a herb–clover mix at approximately 45 days of age grew at similar rates in 2017 and faster in 2018 than lambs reared with their dams on herb–clover mix from birth to conventional weaning. This suggests that farmers could use ryegrass–clover-based pasture for the lambing period and then move ewes and lambs to a herb–clover mix at 45 days of age to improve lamb weaning weights.

Ewes whose lambs were weaned early and offered ryegrass–clover-based pasture until conventional weaning had greater liveweight gains and BCS than unweaned ewes on ryegrass–clover-based pasture. This was likely due to the cessation of lactation, allowing ewes to partition nutrients to gaining live weight and was consistent with previous studies [10,11]. In both years, unweaned ewes on the herb–clover mix also had similar liveweight gains from early weaning to conventional weaning compared to ewes whose lambs were weaned early and offered ryegrass–clover-based pasture, matching previous study results [11]. Combined, these results suggest that ewes can either be weaned early onto ryegrass–clover-based pasture or left unweaned on herb–clover mix and achieve greater live weights than ewes on ryegrass–clover-based pasture with their lambs. Greater live weights at weaning can lead to greater live weights at mating which has been reported to have a positive impact on ewe reproductive performance [41]. Early weaning can be used as a technique to improve future ewe live weights and reproductive performance.

5. Conclusions

Early-weaned lambs on a herb–clover mix had similar liveweight gains to conventional weaning as lambs left unweaned on a ryegrass–clover-based pasture. Prolonged exposure of lambs to the herb–clover mix prior to early weaning, however, had no impact on their rumen development or subsequent growth, therefore, this management option cannot be used to improve performance post-early weaning. Ewes whose lambs were weaned early gained greater live weights and BCS compared to ewes which remained with their lambs on a ryegrass–clover-based pasture. Early weaning of lambs with a herb–clover mix is a management tool that farmers can use to achieve adequate growth in their lambs to a conventional weaning age whilst allowing their ewes to gain additional live weight. Greater lamb performance to conventional weaning, however, was achieved when lambs were left unweaned on the herb–clover mix.

Author Contributions: Conceptualization, L.J.E.; investigation and supervision, R.A.C.-T., L.M.C., P.R.K., S.T.M. and S.J.P.; writing—review and editing, L.J.E., R.A.C.-T., L.M.C., P.R.K. and S.T.M. All authors have read and agreed to the published version of the manuscript.

Acknowledgments: The authors wish to thank D. Burnham and Keeble farm staff for their technical assistance.

References

1. Kenyon, P.R.; Webby, R.W. Pastures and supplements in sheep production systems. In *Pasture and Supplements for Grazing Animals*; Rattray, P., Brookes, I.M., Nicol, A., Eds.; New Zealand Society of Animal Production: Hamilton, New Zealand, 2007; pp. 255–274.
2. Muir, P.D.; Smith, N.B.; Wallace, G.J.; Fugle, C.J.; Bown, M.D. Maximizing lamb growth rates. *Proc. N. Z. Soc. Anim. Prod.* **2000**, *62*, 55–58.
3. Mulvaney, F.J.; Morris, S.T.; Kenyon, P.R.; West, D.M.; Morel, P.C.H. The effect of weaning at 10 or 14 weeks of age on liveweight changes in the hogget and her lambs. *Proc. N. Z. Soc. Anim. Prod.* **2009**, *69*, 68–70.
4. Mulvaney, F.J.; Morris, S.T.; Kenyon, P.R.; Morel, P.C.H.; West, D.M. Is there any advantage of early weaning of twin lambs born to yearlings? *Proc. N. Z. Soc. Anim. Prod.* **2011**, *71*, 79–82.
5. Somasiri, S.C. Effect of Herb-Clover Mixes on Weaned Lamb Growth. Ph.D. Thesis, Massey University, Palmerston North, New Zealand, 2014.
6. Cranston, L.M.; Kenyon, P.R.; Morris, S.T.; Kemp, P.D. A review of the use of chicory, plantain, red clover and white clover in a sward mix for increased sheep and beef production. *J. N. Z. Grassl.* **2015**, *77*, 89–94. [CrossRef]
7. Somasiri, S.C.; Kenyon, P.R.; Kemp, P.D.; Morel, P.C.H.; Morris, S.T. Effect of herb-clover mixes of plantain and chicory on yearling lamb production in the early spring period. *Anim. Prod. Sci.* **2016**, *56*, 1662–1668. [CrossRef]
8. Corner-Thomas, R.; Cranston, L.M.; Kemp, P.; Morris, S.T.; Kenyon, P. Can herb-clover mixes compensate for the lack of milk in the diet of early-weaned lambs? *N. Z. J. Agric. Res.* **2020**, *63*, 233–245. [CrossRef]
9. Ekanayake, W.E.M.L.J.; Corner-Thomas, R.A.; Cranston, L.M.; Kenyon, P.R.; Morris, S.T. The effect of live weight at weaning on liveweight gain of early weaned lambs onto a herb-clover mixed sward. *Proc. N. Z. Soc. Anim. Prod.* **2017**, *77*, 37–42.
10. Ekanayake, W.E.M.L.J.; Corner-Thomas, R.A.; Cranston, L.M.; Kenyon, P.R.; Morris, S.T. A comparison of liveweight gain of lambs weaned early onto a herb-clover mixed sward and weaned conventionally onto a ryegrass-clover pasture and herb-clover mixed sward. *Asian-Australas. J. Anim. Sci.* **2019**, *32*, 201–208. [CrossRef]
11. Ekanayake, L.J.; Corner-Thomas, R.A.; Cranston, L.M.; Kenyon, P.R.; Morris, S.T. Lambs Weaned Early onto a Herb-Clover Mix Have the Potential to Grow at a Similar Rate to Unweaned Lambs on a Grass-Predominant Pasture. *Animals* **2020**, *10*, 613. [CrossRef] [PubMed]
12. Ekanayake, L.J.; Corner-Thomas, R.A.; Cranston, L.M.; Hickson, R.E.; Kenyon, P.R.; Morris, S.T. Characterisation of the nutritional composition of plant components of a herb-clover mix during November to May in New Zealand. *N. Z. J. Anim. Sci. Prod.* **2019**, *79*, 162–167.
13. Jiao, J.; Li, X.; Beauchemin, K.A.; Tan, Z.; Tang, S.; Zhou, C. Rumen development process in goats as affected by supplemental feeding v. grazing: Age-related anatomic development, functional achievement and microbial colonisation. *Br. J. Nutr.* **2015**, *113*, 888–900. [CrossRef] [PubMed]
14. Yanez-Ruiz, D.R.; Abecia, L.; Newbold, C.J. Manipulating rumen microbiome and fermentation through interventions during early life: A review. *Front. Microbiol.* **2015**, *6*, 1–12. [CrossRef] [PubMed]

15. Distel, R.A.; Villalba, J.J.; Laborde, H.E. Effects of early experience on voluntary intake of low-quality roughage by sheep. *J. Anim. Sci.* **1994**, *2*, 1191–1195. [CrossRef] [PubMed]
16. Kittelmann, S.; Pinares-Patiño, C.S.; Seedorf, H.; Kirk, M.R.; Ganesh, S.; McEwan, J.C.; Janssen, P.H. Two different bacterial community types are linked with the low-methane emission trait in sheep. *PLoS ONE* **2014**, *9*, 1–9. [CrossRef] [PubMed]
17. Provenza, F.D.; Balph, D.F. Applicability of five diet selection models to various foraging challenges ruminants encounter. In *Behavioural Mechanisms of Food Selection*; Hughes, R.N., Ed.; Springer: Berlin, Germany, 1990; pp. 423–459.
18. Ramos, A.; Tennessen, T. Effect of previous grazing experience on the grazing behaviour of lambs. *Appl. Anim. Behav. Sci.* **1992**, *33*, 43–52. [CrossRef]
19. Parsons, A.J.; Newman, J.A.; Penning, P.D.; Harvey, A.; Orr, R.J. Diet preference of sheep: Effects of recent diet, physiological state and species abundance. *J. Anim. Ecol.* **1994**, *63*, 465–478. [CrossRef]
20. Provenza, F.D. Post-ingestive feedback as an elementary determinant of food preference and intake in ruminants. *J. Range Manag.* **1995**, *48*, 2–17. [CrossRef]
21. Rutter, S.M.; Orr, R.J.; Rook, A.J. Dietary preference for grass and white clover in sheep and cattle: An overview. In *Grazing Management: The Principles and Practice of Grazing, for Profit and Environmental Gain, Within Temperate Grassland Systems*; Rook, A.J., Penning, P.D., Eds.; British Grassland Society Occasional Symposium No. 34; British Grassland Society: Harrogate, UK, 2000; pp. 73–78.
22. Kyriazakis, I.; Oldham, J.D. Diet selection in sheep: The ability of growing lambs to select a diet that meets their crude protein requirements. *Br. J. Nutr.* **1993**, *69*, 617–629. [CrossRef]
23. Kyriazakis, I.; Oldham, J.D.; Coop, R.L.; Jackson, F. The effect of sub-clinical intestinal nematode infection on the diet selection of growing sheep. *Br. J. Nutr.* **1994**, *72*, 665–677. [CrossRef]
24. Morris, S.T.; Kenyon, P.R. The effect of litter size and sward height on ewe and lamb performance. *N. Z. J. Agric. Res.* **2004**, *47*, 275–286. [CrossRef]
25. Cranston, L.M.; Kenyon, P.R.; Morris, S.T.; Lopez Villalobos, N.; Kemp, P.D. Effect of post grazing height on the productivity, population and morphology of a herb and legume mix. *N. Z. J. Agric. Res.* **2015**, *58*, 397–411. [CrossRef]
26. Nowak, R.; Poindron, P. From birth to colostrum: Early steps leading to lamb survival. *Reprod. Nutr. Dev.* **2006**, *46*, 431–446. [CrossRef] [PubMed]
27. Jefferies, B.C. Body condition scoring and its use in management. *Tasman. J. Agric.* **1961**, *32*, 19–21.
28. McGavin, M.D.; Morrill, J.L. Dissection technique for examination of the bovine rumino-reticulum. *J. Anim. Sci.* **1976**, *42*, 535–538. [CrossRef] [PubMed]
29. Lesmeister, K.E.; Tozer, P.R.; Heinrichs, A.J. Development and analysis of a rumen tissue sampling procedure. *J. Dairy Sci.* **2004**, *87*, 1336–1344. [CrossRef]
30. Frame, J. Herbage Mass. In *Sward Measurement Handbook*; Hodgson, J., Baker, R., Davies, D.A., Laidlaw, A.S., Leaver, J., Eds.; British Grassland Society: Kenilworth, UK, 1993; pp. 39–69.
31. Roughan, P.; Holland, R. Predicting in-vivo digestibility of herbages by exhaustive enzymic hydrolysis of cell walls. *J. Sci. Food Agric.* **1977**, *28*, 1057–1064. [CrossRef]
32. Robertson, J.; Van Soest, P. The detergent system of analysis and its application to human foods. In *The Analysis of Dietary Fiber in Food*; James, W.P.T., Theander, O., Eds.; Marcel Dekker Inc.: New York, NY, USA, 1981; pp. 123–158.
33. Kerr, P. *400 Plus—A Guide to Improved Lamb Growth*; New Zealand Sheep Council: Wellington, New Zealand, 2000; p. 160.
34. Litherland, A.J.; Lambert, M.G. Herbage quality and growth rate of single and twin lambs at foot. *Proc. N. Z. Soc. Anim. Prod.* **2000**, *60*, 55–57.
35. Golding, K.P.; Kemp, P.D.; Kenyon, P.R.; Morris, S.T. High weaned lamb live weight gains on herbs. *Agron. N. Z.* **2008**, *38*, 33–39.
36. Pain, S.J.; Hutton, P.G.; Kenyon, P.R.; Morris, S.T.; Kemp, P.D. Preference of lambs for novel pasture herbs. *Proc. N. Z. Soc. Anim. Prod.* **2010**, *70*, 258–287.
37. Xiao, J.; Alugongo, G.M.; Li, J.; Wang, Y.; Li, S.; Cao, Z. Review: How Forage Feeding Early in Life Influences the Growth Rate, Ruminal Environment, and the Establishment of Feeding Behavior in Pre-Weaned Calves. *Animals* **2020**, *10*, 188. [CrossRef]

38. Xie, B.; Huang, W.; Zhang, C.; Diao, Q.; Cui, J.; Wang, S.; Lv, X.; Zhang, N. Influences of starter NDF level on growth performance and rumen development in lambs fed isocaloric and isonitrogenous diets. *J. Anim. Sci.* **2020**, *98*, 1–8. [CrossRef] [PubMed]
39. Hodgson, J.; Brookes, I.M. Nutrition of grazing animals. In *New Zealand Pasture and Crop Science*; Hodgson, J., White, J., Eds.; Oxford University Press: Auckland, New Zealand, 2002; pp. 133–153.
40. Hutton, P.G.; Kenyon, P.R.; Bedi, M.K.; Kemp, P.D.; Stafford, K.J.; West, D.M.; Morris, S.T. A herb and legume sward mix increased milk production and ewe and lamb live weight gain to weaning compared to a ryegrass dominant sward. *Anim. Feed Sci. Technol.* **2011**, *164*, 1–7. [CrossRef]
41. Corner-Thomas, R.A.; Ridler, A.L.; Morris, S.T.; Kenyon, P.R. Ewe lamb live weight and body condition scores affect reproductive rates in commercial flocks. *N. Z. J. Agric. Res.* **2015**, *58*, 26–34. [CrossRef]

7

Influence of the Fermented Feed and Vaccination, and their Interaction on Parameters of Large White/Norwegian Landrace Piglets

Laurynas Vadopalas [1], Sarunas Badaras [1], Modestas Ruzauskas [2,3], Vita Lele [1,4], Vytaute Starkute [1,4], Paulina Zavistanaviciute [1,4], Egle Zokaityte [1,4], Vadims Bartkevics [5,6], Dovile Klupsaite [1], Erika Mozuriene [1,4], Agila Dauksiene [1,2], Sonata Sidlauskiene [1], Romas Gruzauskas [7] and Elena Bartkiene [1,4,*]

1 Institute of Animal Rearing Technologies, Lithuanian University of Health Sciences, Tilžės g. 18, LT-47181 Kaunas, Lithuania; laurynas.vadopalas@lsmuni.lt (L.V.); sarunas.badaras@gmail.com (S.B.); vita.lele@lsmuni.lt (V.L.); vytaute.starkute@lsmuni.lt (V.S.); paulina.zavistanaviciute@lsmuni.lt (P.Z.); egle.zokaityte@lsmuni.lt (E.Z.); dovile.klupsaite@lsmuni.lt (D.K.); erika.mozuriene@lsmuni.lt (E.M.); agila.dauksiene@lsmuni.lt (A.D.); sonata.sidlauskiene@lsmuni.lt (S.S.)

2 Department of Physiology and Anatomy, Lithuanian University of Health Sciences, Tilžės g. 18, LT-47181 Kaunas, Lithuania; modestas.ruzauskas@lsmuni.lt

3 Microbiology and Virology Institute, Lithuanian University of Health Sciences, Tilžės g. 18, LT-47181 Kaunas, Lithuania

4 Department of Food Safety and Quality, Lithuanian University of Health Sciences, Tilžės g. 18, LT-47181 Kaunas, Lithuania

5 University of Latvia Faculty of Chemistry, Jelgavas iela 1, LV-1004 Riga, Latvia; vadims.bartkevics@bior.gov.lv

6 Institute of Food Safety, Animal Health and Environment BIOR, Lejupesiela 3, LV-1076 Riga, Latvia

7 Department of Food Science and Technology, Kaunas University of Technology, Radvilenu str. 19, LT-50254 Kaunas, Lithuania; romas.gruzauskas@ktu.lt

* Correspondence: elena.bartkiene@lsmuni.lt.

Simple Summary: Farm animals are constantly exposed to pathogenic microorganisms, which can lead to bacterial, as well as secondary infections caused by viruses. For this reason, vaccination is used. However, many discussions have been published about the effectiveness of this prophylactic measure. The aim of this study was to evaluate the influence of fermented feed on non-vaccinated (NV) and vaccinated with Circovac porcine circovirus type 2 vaccine piglets' blood parameters, gut microbial composition, growth performance and ammonia emission. The 36-day experiment was conducted using 25-day-old Large White/Norwegian Landrace piglets, which were randomly divided into four groups each comprising 100 piglets (NV piglets fed with soya meal, vaccinated piglets fed with soya meal, NV piglets fed with fermented rapeseed meal and vaccinated piglets fed with fermented rapeseed meal). The results revealed that vaccination, as a separate factor, did not significantly influence piglets' blood parameters. Finally, rapeseed meal fermented with the selected lactic acid bacteria strains can be used instead of expensive imported soya, because the fermented feed did not cause undesirable changes in piglets' health and growth performance. Furthermore, the process is more sustainable.

Abstract: The aim of this study was to evaluate the influence of fermented with a newly isolated lactic acid bacteria (LAB) strains combination (*Lactobacillus plantarum* LUHS122, *Lactobacillus casei* LUHS210, *Lactobacillus farraginis* LUHS206, *Pediococcus acidilactici* LUHS29, *Lactobacillus plantarum* LUHS135 and *Lactobacillus uvarum* LUHS245) feed on non-vaccinated (NV) and vaccinated with Circovac porcine circovirus type 2 vaccine (QI09AA07, CEVA-PHYLAXIA Co. Ltd. Szállás u. 5. 1107 Budapest, Hungary) piglets' blood parameters, gut microbial composition, growth performance and ammonia emission. The 36-day experiment was conducted using 25-day-old Large White/Norwegian

Landrace (LW/NL) piglets, which were randomly divided into four groups with 100 piglets each: S_{nonV}—non-vaccinated piglets fed with control group compound feed; S_V—vaccinated piglets fed with control group compound feed; RF_{nonV}—non-vaccinated piglets fed with fermented compound feed; RF_V—vaccinated piglets fed with fermented compound feed. Samples from 10 animals per group were collected at the beginning and end of the experiment. Metagenomic analysis showed that fermentation had a positive impact on the *Lactobacillus* prevalence during the post-weaning period of pigs, and vaccination had no negative impact on microbial communities. Although a higher amount of *Lactobacillus* was detected in vaccinated, compared with non-vaccinated groups. At the end of experiment, there was a significantly higher LAB count in the faeces of both vaccinated compared to non-vaccinated groups (26.6% for S_V and 17.2% for RF_V), with the highest LAB count in the S_V group. At the end of experiment, the S_V faeces also had the highest total bacteria count (TBC). The RF_V group had a 13.2% increase in total enterobacteria count (TEC) at the end of experiment, and the S_V group showed a 31.2% higher yeast/mould (Y/M) count. There were no significant differences in the average daily gain (ADG) among the groups; however, there were significant differences in the feed conversion ratios (FCR) between several groups: S_V vs. S_{nonV} (11.5% lower in the S_V group), RF_V vs. RF_{nonV} (10.2% lower in the RFnonV group) and S_V vs. RF_V (21.6% lower in the S_V group). Furthermore, there was a significant, very strong positive correlation between FCR and TEC in piglets' faeces ($R = 0.919$, $p = 0.041$). The lowest ammonia emission was in RF_V group section (58.2, 23.8, and 47.33% lower compared with the S_{nonV}, S_V and RF_{nonV} groups, respectively). Notably, there was lower ammonia emission in vaccinated groups (45.2% lower in S_V vs. S_{nonV} and 47.33% lower in RF_V vs. RF_{nonV}). There was also a significant, very strong positive correlation between ammonia emission and Y/M count in piglets' faeces at the end of the experiment ($R = 0.974$; $p = 0.013$). Vaccination as a separate factor did not significantly influence piglets' blood parameters. Overall, by changing from an extruded soya to cheaper rapeseed meal and applying the fermentation model with the selected LAB combination, it is possible to feed piglets without any undesirable changes in health and growth performance in a more sustainable manner. However, to evaluate the influence of vaccination and its interaction with other parameters (feed, piglets' age, breed, etc.) on piglets' parameters, additional studies should be performed and methods should be standardised to ensure the results may be compared.

Keywords: piglets; fermentation; antimicrobial properties; vaccination; microbiota; ammonia emission

---

1. Introduction

The challenges associated with livestock production become greater each year. This phenomenon is associated with population growth, as well as growing needs for animal-based products. Farms have become bigger, and animals are always exposed to opportunistic and pathogenic microorganisms. Most pathogen infections have a multifunctional influence on the host tissues and cells, and the response to the infection rather than the infection itself results in most of the damage and a general disturbance in physiological processes [1–5]. One consequence of infections on pig production is multisystemic wasting syndrome after weaning. Indeed, it is crucial to ensure piglets health at the earliest stage, because type 2 porcine circovirus (PCV2) leads to multisystemic wasting syndrome after weaning, and co-infection with other pathogens can be lethal [6,7]. Some published findings have indicated the potential participation of PPV2 in clinical diseases. PCV2 can occur as a coinfection with other pathogens in animals afflicted with respiratory disease [8], as well as in the lungs of animals with inflammatory lesions [9]. PCV2 infection leads to metabolic and neurological disorders; the latter cause multisystemic wasting syndrome [10–12]. To improve animal keeping conditions, understanding the interactions between desirable microorganisms and pathogens and reduce the risk of secondary infections is critical to develop preventive measures against various diseases. In this study, we hypothesised that, for vaccinated piglets, incorporating antimicrobial strains for feed fermentation can be a very attractive feeding option. Finally, incorporation and biomodification of local feedstock with antimicrobial lactic acid bacteria (LAB) strains can lead to higher value feedstock for farm animal nutrition and improve their conditions. This use can also increase the sustainability of pig farming and reduce dependence on imported soya stock. In this case, natural antimicrobial treatment, e.g., feed fermentation with antimicrobial LAB strains, is also desirable because the genetic diversity of PCV2 continuously increases, and novel subtypes are still emerging [13–15]. PCV2-related diseases are considered multifactorial in the swine industry [7], and they can promote co-infection with other viruses or bacteria [8,9,16]. In addition, co-infection of different viruses complicates the infection status and makes it more difficult to prevent and control the diseases [17]. Vaccines are used to prevent viral infections and antibiotics are used to treat bacterial diseases, but there is a lack of complex tools for prevention of diseases. Furthermore, viruses and bacteria can mutate very quickly, a phenomenon that remains a huge challenge for the livestock industry. Animals are always exposed to the environment, and thus changes in the microorganisms present in digestive tract should be studied to understand the mechanism of action, which can be associated with vaccination, as well as with the natural antimicrobials, as the microorganisms in the gut strongly influence the immune system and animals' health.

The aim of this study was to evaluate the influence of fermented with a newly isolated LAB strains combination (*Lactobacillus plantarum* LUHS122, *Lactobacillus casei* LUHS210, *Lactobacillus farraginis* LUHS206, *Pediococcus acidilactici* LUHS29, *Lactobacillus plantarum* LUHS135 and *Lactobacillus uvarum* LUHS245) feed on non-vaccinated (NV) and vaccinated with Circovac porcine circovirus type 2 vaccine (QI09AA07, CEVA-PHYLAXIA Co. Ltd. Szállás u. 5. 1107 Budapest, Hungary) piglets' blood parameters, gut microbial composition, growth performance and ammonia emission. This LAB starter combination, with antimicrobial characteristics, was used to ferment local feed stock (rapeseed meal), and the influence of changing from an extruded soya to biomodified rapeseed meal on the parameters of non-vaccinated piglets and those vaccinated with the Circovac PCV2 vaccine was evaluated.

2. Materials and Methods

2.1. LAB Strains Used for Feed Fermentation, Feed Fermentation and Fermented Feed Parameters

The *L. plantarum* LUHS122, *L. casei* LUHS210, *L. farraginis* LUHS206, *P. acidilactici* LUHS29, *L. plantarum* LUHS135 and *L. uvarum* LUHS245 strains were obtained from the Lithuanian University of Health Sciences collection (Kaunas, Lithuania). Our previous studies showed that the above-mentioned strains inhibit various pathogenic and opportunistic microorganisms and are suitable for fermentation of various cereal substrates [18–22], as well for rapeseed meal fermentation [23]. The rapeseed meal, water and LAB strains (equal parts of each strain by volume) suspension (3% from dry matter of feed mass, v/m), containing 8.9 $\log_{10}$ colony forming units CFU/mL, was fermented at 30 ± 2 °C for 12 h. The fermentation process involves wheat and rapeseed meal (1:1 ratio). The crude protein content of the rapeseed meal was 33.58%. Fermented feed was composed of 25% fermented rapeseed meal and 25% fermented wheat. This combination contained 23.24% crude protein. The compound feed for all groups contained 19% crude protein, crude fibre—3.15%, crude oil—6.51%, lysine—1.45%, methionine—0.55%, tryptophan—0.26%, threonine—0.93%, Ca—0.90%, total P—0.59% and Na—0.20%. The entire fermented wheat/rapeseed meal mass was divided in two parts: 70% (by mass) was used for piglet feeding, and 30% (by mass) was used as a starter for additional feed fermentation cycles (Figure 1). The detailed fermented feed technology and characteristics are described by Vadopalas et al. [23].

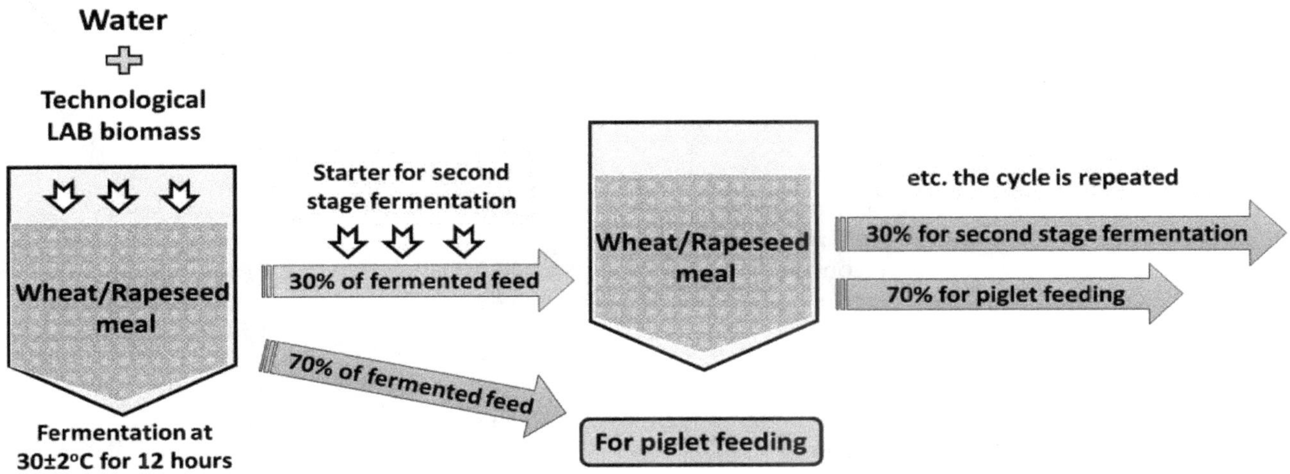

Figure 1. Scheme of fermented feed preparation. The technological lactic acid bacteria (LAB) biomass consisted of *Lactobacillus plantarum* LUHS122, *Lactobacillus casei* LUHS210, *Lactobacillus farraginis* LUHS206, *Pediococcus acidilactici* LUHS29, *Lactobacillus plantarum* LUHS135 and *Lactobacillus uvarum* LUHS245.

2.2. Animals and Housing

All animal procedures were conducted according to the European Union (EU) Directive of the European Parliament and Council, from 22 September, 2010 [24], on the protection of animals used for scientific purposes and the 'Requirements for the Keeping, Maintenance and Use of Animals Intended for Science and Education Purposes', approved by the order of the Lithuanian Director of the State Food and Veterinary Service [25]. The study was conducted at a pig farm in the Klaipeda district (Kontvainiai, Lithuania) and at the Institute of Animal Rearing Technologies, Lithuanian University of Health Sciences (Kaunas, Lithuania). A 36-day experiment was conducted using 25-day-old Large White/Norwegian Landrace (LW/NL) piglets with an initial body weight of 6.9–7.0 kg. Four groups were formed, each with 100 piglets: S_{nonV}—non-vaccinated piglets fed with control group compound feed; S_V—vaccinated piglets fed with control group compound feed; RF_{nonV}—non-vaccinated piglets fed with fermented compound feed; RF_V—vaccinated piglets fed with fermented compound feed. The weaner piglets were kept at the same conditions in a section with two climate zones. The first had a heated concrete floor (36 °C) and roof on it; the second had plastic piglet floors and optimum ventilated air and temperature for the active period. Drinking water and compound liquid feed were available ad libitum throughout the trial. Antibiotic treatment was not applied.

2.3. Experimental Design and Diets

Samples from 10 animals per group were collected at the beginning and end of experiment. Two groups (S_V and RF_V) were vaccinated by intramuscular injection with Circovac PCV2 vaccine (inactivated) (QI09AA07, CEVA-PHYLAXIA Co. Ltd. Szállás u. 5. 1107 Budapest, Hungary), which is used to protect pigs against PCV2. The influence of two dietary treatments on vaccinated and non-vaccinated piglets' parameters were compared. Fermented feed, comprising 500 g/kg of total feed, was included in the diet of treated group beginning at day 25 of life until day 61. Piglets' growth performance was evaluated by testing all 100 piglets in each group. The basal feed was formulated according to the nutritional requirements prescribed in the Nutrient Requirements of Swine [26]. The feed composition and nutritional value are shown in Table 1. The nutritional value of compound feed was determined according to the analytic methods described by Association of Official Analytical Chemists (AOAC) [27].

Table 1. Diet composition (dry matter 87 %).

Ingredients (%)	Control Group	Treated Group
Barley	38.45	33.25
Rapeseed meal	-	25.00
Wheat	32.12	25.00
Soya beans (extruded)	9.30	-
Potato protein	5.00	2.00
Soybean protein concentrate	2.00	-
Whey powder	5.80	5.80
Sunflower oil	2.72	4.51
Limestone	1.48	1.1
NaCl	0.38	0.35
Monocalcium phosphate	0.33	0.41
L-Lysine sulphate	0.87	1.1
DL-Methionine	0.25	0.16
Acidal NC (formic and acetic acids)	0.30	0.30
[1] Vitamins and trace elements (premix)	1.00	1.00
Nutritional value		
ME swine (MJ/kg)	13.86	13.95
Crude protein (%)	19.00	19.00
Crude fat (%)	6.51	6.51
Crude fibre (%)	3.15	5.14
Lysine (%)	1.45	1.45
Methionine (%)	0.55	0.55
Threonine (%)	0.93	0.94
Tryptophan (%)	0.26	0.25
Methionine + Cystine (%)	0.87	0.88
Ca (%)	0.90	0.90
Total P (%)	0.59	0.62
Available P (%)	0.37	0.38
Na (%)	0.20	0.21

ME—metabolisable energy. [1] Composition of premix per 1 kg of feed: Vitamin A—18,180 IU; vitamin D3—2040 IU; vitamin E—161 mg kg^{-1}; vitamin K3—5.03 mg; thiamine—3.64 mg; riboflavin—9.16 mg; choline chloride—404 mg; pyridoxine—4.60 mg; vitamin B12—0.05 mg; niacin—41 mg; pantothenic acid—22.85 mg; folic acid—1.85 mg; biotin—0.21 mg; Fe—152 mg; Cu—100 mg; Zn—91 mg; Mn—80 mg; I—0.81 mg; Co—0.53 mg; Se—0.30 mg.

2.4. Metagenomics and Microbial Profiling Analysis

Before the experiment, faeces from 25-day old piglets from each group (10 piglets per group) were collected. The same procedure, using 10 piglets per group, was performed at the end of the experiment (day 61 of the piglets' life). Faecal samples from each of the group were pooled making 4 samples before and 4 samples after the experiment. Therefore, 8 samples representing 4 tested groups (before and after the experiment) were sequenced for microbial profiling. Specimens were kept in DNA/RNA Shield diluted 1:10 (R1100-250, Zymo Research, Irvine, CA USA) at −80 °C before DNA extraction. DNA was extracted with a faecal DNA MiniPrep kit (D6010, Zymo Research, Irvine, CA USA). Library preparation, metagenomic sequencing and taxonomic characterisation of reads was performed as previously described [28]. ZymoBIOMICS Microbial Community Standard (D6300, Zymo Research, Irvine, CA USA) was used as a microbiome profiling quality control. The results of taxonomic classification were visualised using the interactive online platform https://genome-explorer.com.

2.5. Microbiological Analysis of Faecal Samples

Piglets' faecal samples were collected before and after the experiment, stored in vials (+4 °C) with a transport medium (Faecal Enteric Plus, Oxoid, Basingstoke, UK) and analysed on the same day. Evaluation of the microbiological parameters (LAB, total viable bacteria count [TVC], total enterobacteria count [TEC], and yeast and mould [Y/M] counts) was performed according to methods described by Zavistanaviciute et al. [29].

2.6. Blood Analysis

Piglets were bled from the jugular vein into vacuum blood tubes (BD Vacutainer, Wokingham, UK) before morning feeding. Tubes with clot activator were used for biochemical examination. Blood biochemical variables were evaluated before and after the experiment (on days 25 and 61 of the piglets' life). The parameters included aspartate aminotransferase (AST), alanine aminotransferase (ALT), cholesterol (Chol), high density lipoprotein cholesterol (HDL-C), low-density lipoprotein cholesterol (LDL-C), triglycerides (TG), total protein (TP), albumin (ALB), phosphorus (IP), Mg, K, Na, triiodothyronine (T3), thyroxine (T4), immunoglobulin G (IgG), vitamin B12, albumin (ALB), Fe, glucose (GLU), Ca, creatinine analysed by the Jaffe method (CREA), alkaline phosphatase (AP), thyroid-stimulating hormone (TSH), total bilirubin and urea. Blood parameters were analysed with an automatic biochemistry analyser in the accredited laboratory 'Anteja' (Klaipeda, Lithuania).

2.7. Evaluation of Piglets' Growth Performance and Cases of Mortality and Diarrhoea

Group body weight gain (BWG) was recorded on days 25, 32, 39, 46, 53, and 61 of age using an electronic weighing system (model type: IT1000, SysTec GmbH, Bergheim, Germany) and average daily gain (ADG) was calculated. The feed conversion ratio (FCR) was calculated from feed intake (87% of dry matter) and body weight gain (BWG). Feed consumption was measured using a WEDA (Dammann & Westerkamp GmbH, Goldenstedt, Germany) automated feeding system that has an electronic flowmeter and weighing system. Cases of mortality and diarrhoea were recorded in all tested groups throughout the experiment.

2.8. Analysis of Ammonia Emission

Analysis of ammonia emission was conducted according to the Environmental Protection Document LAND 88-2009 method, approved by the Nr. D1-862 order (31 December, 2009) of the Lithuanian Minister of Environment [30]. Ammonia concentration in the air was analysed by the accredited laboratory "Labtesta" (Kretinga, Lithuania). Air samples were taken on the first and the last day of experiment in four tested farm sectors: S_{nonV}, S_V, RF_{nonV}, and RF_V.

2.9. Statistical Analysis

Data were subjected to multivariate analysis of variance (ANOVA) using the statistical package SPSS for Windows (Ver.15.0, SPSS, Chicago, IL, USA). Baseline measurements were used as covariates to consider the experimental conditions. The mean values were compared using Duncan's multiple range test with a significance level defined at $p \leq 0.05$. In order to evaluate the influence of three different factors (piglets age related differences, the use of fermented feed, vaccination) and their interaction on piglets' parameters, data were subjected to three-way ANOVA and the post hoc Tukey honest significant difference (HSD) test. In the tables, the results are presented as mean values (n = 10). Differences in bacterial genera between the groups at the end of the experiment were assessed using the Z-Test Calculator for two Population Proportions (Social Science Statistics, socscistatistics.com, 2019). The results were considered statistically significant at $p \leq 0.05$.

3. Results and Discussion

3.1. Microbial Profiles of Pig Faeces

The number of bacterial reads in pig faeces before and after the experiment was quite similar: it varied between 35,000 and 40,000 reads among the groups. The number of species with a prevalence of at least 0.01% from the total amount of bacteria in different groups before the experiment was also similar: it varied from 400 in the RF_{nonV} group to 473 in the S_V group. The number of species with the same prevalence rate at the end of experiment varied from 340 to 387. Although the number of bacterial reads and bacterial composition among the groups was similar, we detected some distinct differences, particularly in the number of lactobacilli.

Before experiment, two bacterial genera—*Prevotella* and *Lactobacillus*, had the highest prevalence rates in all groups of pigs; they accounted for more than 40% of the total amount of bacterial composition (Figure 2). The prevalence of *Prevotella* varied from 21.8% to 38.2%, while *Lactobacillus* prevalence ranged from 19.9% to 29.7%. The lowest *Lactobacillus* prevalence was in the RF_V group, while the highest was in S_{nonV} group. In all groups, the most prevalent species among *Lactobacillus* was *Lactobacillus amylovorus* (Tables S1–S4)). The other most prevalent genera included *Barnesiella*, *Clostridium*, *Blautia*, *Faecalibacterium*, *Roseburia* and *Eubacterium*, which ranged from 1.2% (*Eubacterium*) up to 3.4% (*Barnesiella*) in the faeces. Overall, the microbial profiles were similar among all the groups before the experiment (Figure 2), however, it should be mentioned that *Bifidobacterium* was detected only in two groups—RF_{nonV} and RF_V with the prevalence of 4.5% and 1.7% respectively.

At the end of experiment, the microbial profiles had changed depending on the pig group (Figure 3). Overall, the main genera remained similar as before experiment, but there were obvious differences in the *Lactobacillus* prevalence. The lowest prevalence occurred in faeces from the S_{nonV} group, while the highest was in the RF_V group ($p \leq 0.05$). In the latter group, lactobacilli were the most abundant bacteria: their prevalence reached 47.9% of the total bacteria, while in the rest of the groups the highest prevalence was *Prevotella*. When comparing groups with fermented vs. non-fermented feed, pigs that received fermented feed had a notably higher *Lactobacillus* prevalence compared to pigs fed non-fermented feed. The pigs from non-fermented feed groups had higher *Clostridium* and *Terrisporobacter* prevalence ($p \leq 0.05$), while the prevalence of other bacterial genera was low overall in all groups.

When comparing non-vaccinated with vaccinated groups, the main differences were also associated with the *Lactobacillus* prevalence: it was higher in vaccinated compared to non-vaccinated groups ($p \leq 0.05$). At the same time, the amount of *Prevotella* was higher in non-vaccinated groups, although the differences between RF_{nonV} and S_{nonV} groups were small (31.4% vs. 30.2%). The species prevalence and variety at the end of the experiment are presented in Tables S5–S8).

The microbial communities in the pig gut perform a variety of beneficial functions [31]. Previous studies of the gut microbial community have illustrated how populations of constituents are shaped by environmental exposure to microbes, diet, immunological pressures, host genetics and ecological forces within the ecosystem itself [32–34]. Understanding which microbial populations are influenced by fermented feed provides insight into how dietary changes in pigs shape the gut microbiome. Although there have been studies about microbial communities within the gut of healthy pigs, the microbial composition can differ depending on breed, age, place, feed, hygiene conditions and other factors [31,35,36]. This study demonstrated changes in microbial profiles using feed prepared by different technologies within similar groups of animals and the same farm. While the exact mechanism(s) by which vaccination can influence microbial changes within the gut of mammals remains unknown, this study suggests that vaccination has a certain influence, particularly on the number of *Lactobacillus* spp.

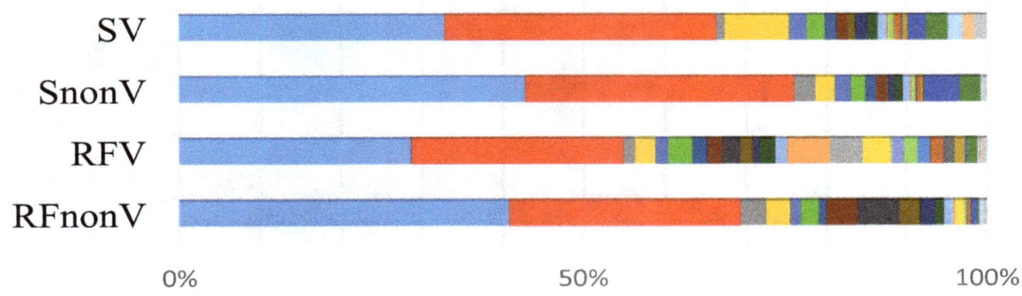

	RFnonV	RFV	SnonV	SV
Prevotella	34.6	21.8	38.2	28.1
Lactobacillus	24.3	19.9	29.7	28.8
Clostridium	2.7	1.1	2.4	0.9
Barnesiella	2.5	1.9	2.2	6.8
Blautia	1.1	1.2	1.8	1.9
Faecalibacterium	1.9	2.3	1.6	1.9
Eubacterium	0.8	1.4	1.2	1.3
Roseburia	3.3	1.4	1.2	1.2
Bifidobacterium	4.5	1.7	0	0
Megasphaera	2.2	1.2	0.1	0.8
Anaerovibrio	1.5	0.6	0.8	1.3
Ruminococcus	1	1.5	0.9	1.1
Alloprevotella	1	1.1	0.8	1
Collinsella	0.2	4.1	0.1	0.1
Olsenella	0	3.1	0.1	0
Selenomonas	1	2.6	0.1	0.1
Oligosphaera	0	1.3	0	0
Oscillibacter	0.2	1.3	0.3	0.5
Denitrobacterium	0.1	1.2	0.1	0
Intestinimonas	0.3	1.2	0.5	0.8
Succinivibrio	0	1.1	0	0.2
Oscillospira	0.1	1	0.3	0.5
Terrisporobacter	0.7	0.2	4.1	2
Gemmiger	0.2	1	2.3	2.3
Catenibacterium	0.5	0.2	0.3	1.6
Anaerobium	0.1	0.2	0.1	1.3
Escherichia	0.2	0.5	0.3	1.3

Figure 2. Microbial profiles (prevalence, %) at the genus level in pig faeces before the experiment.

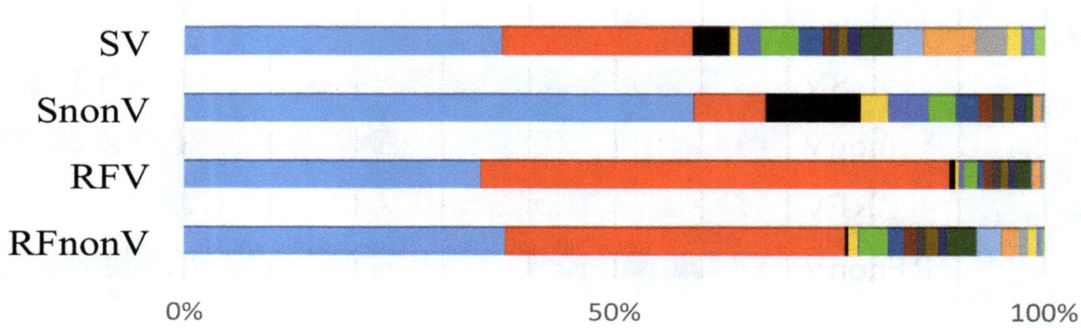

	RFnonV	RFV	SnonV	SV
Prevotella	31.4	30.2	51.4	29.1
Lactobacillus	33.3	47.9	7.2	17.5
Clostridium	0.3	0.6	9.6	3.4
Anaerovibrio	1	0.4	2.8	0.8
Terrisporobacter	0.1	0.5	4.1	2
Faecalibacterium	2.8	1.4	2.8	3.5
Gemmiger	1.6	0.6	2.3	2.3
Alloprevotella	1.2	0.9	1.4	0.6
Eubacterium	0.9	0.9	1.1	0.8
Blautia	1.3	0.8	1.1	0.8
Roseburia	0.8	0.7	1.1	1.2
Megasphaera	3	1.6	0.8	3
Collinsella	2.4	0.1	0.1	2.8
Barnesiella	1.7	0.9	0.7	4.9
Olsenella	0.9	0	0.1	2.9
Denitrobacterium	0.9	0	0.04	1.3
Oscillospira	0.5	0.2	0.2	1.2
Oscillibacter	0.3	0.2	0.1	1

Figure 3. Microbial profiles (prevalence, %) at the genus level in pig faeces after experiment.

Before the experiment, the microbial composition in all groups was very similar, with the highest prevalence for *Prevotella* and *Lactobacillus*, both of which accounted for 56% of the bacterial count. These genera include abundant microorganisms in young weaned healthy pigs [37]. *Prevotella* spp. are usually dominant in pigs' gut and gradually increase in number with age [38,39]. *Prevotella* spp. are key microbial members of the gastrointestinal tracts of adult animals; they are crucial for the degradation of starch and plant polysaccharides but also have a strong capacity for mucoprotein catabolism [40,41]. *Lactobacillus* spp. are common in both the proximal and distal regions of the porcine digestive tract; they colonise soon after birth [42]. This genus influences intestinal physiology, regulates the immune system and balances the intestinal ecology of the host [43]. In addition, *Lactobacillus* spp. have been known to metabolise carbohydrates, including oligosaccharides and starch, which are fermented in the large intestine to short chain fatty acids by lactobacilli for subsequent utilisation by the pigs [44]. Previous studies have indicated that stress greatly affects the gastrointestinal microbiota: it decreases total *Lactobacillus* populations and thus provides an opportunity for pathogen overgrowth [45]. Such stress usually occurs during the weaning period. According to the literature, when compared to diarrhoeic piglets, the gut microbiota of healthy piglets has a higher abundance of *Prevotellaceae*, *Lachnospiraceae*, *Ruminococcaceae* and *Lactobacillaceae* [46]. These data suggest that the gut microbial composition may be used as a biomarker to predict the health status of piglets. The present study demonstrated the positive impact of fermented feed on the porcine microbial composition compared with a conventional feeding regimen in healthy pigs. At the end of the experiment (day 61 of the piglets' life), the *Lactobacillus* prevalence decreased on average 2.4 fold in the groups fed non-fermented feed, whereas the prevalence increased on average 1.8 fold in the groups fed fermented feed. Hence, providing fermented feed prevents the decrease in *Lactobacillus* prevalence after weaning. This effect may help prevent digestive disorders associated with microbial changes because this period is one of the most critical regarding different infections [46]. It is known that during this age, *Lactobacillus* spp. in pigs normally decrease in number [37]. The *Lactobacillus* prevalence was also higher in vaccinated compared with non-vaccinated groups irrespective of the feed type. However, the data do not indicate whether vaccination against PCV2 leads to an increase in *Lactobacillus* prevalence; more experiments are required in this area. Moreover, different types of vaccines and larger experiments should be performed to better understand the possible influence of vaccination on the microbial communities. According to this study, it may be assumed that at the very least vaccination against PCV2 does not negatively affect the microbial composition within the gastrointestinal tract of pigs.

3.2. LAB, TVC, TEC and M/Y Counts in Piglets' Faeces

Microbiological parameters of the piglet faeces (from 25- and 61-day-old piglets) are shown in Table 2 (Table S9—Differences between microbiological parameters of the piglets' faeces between all the tested groups). In most groups (except the S_V group), the LAB count was significantly lower at the end compared with the beginning of the experiment (20.1% lower in the S_{nonV} group; 37.9% lower in the RF_{nonV} group; and 25.3% lower in the RF_V). The LAB count at the end of experiment was significantly higher in both vaccinated compared with non-vaccinated groups (26.6% for S_V and 17.2% for RF_V). The LAB count was 26.3% higher in S_V compared to the RF_V group at the end of the experiment. In a previous study, the faeces of pigs receiving a diet with fermented feed contained significantly fewer total bacteria and fungi, as well as coliform bacteria (including Escherichia coli) and anaerobic *Clostridium perfringens* counts. This phenomenon is due in part to the reduction in pH, an increase in the amount of lactic acid and other volatile fatty acids in the intestinal contents and a reduction in the number of *Enterobacteriaceae* [47,48]. From birth until weaning and then during the post-weaning period, the gut microbiota is dynamic and undergoes major compositional changes that are driven by age, exposure to microbes, environmental conditions and diet [49]. Many authors have described the great impact of early-life events in mammals, and particularly in pigs, on their future health. These experiences shape immune system development through changes in the pattern of microbial intestinal colonisation [50,51]. Colonisation is initiated at birth and is shaped by consumption of the sow's milk, which provides nutritional advantages to the LAB population, building a milk-oriented microbiome that includes *Bacteroidaceae* and *Lactobacillaceae*. This composition rapidly changes after weaning when a (largely) plant-based diet is introduced [35]. The rapidly varying microbiome of young piglets seems to increase in microbial diversity and richness in the suckling phase and gradually stabilises post-weaning [35,52,53]. Our results are in agreement with Bian et al. [54], namely that members of the *Lactobacillaceae* became predominant on days 7, 14 and 28, but had a lower relative abundance again on day 49.

When comparing the TVC in piglets' faeces, in all but the S_V group it was significantly lower at the end of experiment compared with the beginning (9.5% reduction in the S_{nonV} group; 23.6% reduction in the RF_{nonV} group; 3.4% reduction in the RF_V group). TVC was significantly higher in the vaccinated piglets compared to the non-vaccinated piglet faeces at the end of the experiment (22.2% for S_V and 13.8% for RF_V). Between the vaccinated groups, TVC was 10.0% higher in the S_V compared to the RF_V group. During weaning, there is gut microbiota dysbiosis, including a loss of microbial diversity. Studies have noted significant reductions in total bacterial number, including coliform bacteria and anaerobic *C. perfringens*, as well as an increase in *Acetivibrio*, *Dialister*, *Oribacterium*, *Prevotella* and *Proteobacteriaceae*, including *E. coli* [48,49,55,56]. A decrease in *Lactobacillus* spp. can lead to decrease in TVC. Zimmerman et al. [57] demonstrated that *Lactobacillus*, *Bifidobacterium* and *Lactobacillus* spp. have a positive immunomodulatory effect on vaccines in animals.

Table 2. Microbiological parameters in faeces from 25- and 61-day-old piglets.

Microbiological Parameters (log_{10} CFU/g)	Day	Treatments			
		S_{nonV}	S_V	RF_{nonV}	RF_V
LAB	Baseline	7.78 ± 0.03 $^{A;a}$	8.22 ± 0.04 $^{A;b}$	8.33 ± 0.05 $^{A;a}$	8.35 ± 0.06 $^{A;a}$
	61	6.22 ± 0.04$^{B;a}$	8.47 ± 0.02 $^{B;b}$	5.17 ± 0.06 $^{B;a}$	*6.24 ± 0.07* $^{B;b}$
TVC	Baseline	7.08 ± 0.02 $^{A;a}$	8.25 ± 0.05 $^{A;b}$	8.38 ± 0.07 $^{A;b}$	7.68 ± 0.07 $^{A;a}$
	61	6.41 ± 0.04. $^{B;a}$	8.24 ± 0.03. $^{A;b}$	6.40 ± 0.03. $^{B;a}$	7.42 ± 0.08 $^{B;b}$
TEC	Baseline	7.22 ± 0.06 $^{A;a}$	7.79 ± 0.06 $^{A;b}$	7.40 ± 0.08 $^{A;a}$	7.58 ± 0.04 $^{A;b}$
	61	6.36 ± 0.02 $^{B;b}$	6.18 ± 0.05 $^{B;a}$	6.87 ± 0.07 $^{B;a}$	7.12 ± 0.03 $^{B;b}$
Y/M	Baseline	6.73 ± 0.04 $^{A;a}$	7.71 ± 0.08 $^{A;b}$	6.21 ± 0.04 $^{A;a}$	6.63 ± 0.04 $^{A;b}$
	61	6.36 ± 0.08 $^{B;b}$	4.31 ± 0.06 $^{B;a}$	5.72±0.05 $^{B;a}$	6.26±0.04 $^{B;b}$

Tests of between-subject effects: influence of the analysed factors and their interactions on microbiological parameters in piglets' faeces

Dependent Variable	Treatment Duration	Fermented Feed	Vaccination	Treatment Duration × Fermented Feed	Treatment Duration × Vaccination	Fermented Feed × Vaccination	Treatment Duration × Fermented Feed × Vaccination
	Significance of the differences between groups (p)						
LAB	0.0001	0.0001	0.0001	0.0001	0.0001	0.0001	0.0001
TBC	0.0001	0.0001	0.0001	0.0001	0.0001	0.246	0.0001
TVC	0.0001	0.0001	0.001	0.0001	0.021	0.0001	0.0001
Y/M	0.0001	0.0001	0.0001	0.0001	0.0001	0.0001	0.0001

S_{nonV}—non-vaccinated piglets fed with control group compound feed; S_V—vaccinated piglets fed with control group compound feed; RF_{nonV}—non-vaccinated piglets fed with fermented compound feed; RF_V—vaccinated piglets fed with fermented compound feed. CFU—colony forming units; LAB—lactic acid bacteria; TVC—total viable bacteria count; TEC—total enterobacteria count; Y/M—yeast/mould count. Measurements were made at baseline—day 25, before the start of the feeding experiment—and on day 61, the end of the experiment. ^{A,B} Different uppercase letters indicate significant treatment duration-related differences ($p \leq 0.05$). ^{a,b} Different lowercase letters indicate differences among treatments, comparing S_{nonV} and S_V; RF_{nonV} and RF_V ($p \leq 0.05$). Data are presented as mean values ± standard error (n = 10 per group).

In all the groups, the TEC and Y/M counts were lower in piglets' faeces at the end compared with the beginning of the experiment (11.9 and 5.5%, respectively, for the S_{nonV} group; 26.1 and 44.1%, respectively, for the S_V group; 7.2 and 7.9%, respectively, for the RF_{nonV} group; 6.1 and 5.6%, respectively, for the RF_V group). Both TEC and Y/M were significantly lower in the S_V compared to S_{nonV} group faeces at the end of experiment, but the opposite was true for the RF_V compared to the RF_{nonV} group. At the end of the experiment, the TEC and Y/M counts were higher in the RF_V compared to the S_V group (13.2 and 31.2%, respectively). Our results are in agreement with Dowarah et al. [58], who showed that fermented feed improved the number of beneficial microbes (LAB and bifidobacteria) and reduced *E. coli* and clostridia count, the pathogens responsible for diarrhoea, in faces. Arfken et al. [59] investigated the mutualistic relationship between yeast in the piglet gastrointestinal (GI) tract and *Lactobacillus* count. There was a significant reduction in the total number of fungi compared with the control group [48].

The ANOVA results indicated that there were significant treatment duration, fermented feed and vaccination main effects, as well as significant treatment duration × fermented feed, treatment duration × vaccination, fermented feed × vaccination and treatment duration × fermented feed × vaccination interactions on microbiological parameters of the pig faeces. Indeed, only the fermented feed × vaccination interaction was not significant for TVC in pig faeces.

Finally, microbiological parameters of the piglets' faeces were varied and, in most of the cases, influenced by the analysed factors and their interaction. Considering that genera and species as well as the interaction between strains can lead changes of piglets' growth performance, correlation between the above-mentioned parameters were calculated.

3.3. Piglet Blood Parameters

The piglets' blood parameters are shown in Table 3 (Table S10—Differences between blood parameters of the piglets' between all the tested groups). At the end of experiment, there were significantly lower ALT, T3, T4, IP, Mg, K, Ca and vitamin B12 concentrations in the S_V compared with the S_{nonV} group. However, in the S_V group, Chol, HDL-C, TP and GLU concentrations were significantly higher compared with the SnonV group. At the end of experiment, there were significantly lower ALT, HDL-C, LDL-C, TP, ALB and Fe blood concentrations in the RF_V compared to RF_{nonV} group. However, in the RF_V group blood, TG and total bilirubin concentrations were significantly higher compared with the RF_{nonV} group.

When comparing blood parameters between the vaccinated groups (S_V and RF_V) at the end of experiment, ALT, TG, IgG, Mg, K, Fe and total bilirubin were significantly higher in the RF_V group. By contrast, HDL-C, LDL-C, TP, ALB, T3, T4 and vitamin B12 were significantly lower in the RF_V compared with the S_V group.

In general, blood biochemical indices reflect comprehensive functions of piglet's nutritional metabolism, health and welfare [60]. The IgG level reflects immune status, plays important roles in the humoral immune response and controls bacterial infections in the body; it can also function to control diarrhoeal infections by binding multiple pathogenic antigens [46]. Lu et al. [59] reported that IgG and immunoglobulin M (IgM) concentrations were significantly greater and serum ALT and AST concentrations were decreased in pigs that received fermented feed ($p \leq 0.05$). The activities of the hepatic enzymes ALT and AST are key indices that reflect the acute injury of liver. The ALT blood concentration increases when the liver cell membrane is damaged. The AST content significantly increases when the liver mitochondrial membrane is impaired [60]. The blood TP concentration reflects the relationship between protein absorption in vivo and humoral immunity [60]. The hydrolysis products of glucosinolates are known to depress iodine metabolism in the thyroid gland and inhibit the synthesis of thyroid hormones T3 and T4. When these compounds, especially thiocyanates, interfere with iodine uptake, hypothyroidism and thyroid gland enlargement may ensue [46]. However, in this study the utilised rapeseed meal contained very low concentrations of erucic acid and glucosinolates (7.5–11.2 μmol/g). Hence, the influence of the above-mentioned compounds on T3 and T4 should be minimal. The fermentation process can reduce dietary anti-nutrients (phytates, glucosinolates and trypsin inhibitors) and improve the absorption and use of nutrients, including amino acids and minerals (e.g., P, Ca, Zn, Cu and Fe) [48,61]. These changes can lead to haematological, biochemical and mineral profile variations in the blood [61,62]. However, it was published that the TP concentration was increased in the blood of piglets fed fermented feed [63]. In opposition to above-mentioned findings, Min [64] reported that the TP blood concentration was not affected by the fermented feed diets. In serum samples from piglets fed with fermented feed, there were higher ALP, TP, ALB and GLU concentrations [62,65,66]. Liu et al. [15] published that piglets fed fermented feed had lower levels of serum IgG, but there was no difference in the immunoglobulin IgA and IgM serum levels. Furthermore, the serum glucose level was decreased in piglets fed fermented feed; however, TP, Chol, TG, LDL-C, ALT, AST, Ca, P and Mg were not different among the treatments [67]. Satessa et al. [68] found that fermented rapeseed meal reduced the glucose, HDL-C and LDL-C concentrations, but TP, P, AST, ALT concentrations were increased, and significant changes in blood IgG were not established. According to Czech et al. [61], the blood concentrations of Chol and TG AST in piglets fed with fermented rapeseed meal was reduced, and there was better availability of minerals.

Table 3. Piglets' blood parameters.

Blood Parameters	Day	Treatments			
		S_{nonV}	S_V	RF_{nonV}	RF_V
Aspartate aminotransferase (AST), U/L	Baseline	29.4 $^{A;a}$	42.67 $^{A;a}$	51.4 $^{A;a}$	61.8 $^{A;a}$
	61	34.0 $^{B;a}$	41.0. $^{A;a}$	44.0. $^{A;a}$	42.4. $^{A;a}$
Alanine aminotransferase (ALT), U/L	Baseline	48.4 $^{A;a}$	43.67 $^{A;a}$	53.2 $^{A;b}$	60.8 $^{A;a}$
	61	76.2. $^{B;a}$	69.6 $^{B;b}$	87.0 $^{B;a}$	80.8 $^{A;b}$
Cholesterol (Chol), mmol/L	Baseline	1.64 $^{A;a}$	1.64 $^{A;a}$	1.88 $^{A;a}$	2.04 $^{A;a}$
	61	2.06. $^{B;a}$	2.26. $^{B;b}$	2.34. $^{A;a}$	2.25. $^{B;a}$
High-density lipoprotein cholesterol (HDL-C), mmol/L	Baseline	0.714 $^{A;a}$	0.744 $^{A;b}$	0.898 $^{A;a}$	0.846 $^{A;a}$
	61	0.840 $^{B;a}$	0.944 $^{B;b}$	1.028 $^{B;a}$	0.872 $^{A;b}$
Low-density lipoprotein cholesterol (LDL-C), mmol/L	Baseline	0.758 $^{A;a}$	0.726 $^{A;a}$	0.814 $^{A;a}$	0.976 $^{A; a}$
	61	0.980 $^{B;a}$	1.102 $^{B;a}$	1.032 $^{A;a}$	0.860 $^{A; b}$
Triglycerides (TG), mmol/L	Baseline	0.360 $^{A;a}$	0.372 $^{A;a}$	0.366 $^{A;a}$	0.478 $^{A;a}$
	61	0.466 $^{B;a}$	0.466 $^{A;a}$	0.620 $^{B;a}$	0.650 $^{B;b}$
Total protein (TP), g/L	Baseline	46.2 $^{A;a}$	44.9 $^{A;a}$	44.2 $^{A;a}$	45.6 $^{A;a}$
	61	51.8 $^{B;a}$	56.3 $^{B;b}$	52.8 $^{B;a}$	49.5 $^{B;b}$
Albumin (ALB), g/L	Baseline	30.0 $^{A;a}$	29.0 $^{A;a}$	32.6 $^{A;a}$	32.8 $^{A;a}$
	61	35.8 $^{A;a}$	37.4 $^{B;a}$	36.2 $^{A;a}$	33.4 $^{B;b}$
Immunoglobulin IgG, g/L	Baseline	2.65 $^{A;a}$	2.85 $^{A;a}$	2.35 $^{A;a}$	2.96 $^{A;a}$
	61	3.74 $^{B;a}$	3.44 $^{A;a}$	3.05 $^{B;a}$	5.45 $^{A;b}$
Triiodothyronine (T3), nmol/L	Baseline	1.21 $^{A;a}$	1.14 $^{A;b}$	1.29 $^{A;a}$	1.22 $^{A;a}$
	61	2.14$^{B;a}$	2.02 $^{B;b}$	1.59 $^{A;a}$	1.49 $^{A;a}$
Thyroxine (T4), μ d/L	Baseline	4.50 $^{A;a}$	4.82 $^{A;a}$	3.50 $^{A;a}$	5.22 $^{A;a}$
	61	4.84 $^{A;a}$	3.84 $^{A;b}$	2.92 $^{B;a}$	2.90 $^{A;a}$
Glucose (GLU), nmol/L	Baseline	5.84 $^{A;a}$	6.10 $^{A;a}$	6.12 $^{A;a}$	5.66 $^{A;a}$
	61	5.74 $^{A;a}$	6.22 $^{A;b}$	6.08 $^{A;a}$	6.18 $^{B;a}$
Phosphorus (IP), mmol/L	Baseline	2.94 $^{A;a}$	2.87 $^{A;a}$	2.61 $^{A;a}$	2.76 $^{A;a}$
	61	3.50 $^{B;a}$	3.39 $^{A;b}$	3.28 $^{B;a}$	3.39 $^{B;a}$
Magnesium (Mg), mmol/L	Baseline	1.02 $^{A;a}$	0.928 $^{A;b}$	0.996 $^{A;a}$	0.932 $^{A;b}$
	61	1.07 $^{A;a}$	0.936 $^{A;b}$	0.960 $^{A;a}$	0.966 $^{A;b}$
Potassium (K)	Baseline	4.96 $^{A;a}$	5.66 $^{A;b}$	4.65 $^{A;a}$	5.69 $^{A;b}$
	61	5.81 $^{B;a}$	5.34 $^{B;b}$	4.96 $^{A;a}$	4.74 $^{B;a}$
Sodium (Na)	Baseline	143.4 $^{A;a}$	141.0 $^{A;a}$	144.0 $^{A;a}$	142.0 $^{A;b}$
	61	147.2 $^{A;a}$	147.2 $^{B; a}$	146.6 $^{B;a}$	144.4 $^{A;a}$
Iron (Fe), μmol/L	Baseline	23.6 $^{A;a}$	23.8 $^{A;a}$	31.5 $^{A;a}$	23.5 $^{A;b}$
	61	28.1 $^{A;a}$	26.9 $^{B;a}$	47.1 $^{A;a}$	51.4 $^{B;a}$
Calcium (Ca), nmol/L	Baseline	2.60 $^{A;a}$	2.57 $^{A;a}$	2.71 $^{A;a}$	2.55 $^{A;a}$
	61	2.87 $^{A;a}$	2.78 $^{B;b}$	2.79 $^{A;a}$	2.78 $^{B;a}$
Vitamin B12, pmol/L	Baseline	142.2 $^{A;a}$	93.8 $^{A;b}$	78.2 $^{A;a}$	131.0 $^{A;b}$
	61	214.6 $^{A;a}$	122.2 $^{B;b}$	94.2 $^{A; a}$	98.2 $^{B;a}$
Creatinine (CREA), μmol/L	Baseline	64.2 $^{A;a}$	57.4 $^{A;a}$	78.8 $^{A;a}$	63.4 $^{A;a}$
	61	57.4 $^{A;a}$	54.8 $^{A;a}$	48.2 $^{B;a}$	49.4 $^{A;a}$
Alkaline phosphatase (AP), U/L	Baseline	336.2 $^{A;a}$	285.0 $^{A;a}$	408.6 $^{A;a}$	318.0 $^{A;a}$
	61	263.6 $^{A;a}$	220.4 $^{A;a}$	242.6 $^{B;a}$	245.4 $^{A;a}$
Urea, mmol/L	Baseline	2.36 $^{A;a}$	3.26 $^{A;a}$	2.64 $^{A;a}$	2.58 $^{A;a}$
	61	2.02 $^{A;a}$	2.38 $^{A;a}$	3.19 $^{B;a}$	2.63 $^{B;a}$
Thyroid-stimulating hormone (TSH)	Baseline	0.0200 $^{A;a}$	0.0190 $^{A;a}$	0.0208 $^{A;a}$	0.0236 $^{A;a}$
	61	0.0208 $^{A;a}$	0.0228 $^{A;a}$	0.0230 $^{A;a}$	0.0260 $^{B;a}$
Total bilirubin (pmol/L)	Baseline	1.88 $^{A;a}$	1.99 $^{A;a}$	1.99 $^{A;a}$	1.99 $^{A;a}$
	61	1.98 $^{A;a}$	1.99 $^{A;a}$	1.99 $^{A;a}$	2.65 $^{A;b}$

S_{nonV}—non-vaccinated piglets fed with control group compound feed; S_V—vaccinated piglets fed with control group compound feed; RF_{nonV}—non-vaccinated piglets fed with fermented compound feed; RF_V—vaccinated piglets fed with fermented compound feed. Measurements were made at baseline—day 25, before the start of the feeding experiment—and on day 61, the end of the experiment. ^{A,B} Different uppercase letters indicate significant treatment duration-related differences ($p \leq 0.05$). ^{a,b} Different lowercase letters indicate differences among treatments: S_{nonV} and S_V; RF_{nonV} and RF_V ($p \leq 0.05$). Data are presented as mean values (n = 10 per group).

The influence of the analysed factors and their interaction on piglets' blood parameters is shown in Table 4. Vaccination as a separate factor did not significantly influence piglets' blood parameters. However, the treatment duration × fermented feed interaction was significant for T3 ($p = 0.038$), T4 ($p = 0.041$) and Fe ($p = 0.008$) concentrations in piglets' blood. There was a significant treatment duration × vaccination interaction on T4 ($p = 0.008$) and K ($p = 0.007$) concentrations in piglets' blood. Finally, there was a significant fermented feed × vaccination interaction for T4 ($p = 0.033$), vitamin B12 ($p = 0.003$) and urea ($p = 0.014$) concentrations in piglets' blood. All three factors were significant with regard to the TP content in piglets' blood ($p = 0.041$).

Table 4. Tests of between-subject effects: influence of the analysed factors and their interaction on piglets' blood parameters.

Dependent Variable	TD	FF	V	TD × FF	TD × V	FF × V	TD × FF × V
			Significance (p)				
AST, U/L	0.279	**0.025**	0.191	0.182	0.404	0.598	0.791
ALT, U/L	**0.0001**	**0.017**	0.557	0.997	0.358	0.453	0.481
Chol, mmol/L	**0.004**	0.088	0.608	0.467	0.907	0.809	0.396
HDL-C, mmol/L	**0.069**	0.123	0.768	0.501	0.905	0.185	0.482
LDL-C, mmol/L	**0.045**	0.723	0.807	0.143	0.584	0.760	0.149
TG, mmol/L	**0.002**	**0.018**	0.382	0.206	0.591	0.459	0.688
TP, g/L	**0.0001**	0.155	0.773	0.345	0.811	0.310	**0.041**
ALB, g/L	**0.002**	0.581	0.693	0.061	0.937	0.529	0.276
IgG, g/L	0.112	0.699	0.329	0.612	0.660	0.299	0.442
T3, nmol/L	**0.001**	0.120	0.522	0.038	0.909	0.994	0.949
T4, μ d/L	**0.003**	0.004	0.331	0.041	**0.008**	**0.033**	0.685
GLU, nmol/L	0.633	0.716	0.660	0.458	0.300	0.633	0.745
IP, mmol/L	**0.001**	0.179	0.878	0.658	0.851	0.349	0.993
Mg, mmol/L	0.675	0.487	**0.060**	0.655	0.867	0.250	0.422
K	0.885	**0.041**	0.203	0.153	**0.007**	0.462	0.902
Na	**0.0001**	0.548	**0.039**	0.108	0.464	0.548	0.388
Fe, μmol/L	**0.001**	**0.001**	0.694	**0.008**	0.370	0.830	0.267
Ca, nmol/L	**0.001**	0.985	0.176	0.379	0.657	0.816	0.313
B12, pmol/L	0.158	**0.008**	0.158	0.055	0.121	**0.003**	0.934
CREA, μmol/L	0.069	0.831	0.406	0.222	0.463	0.864	0.660
AP, U/L	**0.027**	0.490	0.256	0.521	0.521	0.966	0.588
Urea, mmol/L	0.164	0.797	0.899	0.286	0.075	**0.014**	0.456
TSH	0.604	0.523	0.736	0.854	0.951	0.854	0.691
Total bilirubin (pmol/L)	0.338	0.280	0.280	0.290	0.290	0.349	0.338

AST—Aspartate aminotransferase, ALT—Alanine aminotransferase, Chol—Cholesterol, HDL-C—High-density lipoprotein cholesterol, LDL-C—Low-density lipoprotein cholesterol, TG—Triglycerides, TP—Total protein, ALB—Albumin, T3—Triiodothyronine, T4—Thyroxine, GLU—Glucose, IP—Phosphorus, Mg—Magnesium, K—Potassium, Na—Sodium, Fe—Iron, Ca—Calcium, CREA—Creatinine, AP—Alkaline phosphatase, TSH—Thyroid-stimulating hormone. TD—treatment duration, FF—fermented feed, V—vaccination. Bold values indicate significant differences ($p \leq 0.05$).

3.4. *Piglets' Growth Performance*

The piglets' average FCR, ADG and the influence of analysed factors and their interaction on piglets' growth performance parameters from day 25 to 61 are shown in Table 5. For ADG, there were no significant differences among the groups. In addition, the analysed factors as well as their interactions did not exert significant effects on piglets' ADG. However, there were significant differences in FCR between S_V and S_{nonV} (11.5% lower in the SV group), between RF_V and RF_{nonV} (10.2% lower in the RF_{nonV} group) and between S_V and RF_V (21.6% lower in the S_V group). There was a significant, very strong positive correlation between FCR and TEC in piglets' faeces ($R = 0.919$, $p = 0.041$). There were no significant correlations between the other analysed faecal microbiological parameters (LAB, TBC and Y/M count) and FCR or ADG.

Mortality and diarrhoea cases were similar in all groups throughout the experiment. The mortality of piglets in the non-vaccinated groups (S_{nonV} and RF_{nonV}) was 2%, while mortality in the vaccinated groups (S_V and RF_V) was 2% and 3%, respectively. There was more intense diarrhoea in the S_{nonV} group at day 31 and 36. There were several diarrhoea cases in the S_V, RF_{nonV} and RF_V groups from day 28 to 46.

Table 5. Influence of the analysed factors and their interactions on piglets' growth performance.

Piglets Growth Performance		**Treatments**			
	Average from day 25 to 61	$\mathbf{S_{nonV}}$	$\mathbf{S_V}$	$\mathbf{RF_{nonV}}$	$\mathbf{RF_V}$
FCR		1.56	1.38	1.58	1.76
ADG		0.395	0.392	0.397	0.399

Tests of between-subjects effects: influence of analysed factors and their interactions on FCR and ADG

Dependent Variable	**Fermented Feed**	**Vaccination**	**Fermented Feed × Vaccination**
	Significance (p)		
FCR	0.154	1.000	0.194
ADG	0.962	0.996	0.979

Differences among treatments ($p < 0.05$)

$\mathbf{S_V}$ **vs.** $\mathbf{S_{nonV}}$	$\mathbf{RF_V}$ **vs.** $\mathbf{RF_{nonV}}$	$\mathbf{S_V}$ **vs.** $\mathbf{RF_V}$	$\mathbf{S_{nonV}}$ **vs.** $\mathbf{RF_{nonV}}$	$\mathbf{S_{nonV}}$ **vs.** $\mathbf{RF_V}$	$\mathbf{RF_{nonV}}$ **vs.** $\mathbf{S_V}$
		ADG			
0.901	0.783	0.55	0.943	0.91	0.286
		FCR			
0.001	0.002	0.0001	0.773	0.91	0.074

S_{nonV}—non-vaccinated piglets fed with control group compound feed; S_V—vaccinated piglets fed with control group compound feed; RF_{nonV}—non-vaccinated piglets fed with fermented compound feed; RF_V—vaccinated piglets fed with fermented compound feed. FCR—*feed conversion ratio*; ADG—average daily gain. Measurements were made at baseline—on day 25, before the start of the feeding experiment—and on day 61, the end of the experiment. Bold values indicate significant differences ($p \leq 0.05$). Data are presented as mean values ($n = 100$ per group).

The available published information about the influence of vaccination on piglets' growth performance is varied. According to Oliver-Ferrando et al. [69], the optimal time for PCV2 vaccination is at either 3 or 6 weeks of age. In our study, piglets were vaccinated on day 25 of life. Duivon et al. [70] published that the added PCV2 valence in the vaccination protocol helps counter the negative impact of subclinical PCV2 infection on growth. Jeong et al. [71] indicated that ADG showed improvement in vaccinated compared with non-vaccinated animals. Da Silva et al. [72] compared the ADG between vaccinated and non-vaccinated animals from weaning to finishing for four commercial vaccine products. All products significantly increased ADG values. However, according to Woźniak et al. [73] data from four farms where vaccination was used, the results were similar to those from non-vaccinated farms. The authors suggested revising the vaccination protocol. The above-mentioned findings can be explained by a major PCV2 genotype shift from the predominant PCV2 genotype 2b towards 2d [74]. While the commercial vaccines that were first introduced to the US in 2006 have been highly effective in reducing clinical signs and improving production, recent studies have indicated a declining level of PCV2 prevalence and viraemia in the field. Hence, the efficiency of current vaccines against new and emerging strains, as well as new vaccine development, is crucial [75]. Vaccination against PCV2 is an important co-working agent that confers unintended benefits in the protection against the other agents [76]. Notably, the breed information is frequently omitted from experimental publications of vaccine studies, as well as environmental conditions (temperature, stocking density, etc.). These factors are very important for the further standardisation across studies before logical comparisons can be made among studies [77].

While we found differences in the ADG between different groups, the influence of vaccination on different feed groups was the opposite. Specifically, the FCR was lower in the S_V compared to the S_{nonV} group. However, the effect was the opposite in the rapeseed meal groups: FCR was lower in the R_{nonV} compared to the RF_V group. The lowest FCR (1.38) occurred in the S_V group. Notably, soya-based feed is much more expensive compared with the feed composed of local rapeseed meal. Overall, the data support the use of fermented local feedstock for piglet feeding because it provides similar growth parameters as imported soya.

3.5. *Influence of Analysed Factors on Ammonia Emission*

Table 6 presents the influence of the analysed factors on ammonia emission. At the end of experiment, the RF_V group had the lowest ammonia emission: 58.2, 23.8 and 47.33% lower than the S_{nonV}, S_V, and RF_{nonV} groups, respectively. There was lower ammonia emission in vaccinated groups at the end of the experiment: 45.2% lower in the S_V compared to S_{nonV} group and 47.33% lower in the RF_V compared to the RF_{nonV} group. ANOVA revealed a significant effect for all the analysed factors and their interaction on ammonia emission. Furthermore, there was a significant, very strong positive correlation between ammonia emission and Y/M count in piglets' faeces at the end of experiment ($R = 0.974$; $p = 0.013$).

Table 6. Influence of the analysed factors on ammonia emission.

Ammonia Emission	Day	Treatments			
		S_{nonV}	S_V	RF_{nonV}	RF_V
	Baseline	0.837 $^{A;a}$	1.172 $^{A;b}$	1.584 $^{A;a}$	1.252 $^{A;b}$
	61	1.938 $^{B;a}$	1.063 $^{A;b}$	1.538 $^{A;a}$	0.810 $^{B;b}$

Tests of between-subject effects: influence of the analysed factors and their interactions on ammonia emission

Treatment Duration	Fermented Feed	Vaccination	Treatment Duration × Fermented Feed	Treatment Duration × Vaccination	Fermented Feed × Vaccination	Treatment Duration × Fermented Feed × Vaccination
Significance (p)						
0.0001	**0.003**	**0.0001**	**0.0001**	**0.0001**	**0.0001**	**0.0001**

S_{nonV}—non-vaccinated piglets fed with soya meal; S_V—vaccinated piglets fed with soya meal; RF_{nonV}—non-vaccinated piglets fed with fermented rapeseed meal; RF_V—vaccinated piglets fed with fermented rapeseed meal. Measurements were done at baseline, on day 25 before the start of the feeding experiment, and on day 61, the end of the experiment. ^{A,B} Different uppercase letters indicate significant treatment duration-related differences ($p \leq 0.05$). ^{a,b} Different lowercase letters indicate differences among treatments ($p \leq 0.05$). Bold values indicate significant differences ($p \leq 0.05$). Data are presented as the mean values ($n = 3$).

There has been no information published about the influence of vaccination on ammonia emission; most of the published studies have focussed on dietary aspects. Yi et al. [37] and Lee et al. [78] reported that protected organic acids used for weanling pigs reduce faecal ammonia. Nguyen et al. [79] concluded that diets with probiotics mixture decreased pigs' ammonia emission. According to Wang et al. [51], diets with fermented feed ingredients tended to decrease ammonia emissions, but this effect was not statistically significant. According to Bindas et al. [80], ammonia in the faeces of experimental animals fed with fermented feed was significantly lower compared with the control animals. Reducing ammonia emissions by changing the dietary composition is considered economical and feasible because it can improve nitrogen utilisation and, consequently, reduce ammonia emissions [53]. Lactobacilli-based feed fermentation decreases the emissions of total organic carbon and ammonium [67]. Finally, many factors can contribute to ammonia emission on pig farms. Our data indicate that vaccination as a factor should be considered, as well as microbiological parameters of the piglets' faeces. Indeed, we found a very strong correlation between ammonia emission and Y/M count in piglets' faeces.

4. Conclusions

Fermented feed had a positive impact on the amount of *Lactobacillus* during the post-weaning period of pigs. In addition, vaccination had no negative impact on microbial communities. There was a higher *Lactobacillus* prevalence in vaccinated compared with non-vaccinated groups, as well as a higher viable LAB count in faeces from both vaccinated groups (S_V and RF_V) at the end of experiment. There were no significant differences in ADG between piglets' groups. There were differences in FCR: 11.5% lower in the S_V compared with the S_{nonV} group, 10.2% lower in the RF_{nonV} compared with the RF_V group, and 21.6% lower in the S_V compared with the RF_V group. There were a significant, very strong positive correlations between FCR and TEC in piglets' faeces (R = 0.919, p = 0.041) and between ammonia emission and Y/M count in piglets' faeces at the end of experiment (R = 0.974, p = 0.013). The lowest ammonia emission was in the RF_V group section. Vaccination as a separate factor did not significantly influence piglets' blood parameters. Overall, by changing from an extruded soya to cheaper rapeseed meal and applying the fermentation model with the selected LAB combination, it is possible to feed piglets without any undesirable changes in health and growth performance, as well as in a more sustainable manner. However, to evaluate the influence of vaccination and its interaction with other parameters (feed, piglet age, breed, etc.) on piglets' parameters, additional studies should be performed and methods should be standardised so that the results may be compared.

Supplementary Materials:
Table S1: Species S_{nonV} group before experiment; Table S2: Species S_V group before experiment Species S_V group before experiment; Table S3: Species RF_{nonV} group before experiment; Table S4: Species RFV group before experiment; Table S5: Species S_{nonV} group after experiment; Table S6: Species SV group after experiment; Table S7: Species RF_{nonV} group after experiment; Table S8: Species RFV group after experiment; Table S9: Differences between microbiological parameters of the piglets' faeces between all the tested groups; Table S10: Differences between blood parameters of the piglets between all the tested groups.

Author Contributions: Conceptualisation, methodology, E.B., R.G., L.V., S.B. and M.R.; investigation, L.V., S.B., S.S., A.D., V.S., P.Z., E.Z. and V.L; writing—original draft preparation, L.V., D.K., P.Z., V.S., E.M. and V.L.; writing—review and editing, E.B., V.B. and R.G.; visualisation, P.Z., V.S., V.L., D.K. and E.M.; supervision, E.B. All authors have read and agreed to the published version of the manuscript.

Acknowledgments: The authors gratefully acknowledge the EUREKA Network Project E!13309 "SUSFEETECH" (Nr. 01.2.2-MITA-K-702-05-0001) and COST Action CA18101 'SOURDOugh biotechnology network towards novel, healthier and sustainable food and bIoproCesseS'.

References

1. Tischer, I.; Mields, W.; Wolff, D.; Vagt, M.; Griem, W. Studies on epidemiology and pathogenicity of porcine circovirus. *Arch. Virol.* **1986**, *91*, 271–276. [CrossRef]
2. Ellis, J.; Hassard, L.; Clark, E.; Harding, J.; Allan, G.; Willson, P.; Strokappe, J.; Martin, K.; McNeilly, F.; Meehan, B.; et al. Isolation of circovirus from lesions of pigs with postweaning multisystemic wasting syndrome. *Can. Vet. J.* **1998**, *39*, 44–51.
3. Argilés, J.M.; Busquets, S.; Toledo, M.; López-Soriano, F.J. The role of cytokines in cancer cachexia. *Curr. Opin. Support Palliat. Care* **2009**, *3*, 263–268. [CrossRef] [PubMed]
4. Ottenhoff, T.H.M. The knowns and unknowns of the immunopathogenesis of tuberculosis. *Int. J. Tuberc. Lung Dis.* **2012**, *16*, 1424–1432. [CrossRef] [PubMed]
5. Szabo, G.; Csak, T. Inflammasomes in liver diseases. *J. Hepatol.* **2012**, *57*, 642–654. [CrossRef]
6. Ramamoorthy, S.; Meng, X.-J. Porcine circoviruses: A minuscule yet mammoth paradox. *Anim. Health Res. Rev.* **2009**, *10*, 1–20. [CrossRef]
7. Gillespie, J.; Opriessnig, T.; Meng, X.J.; Pelzer, K.; Buechner-Maxwell, V. Porcine circovirus type 2 and porcine circovirus-associated disease. *J. Vet. Intern. Med.* **2009**, *23*, 1151–1163. [CrossRef]
8. Cságola, A.; Zádori, Z.; Mészáros, I.; Tuboly, T. Detection of porcine parvovirus 2 (*Ungulate Tetraparvovirus* 3) Specific antibodies and examination of the serological profile of an infected swine herd. *PLoS ONE* **2016**, *11*, e0151036. [CrossRef]

9. Novosel, D.; Cadar, D.; Tuboly, T.; Jungic, A.; Stadejek, T.; Ait-Ali, T.; Cságola, A. Investigating porcine parvoviruses genogroup 2 infection using in situ polymerase chain reaction. *BMC Vet. Res.* **2018**, *14*, 163. [CrossRef] [PubMed]
10. Thomson, J.; Smith, B.; Alan, G.; McNeilly, F.; McVicar, C. PDNS, PMWS and porcine circovirus type 2 in Scotland. *Vet. Rec.* **2000**, *146*, 651–652.
11. Dán, Á.; Molnár, T.; Biksi, I.; Glávits, R.; Shaheim, M.; Harrach, B. Characterisation of Hungarian porcine circovirus 2 genomes associated with PMWS and PDNS cases. *Acta Vet. Hung.* **2003**, *51*, 551–562. [CrossRef]
12. An, D.; Roh, I.; Song, D.; Park, C.; Park, B. Phylogenetic characterization of porcine circovirus type 2 in PMWS and PDNS Korean pigs between 1999 and 2006. *Virus Res.* **2007**, *129*, 115–122. [CrossRef] [PubMed]
13. Zheng, G.; Lu, Q.; Wang, F.; Xing, G.; Feng, H.; Jin, Q.; Guo, Z.; Teng, M.; Hao, H.; Li, D.; et al. Phylogenetic analysis of porcine circovirus type 2 (PCV2) between 2015 and 2018 in Henan Province, China. *BMC Vet. Res.* **2020**, *16*, 6. [CrossRef] [PubMed]
14. Xiao, C.-T.; Halbur, P.G.; Opriessnig, T. Global molecular genetic analysis of porcine circovirus type 2 (PCV2) sequences confirms the presence of four main PCV2 genotypes and reveals a rapid increase of PCV2d. *J. Gen. Virol.* **2015**, *96*, 1830–1841. [CrossRef]
15. Liu, J.; Wei, C.; Dai, A.; Lin, Z.; Fan, K.; Fan, J.; Liu, J.; Luo, M.; Yang, X. Detection of PCV2e strains in Southeast China. *PeerJ* **2018**, *6*, e4476. [CrossRef]
16. Opriessnig, T.; Xiao, C.-T.; Gerber, P.F.; Halbur, P.G. Identification of recently described porcine parvoviruses in archived North American samples from 1996 and association with porcine circovirus associated disease. *Vet. Microbiol.* **2014**, *173*, 9–16. [CrossRef]
17. Chen, N.; Huang, Y.; Ye, M.; Li, S.; Xiao, Y.; Cui, B.; Zhu, J. Co-infection status of classical swine fever virus (CSFV), porcine reproductive and respiratory syndrome virus (PRRSV) and porcine circoviruses (PCV2 and PCV3) in eight regions of China from 2016 to 2018. *Infect. Genet. Evol.* **2019**, *68*, 127–135. [CrossRef] [PubMed]
18. Bartkiene, E.; Bartkevics, V.; Krungleviciute, V.; Pugajeva, I.; Zadeike, D.; Juodeikiene, G. Lactic acid bacteria combinations for wheat sourdough preparation and their influence on wheat bread quality and acrylamide formation. *J. Food Sci.* **2017**, *82*, 2371–2378. [CrossRef]
19. Bartkiene, E.; Bartkevics, V.; Lele, V.; Pugajeva, I.; Zavistanaviciute, P.; Mickiene, R.; Zadeike, D.; Juodeikiene, G. A concept of mould spoilage prevention and acrylamide reduction in wheat bread: Application of lactobacilli in combination with a cranberry coating. *Food Control* **2018**, *91*, 284–293. [CrossRef]
20. Bartkiene, E.; Sakiene, V.; Bartkevics, V.; Juodeikiene, G.; Lele, V.; Wiacek, C.; Braun, P.G. Modulation of the nutritional value of lupine wholemeal and protein isolates using submerged and solid-state fermentation with *Pediococcus pentosaceus* strains. *Int. J. Food Sci. Technol.* **2018**, *53*, 1896–1905. [CrossRef]
21. Bartkiene, E.; Sakiene, V.; Lele, V.; Bartkevics, V.; Rusko, J.; Wiacek, C.; Ruzauskas, M.; Braun, P.G.; Matusevicius, P.; Zdunczyk, Z.; et al. Perspectives of lupine wholemeal protein and protein isolates biodegradation. *Int. J. Food Sci. Technol.* **2019**, *54*, 1989–2001. [CrossRef]
22. Bartkiene, E.; Mozuriene, E.; Lele, V.; Zokaityte, E.; Gruzauskas, R.; Jakobsone, I.; Juodeikiene, G.; Ruibys, R.; Bartkevics, V. Changes of bioactive compounds in barley industry by-products during submerged and solid state fermentation with antimicrobial *Pediococcus acidilactici* strain LUHS29. *Food Sci. Nutr.* **2020**, *8*, 340–350. [CrossRef] [PubMed]
23. Vadopalas, L.; Ruzauskas, M.; Lele, V.; Starkute, V.; Zavistanaviciute, P.; Zokaityte, E.; Bartkevics, V.; Badaras, S.; Klupsaite, D.; Mozuriene, E.; et al. Pigs' feed fermentation model with antimicrobial lactic acid bacteria strains combination by changing extruded soya to biomodified local feed stock. *Animals* **2020**, *10*, 783. [CrossRef] [PubMed]
24. European Union Law. European Parliament Directive 2010/63/EU of the European Parliament and of the Council of 22 September 2010 on the Protection of Animals Used for Scientific Purposes Text with EEA Relevance. Available online: https://eur-lex.europa.eu/legal-content/EN/TXT/?uri=celex%3A32010L0063 (accessed on 13 July 2020).
25. Lithuanian Director of the State Food and Veterinary Service. B1-866 Dėl Mokslo ir mokymo tikslais naudojamų gyvūnų laikymo, priežiūros ir naudojimo reikalavimų patvirtinimo. Available online: https://e-seimas.lrs.lt/portal/legalAct/lt/T

AD/TAIS.437081?positionInSearchResults=0&searchModelUUID=64a1f51f-6356-4b60-a21e-3a67ac3b1107 (accessed on 10 March 2020).

26. National Research Council. *Nutrient Requirements of Swine: Eleventh Revised Edition*; The National Academies Press: Washington, DC, USA, 2012.
27. AOAC. *Official Methods of Analysis*, 21st ed.; AOAC International: Rockland, MD, USA, 2019.
28. Merkeviciene, L.; Ruzauskaite, N.; Klimiene, I.; Siugzdiniene, R.; Dailidaviciene, J.; Virgailis, M.; Mockeliunas, R.; Ruzauskas, M. Microbiome and antimicrobial resistance genes in microbiota of cloacal samples from European herring gulls (*Larus Argentatus*). *J. Vet. Res.* **2017**, *4*, 27–35. [CrossRef]
29. Zavistanaviciute, P.; Poskiene, I.; Lele, V.; Antanaitis, R.; Kantautaite, J.; Bartkiene, E. The influence of the newly isolated Lactobacillus plantarum LUHS135 and *Lactobacillus paracasei* LUHS244 strains on blood and faeces parameters in endurance horses. *Pol. J. Vet. Sci.* **2019**, *22*, 513–521. [CrossRef]
30. The Lithuanian Minister of Environment. D1-862. Dėl Lietuvos Respublikos Aplinkos Apsaugos Normatyvinio Dokumento LAND 88-2009 "Amoniako Koncentracijos Nustatymas Aplinkos Ore Spektrometriniu Metodu" Patvirtinimo. Available online: https://e-seimas.lrs.lt/portal/legalAct/lt/TAD/TAIS.363222 (accessed on 13 July 2020).
31. Pajarillo, E.A.B.; Chae, J.P.; Kim, H.B.; Kim, I.H.; Kang, D.-K. Barcoded pyrosequencing-based metagenomic analysis of the faecal microbiome of three purebred pig lines after cohabitation. *Appl. Microbiol. Biotechnol.* **2015**, *99*, 5647–5656. [CrossRef]
32. Round, J.L.; Mazmanian, S.K. The gut microbiota shapes intestinal immune responses during health and disease. *Nat. Rev. Immunol.* **2009**, *9*, 313–323. [CrossRef]
33. Benson, A.K.; Kelly, S.A.; Legge, R.; Ma, F.; Low, S.J.; Kim, J.; Zhang, M.; Oh, P.L.; Nehrenberg, D.; Hua, K.; et al. Individuality in gut microbiota composition is a complex polygenic trait shaped by multiple environmental and host genetic factors. *Proc. Natl. Acad. Sci. USA* **2010**, *107*, 18933–18938. [CrossRef]
34. Leamy, L.J.; Kelly, S.A.; Nietfeldt, J.; Legge, R.M.; Ma, F.; Hua, K.; Sinha, R.; Peterson, D.A.; Walter, J.; Benson, A.K.; et al. Host genetics and diet, but not immunoglobulin A expression, converge to shape compositional features of the gut microbiome in an advanced intercross population of mice. *Genome Biol.* **2014**, *15*, 552. [CrossRef]
35. Frese, S.A.; Parker, K.; Calvert, C.C.; Mills, D.A. Diet shapes the gut microbiome of pigs during nursing and weaning. *Microbiome* **2015**, *3*, 28. [CrossRef] [PubMed]
36. Le Floc'h, N.; Knudsen, C.; Gidenne, T.; Montagne, L.; Merlot, E. Impact of feed restriction on health, digestion and faecal microbiota of growing pigs housed in good or poor hygiene conditions. *Animal* **2014**, *8*, 1632–1642. [CrossRef] [PubMed]
37. Yang, Q.; Huang, X.; Wang, P.; Yan, Z.; Sun, W.; Zhao, S. Longitudinal development of the gut microbiota in healthy and diarrheic piglets induced by age-related dietary changes. *Microbiologyopen* **2019**, *8*, e923. [CrossRef] [PubMed]
38. Lamendella, R.; Domingo, J.W.; Ghosh, S.; Martinson, J.; Oerther, D.B. Comparative fecal metagenomics unveils unique functional capacity of the swine gut. *BMC Microbiol.* **2011**, *15*, 103. [CrossRef] [PubMed]
39. Pajarillo, E.A.B.; Chae, J.-P.; Balolong, M.P.; Kim, H.B.; Kang, D.-K. Assessment of fecal bacterial diversity among healthy piglets during the weaning transition. *J. Gen. Appl. Microbiol.* **2014**, *60*, 140–146. [CrossRef]
40. Ivarsson, E.; Roos, S.; Liu, H.Y.; Lindberg, J.E. Fermentable non-starch polysaccharides increase the abundance of *Bacteroides-Prevotella-Porphyromonas* in ileal microbial community of growing pigs. *Animal* **2014**, *8*, 1777–1787. [CrossRef] [PubMed]
41. Rho, J.; Wright, D.P.; Christie, D.L.; Clinch, K.; Furneaux, R.H.; Roberton, A.M. A novel mechanism for desulfation of mucin: Identification and cloning of a mucin-desulfating glycosidase (*sulfoglycosidase*) from Prevotella strain RS2. *J. Bacteriol.* **2005**, *187*, 1543–1551. [CrossRef] [PubMed]
42. Delia, E.; Tafaj, M.; Männer, K. Efficiency of Probiotics in Farm Animals. Available online: https://www.intechopen.com/books/probiotic-in-animals/efficiency-of-probiotics-in-farm-animals (accessed on 17 May 2020).
43. Valeriano, V.D.V.; Balolong, M.P.; Kang, D.-K. Probiotic roles of Lactobacillus sp. in swine: Insights from gut microbiota. *J. Appl. Microbiol.* **2017**, *122*, 554–567. [CrossRef]
44. Gänzle, M.G.; Follador, R. Metabolism of oligosaccharides and starch in lactobacilli: A review. *Front. Microbiol.* **2012**, *3*, 340. [CrossRef]

45. Konstantinov, S.R.; Awati, A.A.; Williams, B.A.; Miller, B.G.; Jones, P.; Stokes, C.R.; Akkermans, A.D.L.; Smidt, H.; de Vos, W.M. Post-natal development of the porcine microbiota composition and activities. *Environ. Microbiol.* **2006**, *8*, 1191–1199. [CrossRef]
46. Dou, S.; Gadonna-Widehem, P.; Rome, V.; Hamoudi, D.; Rhazi, L.; Lakhal, L.; Larcher, T.; Bahi-Jaber, N.; Pinon-Quintana, A.; Guyonvarch, A.; et al. Characterisation of early-life fecal microbiota in susceptible and healthy pigs to post-weaning diarrhoea. *PLoS ONE* **2017**, *12*, e0169851. [CrossRef]
47. Canibe, N.; Jensen, B.B. Fermented liquid feed—Microbial and nutritional aspects and impact on enteric diseases in pigs. *Anim. Feed Sci. Technol.* **2012**, *173*, 17–40. [CrossRef]
48. Grela, E.R.; Czech, A.; Kiesz, M.; Wlazło, Ł.; Nowakowicz-Dębek, B. A fermented rapeseed meal additive: Effects on production performance, nutrient digestibility, colostrum immunoglobulin content and microbial flora in sows. *Anim. Nutr.* **2019**, *5*, 373–379. [CrossRef] [PubMed]
49. Mach, N.; Berri, M.; Estellé, J.; Levenez, F.; Lemonnier, G.; Denis, C.; Leplat, J.-J.; Chevaleyre, C.; Billon, Y.; Doré, J.; et al. Early-life establishment of the swine gut microbiome and impact on host phenotypes. *Environ. Microbiol. Rep.* **2015**, *7*, 554–569. [CrossRef]
50. Shin, D.; Chang, S.Y.; Bogere, P.; Won, K.; Choi, J.-Y.; Choi, Y.-J.; Lee, H.K.; Hur, J.; Park, B.-Y.; Kim, Y.; et al. Beneficial roles of probiotics on the modulation of gut microbiota and immune response in pigs. *PLoS ONE* **2019**, *14*, e0220843. [CrossRef]
51. Wang, L.; Zhu, L.; Qin, S. Gut microbiota modulation on intestinal mucosal adaptive immunity. *J. Immunol. Res.* **2019**, *2019*, 4735040. [CrossRef] [PubMed]
52. Slifierz, M.J.; Friendship, R.; Weese, J.S. Zinc oxide therapy increases prevalence and persistence of methicillin-resistant Staphylococcus aureus in pigs: A randomized controlled trial. *Zoonoses Public Health* **2015**, *62*, 301–308. [CrossRef]
53. Chen, L.; Xu, Y.; Chen, X.; Fang, C.; Zhao, L.; Chen, F. The maturing development of gut microbiota in commercial piglets during the weaning transition. *Front. Microbiol.* **2017**, *8*, 1688. [CrossRef]
54. Bian, G.; Ma, S.; Zhu, Z.; Su, Y.; Zoetendal, E.G.; Mackie, R.; Liu, J.; Mu, C.; Huang, R.; Smidt, H.; et al. Age, introduction of solid feed and weaning are more important determinants of gut bacterial succession in piglets than breed and nursing mother as revealed by a reciprocal cross-fostering model. *Environ. Microbiol.* **2016**, *18*, 1566–1577. [CrossRef]
55. Levesque, C.L.; Hooda, S.; Swanson, K.S.; de Lange, K. Alterations in ileal mucosa bacteria related to diet complexity and growth performance in young pigs. *PLoS ONE* **2014**, *9*, e108472. [CrossRef]
56. Zhu, L.; Liao, R.; Tu, W.; Lu, Y.; Cai, X. Pyrodextrin enhances intestinal function through changing the intestinal microbiota composition and metabolism in early weaned piglets. *Appl. Microbiol. Biotechnol.* **2020**, *104*, 4141–4154. [CrossRef]
57. Zimmermann, P.; Curtis, N. The influence of probiotics on vaccine responses—A systematic review. *Vaccine* **2018**, *36*, 207–213. [CrossRef] [PubMed]
58. Dowarah, R.; Verma, A.K.; Agarwal, N.; Singh, P.; Singh, B.R. Selection and characterization of probiotic lactic acid bacteria and its impact on growth, nutrient digestibility, health and antioxidant status in weaned piglets. *PLoS ONE* **2018**, *13*, e0192978. [CrossRef] [PubMed]
59. Arfken, A.M.; Frey, J.F.; Ramsay, T.G.; Summers, K.L. Yeasts of burden: Exploring the mycobiome–bacteriome of the piglet GI tract. *Front. Microbiol.* **2019**, *10*, 2286. [CrossRef] [PubMed]
60. Xu, X.; Li, L.-M.; Li, B.; Guo, W.-J.; Ding, X.-L.; Xu, F.-Z. Effect of fermented biogas residue on growth performance, serum biochemical parameters, and meat quality in pigs. *Asian Austral. J. Anim. Sci.* **2017**, *30*, 1464–1470. [CrossRef]
61. Czech, A.; Grela, E.; Kiesz, M.; Kłys, S. Biochemical and haematological blood parameters of sows and piglets fed a diet with a dried fermented rapeseed meal. *Ann. Anim. Sci.* **2020**, *20*, 535–550. [CrossRef]
62. Shi, C.; He, J.; Wang, J.; Yu, J.; Yu, B.; Mao, X.; Zheng, P.; Huang, Z.; Chen, D. Effects of Aspergillus niger fermented rapeseed meal on nutrient digestibility, growth performance and serum parameters in growing pigs. *Anim. Sci. J.* **2016**, *87*, 557–563. [CrossRef]
63. Cho, J.; Min, B.; Chen, Y.; Yoo, J.S.; Wang, Q. Evaluation of FSP (fermented soy protein) to replace soybean meal in weaned pigs: Growth performance, blood urea nitrogen and total protein concentrations in serum and nutrient digestibility. *Asian Austral. J. Anim. Sci.* **2007**, *20*, 1874–1879. [CrossRef]
64. Min, B. Effects of diet complexity and fermented soy protein on growth performance and apparent ileal amino acid digestibility in weanling pigs. *Asian Austral. J. Anim. Sci.* **2010**, *23*, 1496–1502. [CrossRef]

65. Zhu, J.; Gao, M.; Zhang, R.; Sun, Z.; Wang, C.; Yang, F.; Huang, T.; Qu, S.; Zhao, L.; Li, Y.; et al. Effects of soybean meal fermented by *L. plantarum*, *B. subtilis* and *S. cerevisieae* on growth, immune function and intestinal morphology in weaned piglets. *Microb. Cell Fact.* **2017**, *16*, 191. [CrossRef]
66. Liu, X.; Feng, J.; Xu, Z. The effects of fermented soybean meal on growth performance and immune characteristics in weaned piglets. *Turk. J. Vet. Anim. Sci.* **2007**, *31*, 341–345.
67. Jeong, Y.D.; Ko, H.S.; Hosseindoust, A.; Choi, Y.H.; Chae, B.J.; Yu, D.J.; Cho, E.S.; Cho, K.H.; Shim, S.M.; Ra, C.S.; et al. Lactobacillus-based fermentation product and lactose level in the feed for weanling pigs: Effects on intestinal morphology, microbiota, gas emission, and targeted intestinal coliforms. *Livest. Sci.* **2019**, *227*, 90–96. [CrossRef]
68. Satessa, G.D.; Tamez-Hidalgo, P.; Hui, Y.; Cieplak, T.; Krych, L.; Kjærulff, S.; Brunsgaard, G.; Nielsen, D.S.; Nielsen, M.O. Impact of dietary supplementation of lactic acid bacteria fermented rapeseed with or without macroalgae on performance and health of piglets following omission of medicinal zinc from weaner diets. *Animals* **2020**, *10*, 137. [CrossRef] [PubMed]
69. Oliver-Ferrando, S.; Segalés, J.; López-Soria, S.; Callén, A.; Merdy, O.; Joisel, F.; Sibila, M. Evaluation of natural porcine circovirus type 2 (PCV2) subclinical infection and seroconversion dynamics in piglets vaccinated at different ages. *Vet. Res.* **2016**, *47*, 121. [CrossRef]
70. Duivon, D.; Corrégé, I.; Hémonic, A.; Rigaut, M.; Roudaut, D.; Jolie, R. Field evaluation of piglet vaccination with a Mycoplasma hyopneumoniae bacterin as compared to a ready-to-use product including porcine circovirus 2 and M. hyopneumoniae in a conventional French farrow-to-finish farm. *Porc. Health Manag.* **2018**, *4*, 4. [CrossRef] [PubMed]
71. Jeong, J.; Park, C.; Choi, K.; Chae, C. Comparison of three commercial one-dose porcine circovirus type 2 (PCV2) vaccines in a herd with concurrent circulation of PCV2b and mutant PCV2b. *Vet. Microbiol.* **2015**, *177*, 43–52. [CrossRef] [PubMed]
72. da Silva, N.; Carriquiry, A.; O'Neill, K.; Opriessnig, T.; O'Connor, A.M. Mixed treatment comparison meta-analysis of porcine circovirus type 2 (PCV2) vaccines used in piglets. *Prev. Vet. Med.* **2014**, *117*, 413–424. [CrossRef]
73. Woźniak, A.; Miłek, D.; Matyba, P.; Stadejek, T. Real-time PCR detection patterns of porcine circovirus type 2 (PCV2) in Polish farms with different statuses of vaccination against PCV2. *Viruses* **2019**, *11*, 1135. [CrossRef]
74. Opriessnig, T.; Xiao, C.-T.; Halbur, P.G.; Gerber, P.F.; Matzinger, S.R.; Meng, X.-J. A commercial porcine circovirus (PCV) type 2a-based vaccine reduces PCV2d viremia and shedding and prevents PCV2d transmission to naïve pigs under experimental conditions. *Vaccine* **2017**, *35*, 248–254. [CrossRef]
75. Afghah, Z.; Webb, B.; Meng, X.-J.; Ramamoorthy, S. Ten years of PCV2 vaccines and vaccination: Is eradication a possibility? *Vet. Microbiol.* **2017**, *206*, 21–28. [CrossRef]
76. Chae, C. Porcine respiratory disease complex: Interaction of vaccination and porcine circovirus type 2, porcine reproductive and respiratory syndrome virus, and Mycoplasma hyopneumoniae. *Vet. J.* **2016**, *212*, 1–6. [CrossRef]
77. Patterson, R.; Nevel, A.; Diaz, A.V.; Martineau, H.M.; Demmers, T.; Browne, C.; Mavrommatis, B.; Werling, D. Exposure to environmental stressors result in increased viral load and further reduction of production parameters in pigs experimentally infected with PCV2b. *Vet. Microbiol.* **2015**, *177*, 261–269. [CrossRef] [PubMed]
78. Lee, D.J.; Yang, Y.; Jung, H.I.; Nguyen, D.H.; Kim, I.H. Effect of a protected blend of organic acids and medium-chain fatty acids on growth performance, nutrient digestibility, blood profiles, meat quality, faecal microflora, and faecal gas emission in finishing pigs. *Can. J. Anim. Sci.* **2019**, *99*, 448–455.
79. Nguyen, D.H.; Nyachoti, C.; Kim, I.H. Evaluation of effect of probiotics mixture supplementation on growth performance, nutrient digestibility, faecal bacterial enumeration, and noxious gas emission in weaning pigs. *Ital. J. Anim. Sci.* **2019**, *18*, 466–473. [CrossRef]
80. Bindas, L.; Bujnak, L.; Maskaľová, I.; Mihok, T.; Lacková, P.; Nad, P. Effects of feeding low protein diets on serum and faeces parameters in weaned piglets. *Folia Vet.* **2019**, *63*, 37–44. [CrossRef]

8

Former Foodstuff Products in *Tenebrio Molitor* Rearing: Effects on Growth, Chemical Composition, Microbiological Load, and Antioxidant Status

Simone Mancini [1,*], Filippo Fratini [1,2], Barbara Turchi [1], Simona Mattioli [3], Alessandro Dal Bosco [3], Tiziano Tuccinardi [4], Sanjin Nozic [1,4] and Gisella Paci [1,2]

1 Department of Veterinary Sciences, University of Pisa, Viale delle Piagge 2, 56124 Pisa, Italy
2 Interdepartmental Research Center "Nutraceuticals and Food for Health", University of Pisa, Via del Borghetto 80, 56124 Pisa, Italy
3 Department of Agricultural, Food and Environmental Science, University of Perugia, Borgo XX Giugno 74, 06100 Perugia, Italy
4 Department of Pharmacy, University of Pisa, Via Bonanno 6, 56124 Pisa, Italy
* Correspondence: simone.mancini@for.unipi.it or simafo@gmail.com.

Simple Summary: Insects represent a possible alternative nutrient source for food and feed purposes. Insects could be reared on a feed basis alternative to conventional ones of animal origin and could help to face the rising demand of proteins. Mealworm could be reared directly on former foodstuff products allowing to reduce waste materials and enhance profits in several sectors. This study demonstrates that *Tenebrio molitor* rearing can be done on leftovers and by-products with proficient outcomes and high-quality final products. However, rearing substrates must be carefully selected in order to maximize the outcomes in relation to the prefixed goals.

Abstract: *Tenebrio molitor* (mealworm) larvae represent one of the most interesting edible insects and could be reared on alternative feeds, such as former foodstuff products (FFPs). In the present work, five different FFPs (brewery spent grains, bread and cookie leftovers, and mixes of brewer's spent grain or bread with cookies) were employed as feeding substrates. Larvae's growth performances, chemical composition, microbial loads, and antioxidant status were determined. Chemical compositions of the substrates affected all the tested parameters. Brewery spent grains-fed larvae showed a faster growth period and higher crude protein and carbohydrate contents. The use of cookies as a single substrate or their addition to spent grains or bread increased the lipids contents, while growth was delayed. Microbial loads were partially affected by the fed diet. The antioxidant status of larvae showed different concentrations of tocopherols isoforms (δ, γ, α) in relation to the diet; however, no differences were detected in relation to the global antioxidant capacity (2,2-azinobis-(3 ethylbenzothiazoline-6-sulfonic acid), ABTS reducing activity; 1,1-diphenyl-2-pircydrazyl, DPPH radical scavenging activity; ferric reducing ability, FRAP). Results point out a high plasticity of mealworm larvae and the potential to tailor the final outcomes in relation to the substrate employed. Mealworms could be practically reared on FFPs to produce food-feed with high nutrient values.

Keywords: mealworm; edible insects; animal protein; by-product; entomophagy

1. Introduction

In the light of the increasing world population and the demand of proteins from the developing countries, insects could represent one of the more suitable answers [1]. Insects, compared to conventional production animals, have a lower environmental impact (mostly linked to greenhouse gas production, use of water, and use of arable land); moreover, due to their higher reproductive capacity, nutritional quality, and feed conversion efficiency, they could be taken into account for both food and feed purposes.

Despite researchers' interests in edible insects, western consumers seem to be cautious to practice entomophagy, mainly in relation to disgust related to western socio-cultural ideas of these animals [2,3]. During the last years, several research studies were carried out to determine which items could be positively accepted by consumers in order to introduce edible insects on the market and increase their use both as food and feed. The main drivers which induced an increase of familiarity and a decrease in the awareness in relation to entomophagy have been identified with the visual presence of insects (visible/invisible, powder/whole), the familiarity with the products containing insects, the correct information about nutritional values, and possible positive effects of edible insects from sustainability and environmental perspectives [4–6]. With respect to these last perspectives, the rearing conditions could play a major role, as insects can be reared on sustainable feeds, such as waste or by-products that do not meet the nutritional values needed for the rearing of other farmed animals [7]. Former foodstuff products (FFPs) are defined by the European Commission in the Commission Regulation (EU) No 68/2013 of the catalogue of feed materials as "foodstuffs other than catering reflux, which are manufactured in full compliance with EU food law but are no longer intended for human consumption for practical and logistical reasons or due to problems in manufacturing or packaging which are unlikely to cause any health risks when used as feed". Therefore, in this category, both leftovers from the food industry mainly composed of bakery products, such as bread, and leftovers composed of confectionery products, such as cookies, are listed. The use of FFPs as feed leads to a decrease in waste material and pollution and represents an interesting opportunity in terms of the circular economy [8]. Thus, FFPs represent a valid and cheap source of energy, and they could be used in insect feeding concerning their essential nutrients [9].

Based on the above-mentioned considerations, the use of FFPs in the rearing of edible insects seems a good compromise to reduce food waste materials, produce valuable new food/feed products, and have a positive impact on the consumers' perception. However, it has to be mentioned that, even in the insect production sector, there are some limitations for the employment of certain feeds as the diet composition could affect the insect's development rate and body nutrient composition [7,10].

The mealworm (*Tenebrio molitor* L. 1758; Coleoptera Tenebrionidae) represents one of the most studied edible insects, both as food and feed. The mealworm is a holometabolic insect (complete metamorphosis) and probably originated in the Mediterranean region, but it is nowadays cosmopolitan in its distribution. Due to its low rearing request, the mealworm is one of the promising candidates for an industrial scale production. Indeed, in the last years, several farms and food-feed companies worldwide have started to employ this insect. Nutritional composition of mealworm larvae revealed a relatively high content of protein, 50% on a dry-matter basis [11,12], and lipids, about 30%–34% on a dry-matter basis [12–14], with a good composition in amino acids (good source of the essential amino acids), vitamins (i.e., vitamin E, vitamin B12, niacin, riboflavin, pantothenic acid, and biotin), and minerals (P, K, Mg, Zn, and Mn) [15].

Few research studies were carried out on the rearing conditions and on the effects of different substrates on the development and chemical-biological characteristics of mealworm larvae [7,10,13,14,16,17]. Most of the cited studies highlighted a plasticity of *T. molitor* in relation to the substrate, with variations of both development times and nutritional values of the larvae. However, researchers employed mixes of several ingredients as substrates in their studies, since the main focus was to determine the effects on combinations of high/low amounts of proteins, lipids, or carbohydrates. In this scenario, it is very difficult to assess if mealworms could be reared on FFPs directly gathered from the producers (one single type of FFP or a mix of a few). Indeed, as insect rearing does not request specific features in geographical or natural environmental conditions (as carried out in indoor controlled conditions), it is possible to speculate that new farms could be located near the substrate suppliers, and feed manipulation will be kept at a minimum in order to maintain low costs.

Hence, the main aim of this research study was to evaluate the use of three FFPs in *T. molitor* rearing in order to produce larvae for food/feed purposes. In particular, larval development, chemical characteristics, and the microbiological and antioxidant status were determined in mealworm larvae fed with brewery spent grains, bread and cookie leftovers, and on mixes of brewer's spent grain or bread with cookies (mixed to increase brewer's spent grain or bread diets in lipid content).

2. Materials and Methods

2.1. Diet Preparation

The mealworms were fed with five different diets: brewery spent grains (SG), bread (B), cookies (C), 50% brewer's spent grain and 50% cookies (SG-C), and 50% bread and 50% cookies (B-C). Spent grains were directly collected from a local brewery and immediately frozen at −20 °C. Bread and cookies were collected from a market shop as daily remains (bread) or soon to expire products (cookies). Excessive humidity of spent grains, bread, and cookies was removed in an oven at 90 °C until the product was completely dry. Spent grains were previously thawed for 18 h at 4 °C. The proximate composition of SG, B, and C is reported in Table 1. The three substrates were then finely ground, and the five diets were formulated.

Table 1. Proximate composition of brewery spent grains (SG), bread (B) and cookies (C).

Item	Unit	SG	B	C
Dry matter (DM)	%	94.81	97.09	99.99
Ether extract	%	3.29	0.31	10.44
Crude protein [1]	%	17.98	11.15	6.55
Ash	%	3.43	1.88	0.70
Carbohydrates	%	70.10	83.75	82.29
Ether extract	% on DM	3.47	0.32	10.45
Crude protein [1]	% on DM	18.97	11.49	6.55
Ash	% on DM	3.62	1.93	0.70
Carbohydrates	% on DM	73.94	86.26	82.30

[1] Conversion factors of 5.83 for SG [18], of 5.70 for B, and 6.25 for C [19] (protocol numbers 950.36 and 935.39, respectively).

2.2. Insect Rearing and Growth Performances

The mealworms were reared in plastic containers (39 × 28 × 14 cm) at the Department of Veterinary Sciences (University of Pisa, Pisa, Italy) under a laboratory scale production. The temperature was maintained at 25 ± 1 °C with 50%–60% of relative humidity. In the first stage of the trial, 15 boxes were employed (three per diet), in which adult beetles (one to two weeks old) were placed for one week to deposit eggs. Then beetles were removed, and the larvae were left to grow in the substrate. Experimental diets were added weekly if needed (*ad libitum*, weighed before adding), and in order to provide moisture, potatoes slices were deposited once a week. Boxes were visually evaluated three times per week to check larvae health and to remove dead ones.

The mealworms were harvested when the first pupa was observed or after one year of rearing time (in order to maintain a profitable low cost). The development time was calculated between the first day of the experiment and the day of harvesting. Every week, representative samples of larvae (100 per box) were weighed to quantify the growth performance. After harvesting, the larvae were starved for 24 h (apart from samples for microbiological analyses performed with and without starving, see below) in sterile plastic containers with plastic web on the base in order to separate feces and to avoid fecal contact.

The feed conversion ratio (FCR) and efficiency of conversion of ingested food (ECI) were calculated assuming that all the provided feed was consumed as reported by Oonincx et al. [7]. Furthermore, the nitrogen conversion efficiency (N-ECI) was also calculated as reported by Oonincx et al. [7].

2.3. Proximate Composition of Feed and Larvae

Proximate composition analyses were performed in triplicate for each sample. Larvae were ground in a blender before being analyzed.

The dry matter content was determined by dehydration in a drying oven at 105 °C until constant weight. The lipids content was quantified by the Soxhlet method using petroleum ether as a solvent. The ash content was determined by incineration in a muffle furnace at 550 °C. The crude protein content was determined by the Kjeldahl method, two protein-to-nitrogen conversion factors were used: 6.25 as normally calculated for meat samples and 4.76 as suggested by Janssen et al. [20] for insects. Carbohydrates were calculated as: 100 − crude protein − lipid content − moisture − ash.

2.4. Microbiological Analyses

For microbiological analyses both un-starved and starved larvae were employed in order to quantify the effect of fasting. Substrates were also analyzed, and microbiological amounts lower than the detection limits were reported.

Ten grams of larvae were weighed in sterile stomacher bags, and then 90 mL of sterile saline solution was added. The mixture was homogenized for 60 s in a stomacher (Stomacher® 400 Circulator, VWR International Sr, Milan, Italy). Ten-fold serial dilution series were performed and plated on different media.

Plate count agar (PCA) was employed for the quantification of the total viable aerobic count (incubated at 30 °C for 72 h) and aerobic bacterial endospores after heating the 10:1 dilution at 80 °C for 10 min and performing ten-fold serial dilutions (incubated at 30 °C for 72 h); violet red bile glucose agar (VRBGA) was used to enumerate Enterobacteriaceae (incubation at 37 °C for 24 h); Tryptone Bile X-Glucuronide medium (TBX) was employed to quantify *Escherichia coli* (42 °C for 24 h); Baird Parker medium (BP) was employed for the enumeration of *staphylococci* both presumptive coagulase-positive and -negative (incubated at 37 °C for 24–48 h); yeast extract, dextrose, chloramphenicol medium (YEDC) was used for yeast and mold counts (incubation at 25 °C for 120 h); de Man–Rogosa–Sharpe agar (MRS) was employed to count lactic acid bacteria (37 °C for 48 h in anaerobic conditions); Bacillus Cereus MYP agar was used for the evaluation of the presence of presumptive *Bacillus cereus* (37 °C for 24 h). Absence of *Listeria monocytogenes* and *Salmonella* spp. in 25 g was assessed according to ISO 11290 and ISO 6579, respectively.

All culture media and supplements were purchased from ThermoFisher Scientific, (Milan, Italy) except for Bacillus Cereus MYP agar, which was purchased from Biolife (Milan, Italy).

Microbial counts were expressed in log colony-forming unit (CFU)/g as mean of three replicates (when one or more samples showed values below the detection threshold, 1.0 log CFU/g, the detection threshold divided by two, 0.5 log CFU/g, was used to calculate the mean and standard deviation as reported by Stoops et al. [21]).

2.5. Antioxidant Status

Antioxidant activity of the larvae was quantified as reported by Mancini et al. [22] for meat samples with minor modifications. Three grams of larvae were homogenized in 10 mL ethanol, followed by a centrifugation and filtration (Whatman number 4 filter paper) steps. The filtrate was used to measure 2,2-azinobis-(3 ethylbenzothiazoline-6-sulfonic acid) (ABTS) reducing activity, 1,1-diphenyl-2-pircydrazyl (DPPH) radical scavenging activity, and ferric reducing ability (FRAP). The results of antioxidant capacity were expressed as mmol of Trolox equivalent per kilogram of sample (calibration curves obtained with Trolox at 0–2000 μM, final concentrations).

The tocopherol (δ-, γ-, α-isoforms) contents of the samples were quantified by a high performance liquid chromatography (HPLC) system, according to Zaspel and Csallany [23]. About 3 g of larvae were saponified in 60 g/100 mL KOH in ethanol in a thermostat bath at 70 °C for 30 min. Then the content was sonicated and extracted twice with 15 mL of n-hexane/ethyl acetate (9:1, v/v). After collecting the upper phase, the samples were nitrogen dried and then reconstituted in 200 μL of acetonitrile. Fifty 50 μL was injected into the HPLC system (Perkin Elmer series 200, equipped with an autosampler system, model AS 950–10, Tokyo, Japan) on a Synergy Hydro-RP column (4 μm, 4.6 × 100 mm; Phenomenex, Bologna, Italy). An Fluorometric detector (FD) (model, FP-1525; Jasco, Tokyo, Japan; excitation and emission wavelengths of 295 and 328 nm, respectively) were used to identify the different isoforms. External calibration curves were used to quantify isoforms by increasing amounts of pure tocopherols in ethanol.

2.6. Statistical Analysis

The data obtained from proximate composition, microbial determinations, antioxidant capacity, and antioxidant compounds were statistically analyzed using a one-way ANOVA with regards to the different diets (SG, B, C, SG-C, and B-C). The effect of starvation (starved and un-starved) was tested against each different microbiological quantification by a Student's *t*-test. Statistical significance was set at 0.05 and differences were assessed using Tukey's test. Linear regression and second-order polynomial quadratic equation were performed to evaluate the effect of each diet on the growth performance. Principal components analysis (PCA) was performed on proximate composition (protein conversion factor of 4.76), microbial loads (fasted samples), growth performance (g weight gain per day), antioxidant capacity (ABTS, DPPH, and FRAP), and antioxidant compounds (tocopherol δ-, γ-, α-isoforms); all the data were mean centered and scaled to a unit standard deviation before analysis. R free statistical software was used [24].

3. Results and Discussion

3.1. Growth Performances and Proximate Compositions

Proximate compositions of the larvae reared on the different substrates are reported in Table 2.

Table 2. Proximate compositions of the larvae reared on the different substrates.

Item	Unit	Rearing Substrates					RMSE	*p*-Value
		SG	B	C	SG-C	B-C		
Dry matter (DM)	%	33.33 ^{c}	32.62 ^{c}	35.55 ^{b}	37.53 ^{a}	35.33 ^{b}	0.733	<0.001
Ether extract	%	6.46 ^{d}	14.82 ^{b}	17.77 ^{a}	11.77 ^{c}	17.48 ^{a}	1.189	<0.001
Crude protein [1]	%	17.36 ^{a}	14.09 ^{ab}	13.33 ^{b}	17.65 ^{a}	14.07 ^{ab}	1.392	0.020
Crude protein [2]	%	13.22 ^{a}	10.73 ^{ab}	10.15 ^{b}	13.44 ^{a}	10.72 ^{ab}	1.392	0.020
Ash	%	1.10	0.98	0.95	1.07	1.01	0.097	0.371
Carbohydrates [1]	%	8.40 ^{a}	2.73 ^{b}	3.50 ^{b}	7.05 ^{a}	2.76 ^{b}	1.620	<0.001
Carbohydrates [2]	%	12.54 ^{a}	6.09 ^{b}	6.72 ^{b}	11.26 ^{a}	6.12 ^{b}	1.523	<0.001
Ether extract	% on DM	19.38 ^{c}	45.43 ^{a}	50.00 ^{a}	31.35 ^{b}	49.48 ^{a}	3.389	<0.001
Crude protein [1]	% on DM	51.34 ^{a}	42.28 ^{ab}	37.31 ^{b}	46.40 ^{ab}	40.96 ^{ab}	3.898	0.022
Crude protein [2]	% on DM	39.10 ^{a}	32.20 ^{ab}	28.41 ^{b}	35.33 ^{ab}	31.20 ^{ab}	3.898	0.022
Ash	% on DM	3.36	3.05	2.69	2.88	2.81	0.268	0.084
Carbohydrates [1]	% on DM	25.15 ^{a}	8.28 ^{b}	9.83 ^{b}	18.76 ^{a}	7.81 ^{b}	4.523	<0.001
Carbohydrates [2]	% on DM	37.58 ^{a}	18.58 ^{c}	18.77 ^{c}	29.97 ^{b}	17.31 ^{c}	4.123	<0.001

SG: brewery spent grains; B: bread; C: cookies; SG-C: 50% spent grains, 50% cookies; B-C: 50% bread, 50% cookies. [1] Conversion factors of 6.25 for crude protein. [2] Conversion factors of 4.76 for crude protein. Means in the same row with no common superscripts (a–d) differ significantly ($p < 0.05$).

Dry matter, ether extract, and crude protein contents were affected by the diet. Dry matter of the SG-C larvae was the highest, followed by that of the C and B-C diets. From these data, it would seem that the presence of the cookies induced a decrease of humidity in the larval body, as the larvae fed with spent grains and bread alone (SG and B, respectively) showed the lowest values of dry matter. Cookies also played a major role in the lipid content of the larvae. Diets C and B-C showed the highest contents, followed by B. Spent grains contained low amounts of lipids (3.29%), and consequently larvae fed with SG and SG-C showed lower values compared to those fed other diets.

The highest crude protein contents were shown by diets which included spent grains (SG and SG-C) with minor differences. This value was lower in diets with bread (B and B-C). In this case, the low amount of crude proteins of the cookies (6.55%) negatively affected the chemical composition of the larvae with a consequent lowest body protein content.

As the dry matter analyses showed to be significantly affected by the diet, the proximate composition was also reported as % of dry matter (Table 2). Expressing the data as % of dry matter, the ether extract content showed that B, C, and B-C had the highest lipid contents, followed by SG-C and SG, respectively. Thus, the effect of the cookie fat content was confirmed but the bread diet (B) also played a key role.

Crude protein contents, as % of dry matter, showed only minor differences in the statistical analyses if compared to the expression as % of fresh sample. The conversion factors of 6.25 or 4.76, as well as the expression as % of fresh sample or % of dry matter, drastically affected the final numerical expression; nevertheless, as no standardization is currently used in the insect research field, we choose to report all the data expressions.

Carbohydrates were higher in larvae fed SG and SG-C diets than those fed the other diets. Interestingly, SG had the lowest amount of carbohydrates within the employed substrates. Particularly, moisture and lipids content seemed to affect the calculations of carbohydrates in larvae reared in SG and SG-C.

Chemical compositions of mealworm confirm the data already published [7,10,20] and in particular that body composition showed a considerable plasticity in relation to the diet [7,10]. Notably, in our trial, mealworms fed cookies (C) showed a high amount of lipids (50% on DM base) mostly related to the detriment of the protein content. This data could be interesting as edible insects are almost exclusively recommended as a protein source even though they could contain a high amount of fats.

The larvae weights are reported in Figure 1; larvae fed SG and SG-C reached the harvesting day approximately after 5 to 6.5 months. After one year of rearing, larvae fed diets with lower content of protein, C, B, and B-C did not reach the pupae stage; therefore, in order to not exceed in the rearing time, their growth was stopped. Larvae fed diets C, B, and B-C reached the final weights of 87, 95, and 112.5 mg, respectively. In linear regression and second-order polynomial quadratic equations (Figure 1), the coefficient of determination indicates that the models explain a high percentage of the variability (R2 between 0.94 and 0.99). Between linear regression and second-order polynomial quadratic equation, the second seems to better fit the larval growth rate.

The FCR (calculated on fresh basis), the ECI, and the N-ECI were severely affected by the dietary treatments. Dietary efficiency of mealworm larvae was strictly related to the chemical composition of the substrate; indeed, larvae fed SG and SG-C showed the lowest values of FCR, 2.22 and 2.76 respectively, confirming the extremely positive potential of this species to convert feed into body weight. On the other hand, larvae fed B, C, and the mix B-C showed FCR values about 8.86, 7.31, and 4.02, which are similar to those of conventional production animals (4.0 for pork and 8.8 for cereal beef; [9]). The variability in FCR values was related to the protein contents of the diets; indeed, SG and SG-C showed the highest amounts of protein. A large variability of the FCR value in relation to the protein content was also reported by Oonincx et al. [7] and van Broekhoven et al. [10] in mealworms fed different diets (from 4.1 to 19.1 and from 2.62 to 6.05, respectively).

The ECI values were higher for SG and SG-C diets (15.85 and 13.94, respectively) than for B, C, and B-C ones (3.79, 4.87, and 8.92, respectively). These data are comparable with those reported by Oonincx et al. [7] and van Broekhoven et al. [10]. As reported before for FCR, the ECI values were affected by the protein content of the diets, which confirms that mealworm efficiency could be modelled through the rearing substrate's chemical composition.

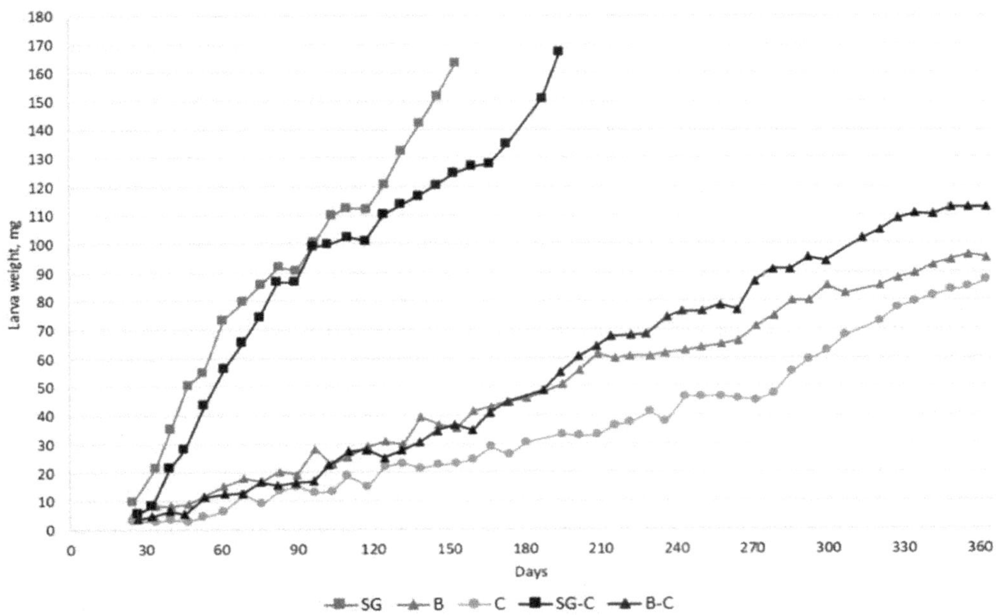

Rearing substrate	Linear regression	R^2	Second order polynomial quadratic equation	R^2
SG	$y = 1.0665x - 4.0790$	0.972	$y = -0.0024x^2 + 1.4989x - 19.763$	0.978
B	$y = 0.2774x - 3.0232$	0.993	$y = 0.00001x^2 + 0.2734x - 2.7408$	0.993
C	$y = 0.2434x - 10.839$	0.952	$y = 0.0005x^2 + 0.0449x + 3.3146$	0.985
SG-C	$y = 0.8480x - 0.5992$	0.943	$y = -0.0033x^2 + 1.5648x - 31.4230$	0.973
B-C	$y = 0.3583x - 13.1800$	0.989	$y = 0.0002x^2 + 0.2820x - 7.7445$	0.991

Figure 1. *Tenebrio molitor* larval growth in relation to the diet (SG: brewery spent grains; B: bread; C: cookies; SG-C: 50% spent grains, 50% cookies; B-C: 50% bread, 50% cookies).

Nitrogen was more efficiently converted than the other diet components in almost all diets, as in the SG, SG-C, C, and B-C diets, the N-ECIs were higher than the ECI (59.11, 71.75, 13.83, and 14.05, respectively). The diet exclusively composed of bread (B) showed the lowest value of N-ECI (2.77), which is lower than the ECI value. This means that mealworm fed only bread did not convert N of the substrate efficiently into body mass [7]. Indeed, larvae fed only cookies (C) reached approximately the same amount of crude protein content (Table 2) even though the substrate showed to contain half the protein (6.55% in C vs. 11.15% in B, Table 1). The high N-ECI showed by SG and SG-C larvae represents a starting point to reach a relevant benefit of insects over conventional production animals [25].

3.2. *Microbiological Analyses*

Microbial determinations of starved and un-starved larvae are reported in Table 3.

In the un-starved larvae, staphylococci, yeast-molds, and bacterial endospores were significantly different in relation to the diet. Staphylococci and yeast-molds loads were higher in larvae fed B, C, and B-C, followed by SG-C and SG. Minor differences in the staphylococci amount was highlighted between SG and SG-C in relation to the presence of the cookies. Bacterial endospores were absent in SG, and their presence was related to B and C diets. Indeed, feeding the larvae with a mix of SG and C induced an increase of these bacteria. The highest number of bacterial endospores was detected in the B-C diet, followed by B and C substrates.

Microbial analyses highlighted the total absence of *Escherichia coli* and *Bacillus cereus*, as well as the absence in 25 g of *Listeria monocytogenes* and *Salmonella* spp. Other research studies reported the absence of these bacteria in edible insects reared for human consumption [26–29].

Table 3. Microbiological analyses of the larvae (un-starved and starved) reared on the different substrates.

Item	Rearing Substrates					RMSE	p-Value
	SG	B	C	SG-C	B-C		
No starvation							
Total viable aerobic counts	7.08	7.63	6.46	6.84	7.77	0.776	0.266
Enterobacteriaceae	6.31	6.44	6.33	5.87	6.30	0.502	0.688
Staphylococci	3.85 ^{c}	5.96 ^{a}	5.29 ^{ab}	4.38 ^{b}	5.19 ^{ab}	0.579	0.005
Yeast and molds	3.30 ^{b}	5.60 ^{a}	6.31 ^{a}	3.13 ^{b}	5.34 ^{ab}	0.944	0.001
Lactic acid bacteria	6.20	5.18	5.57	5.44	5.31	0.685	0.343
Bacterial endospores	0.00 ^{d}	4.44 ^{ab}	3.44 ^{b}	2.05 ^{c}	5.32 ^{a}	0.240	<0.001
Starvation, 24 h							
Total viable aerobic counts	7.04	7.63	6.41	7.17	6.91	0.689	0.295
Enterobacteriaceae	6.36	6.01	4.96	5.52	5.55	0.744	0.211
Staphylococci	4.87	4.20	3.92	4.40	3.73	0.705	0.370
Yeast and molds	2.56 ^{b}	4.45 ^{a}	5.11 ^{a}	3.54 ^{ab}	4.44 ^{a}	0.561	0.002
Lactic acid bacteria	6.10	5.91	4.91	5.73	5.13	0.666	0.168
Bacterial endospores	0.00 ^{c}	3.62 ^{a}	2.44 ^{b}	2.36 ^{b}	3.60 ^{a}	0.243	0.038
Effect of starvation, p-Value							
Total viable aerobic counts	0.954	0.988	0.915	0.658	0.187		
Enterobacteriaceae	0.913	0.364	0.006	0.639	0.285		
Staphylococci	0.115	0.031	0.012	0.968	0.039		
Yeast and molds	0.506	0.014	< 0.001	0.598	0.021		
Lactic acid bacteria	0.835	0.271	0.250	0.653	0.629		
Bacterial endospores	-	0.120	0.003	0.312	0.009		

SG: brewery spent grains; B: bread; C: cookies; SG-C: 50% spent grains, 50% cookies; B-C: 50% bread, 50% cookies. Data were reported as log CFU/g. Means in the same row with no common superscripts (a–d) differ significantly ($p < 0.05$).

In starved larvae, diets affected only yeast-molds and bacterial endospores, showing a similar trend of that reported for the un-starved larvae with minor differences. Starvation was not effective in larvae fed SG and SG-C diets, while in larvae fed B, C, and B-C it partially affected the microbial flora. Total viable aerobic counts and lactic acid bacteria were not affected by the starvation. Staphylococci and yeast-molds amounts were significantly decreased in B, C, and B-C fed larvae. Starvation in larvae fed C induced a significant decrease in the Enterobacteriaceae amount, as well as in bacterial endospores of larvae fed C and B-C.

Studying larvae fed wheat bran supplemented with carrots, Wynants et al. [30] reported that fasting for 24 or 48 h, both with and without fecal contact of the larvae, did not significantly affect total viable aerobic bacteria, Enterobacteriaceae, aerobic bacterial endospores, psychrotrophic aerobic bacteria, and yeast and mold amounts.

Contrarily, Mancini et al. [31] reported that a starvation treatment for 24 h affected the bacterial endospores amount in mealworm larvae reared on wheat bran supplemented with potato slices as a water source without influencing the other microorganisms.

Moreover, starvation also resulted effective in the reduction of *Salmonella enterica* and *Listeria monocytogenes* in mealworm larvae reared in artificially contaminated substrates [32,33]. Thus, the effect of the starvation process seems to vary in relation to the diet as well as the response of different tested bacteria.

3.3. Antioxidant Status

Antioxidant capacity of mealworm larvae is reported in Table 4.

Table 4. Antioxidant capacity (ABTS, DPPH, and FRAP) and antioxidant compounds (tocopherols) of *Tenebrio molitor* larvae reared on different substrates.

Item	Rearing Substrates					RMSE	*p*-Value
	SG	B	C	SG-C	B-C		
Antioxidant Capacity							
ABTS	1.73	2.17	2.45	2.01	1.70	0.407	0.415
DPPH	0.35	0.30	0.34	0.30	0.28	0.042	0.504
FRAP	0.75	0.80	1.04	0.86	0.75	0.139	0.332
Antioxidant Compounds							
δ-tocopherol	0.12 ^{c}	0.13 ^{c}	0.18 ^{c}	0.24 ^{b}	0.33 ^{a}	0.018	<0.001
γ-tocopherol	0.07 ^{b}	0.02 ^{b}	0.15 ^{a}	0.17 ^{a}	0.04 ^{b}	0.017	<0.001
α-tocopherol	0.48 ^{c}	0.35 ^{c}	4.95 ^{a}	2.86 ^{b}	4.59 ^{a}	0.055	<0.001
Total tocopherols	0.79 ^{c}	0.51 ^{c}	5.28 ^{a}	3.16 ^{b}	4.96 ^{ab}	0.601	<0.001

SG: brewery spent grains; B: bread; C: cookies; SG-C: 50% spent grains, 50% cookies; B-C: 50% bread, 50% cookies. ABTS, DPPH and FRAP expressed as mmol of Trolox equivalent per kilogram of sample. Tocopherols and carotenes expressed as mg per kilogram of samples. Means in the same row with no common superscripts (a–c) differ significantly ($p < 0.05$).

No significant differences were found among the experimental diets. However, the C diet seems to slightly improve the antioxidant capacity of the larvae with respect to the other groups (as highlighted by ABTS and FRAP values, even if no statistical significances were determined). Such a trend is justified also by the significantly higher concentration of tocopherols (mainly due to the α-isoform amount) found in the cookies-supplemented groups (C, SG-C, and B-C larvae). Nevertheless, vitamin E being a fat-soluble vitamin, it is not surprising that its concentration was higher in the mealworms with a higher fat content (ether extract values). To the best of our knowledge, this is the first evaluation of the antioxidant capacity of *T. molitor* larvae.

The antioxidant capacity of mealworms could be very interesting in order to enhance dietary antioxidant ingestion; however, as no thermal process was applied to the samples (cooking or drying), enzymes present in the larvae body may have altered the final outcomes. Few published data reported the antioxidant concentration of insects: Finke [15] summarized that the vitamin E contents ranged from 3.3–24.0 mg/kg for mealworms, 5.3–9.1 mg/kg for superworms (*Zophobas morio*), and 8.6–69.2 mg/kg for waxworms (*Galleria mellonella*). In the present study, the total vitamin E concentration of mealworms ranged from 0.51 to 5.28 mg/kg, close to the lower values reported in the literature. However, a higher concentration (>10 mg/kg) of vitamin E is reported to be common in the wild-caught insects with respect to the reared ones [34].

3.4. Principal Component Analysis

A principal component analysis of proximate composition, microbial loads, growth performance, antioxidant capacity, and antioxidant compounds was performed in order to detect the principal components that better describe the modifications highlighted (Figure 2).

Eigenvalues, eigenvectors, and cumulative % of the first three principal components are reported in Table 5. The first two principal components (PC1: 41.07% and PC2: 17.91%) well differentiate the samples in relation to the diet.

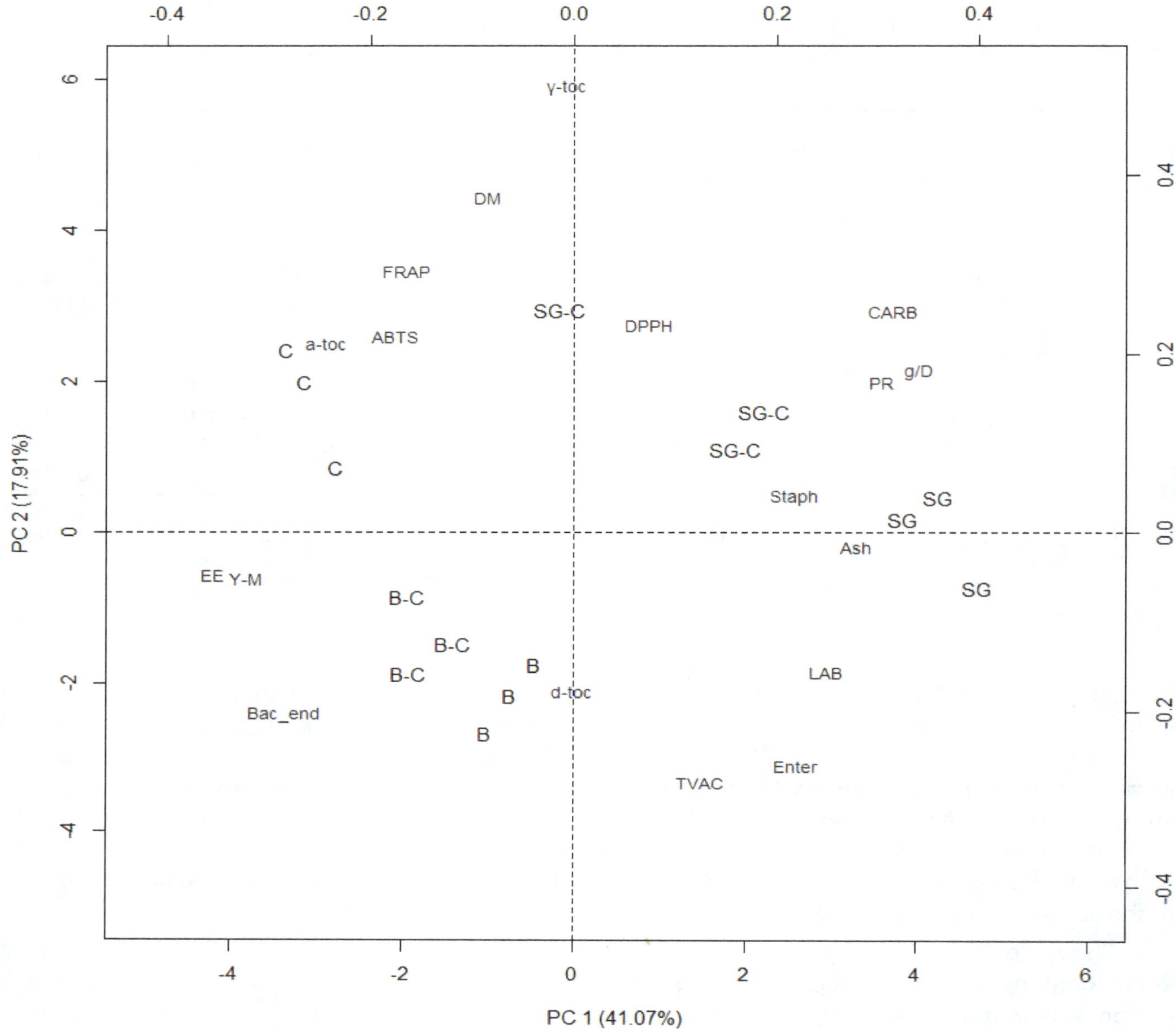

Figure 2. Biplot of the principal component (PC) analysis of mealworm larvae fed different diets (SG: brewery spent grains; B: bread; C: cookies; SG-C: 50% spent grains, 50% cookies; B-C: 50% bread, 50% cookies).

Positives PC1 eigenvectors collocate crude proteins on the right side of the biplot, near to SG and SG-C samples, as well as the growth performance. This relation highlights that diets rich in proteins (SG and SG-C) could promote growth and increase the protein content of the larvae. On the other hand, negative PC1 eigenvectors collocated ether extract content on the left side of the biplot, in contraposition with the protein content and the growth performances. The presence of the cookies in B-C and SG-C diets induced a shift on the left side of the plotted samples and then evidence a worsening of the growth performances and a higher deposition of fats.

Almost all the parameters that showed a statistical significance were plotted in the upper right or the lower left squares (both eigenvectors, positives or negatives) except for dry matter and tocopherol γ- and α-isoforms, which were plotted in the upper left square highlighting that another undetermined effect plays a role. As PC2 differentiates the antioxidant capacity and the tocopherol γ- and α-isoforms in the upper part of the plot and the microbial loads in the lower part, it could possible to hypothesize that a higher content of antioxidants played a role against microbial loads.

The observations reported above suggest that PC1 and PC2 were both a part of the diet effect related to the chemical composition of the substrates.

A residual percentage of variance (13.01%) was expressed also by PC3; this principal component was mostly related to (as absolute values) the antioxidant capacity and tocopherols δ- and α-isoforms.

Table 5. Eigenvalues and eigenvectors of the first three principal components (PC) of principal components analysis performed on proximate composition, microbial loads, growth performance, antioxidant capacity, and antioxidant compounds of mealworm larvae fed experimental diets.

Item	PC		
	PC 1	**PC 2**	**PC 3**
Eigenvalues	7.392	3.223	2.341
Eigenvectors			
Dry matter	−0.085	0.377	−0.269
Ether extract	−0.356	−0.049	−0.052
Crude protein	0.304	0.170	−0.133
Ash	0.279	−0.016	−0.204
Carbohydrates	0.314	0.250	−0.020
Weight gain (g/D)	0.340	0.182	−0.093
Total viable aerobic counts	0.127	−0.279	0.198
Enterobacteriaceae	0.220	−0.260	0.117
Staphylococci	0.219	0.042	0.156
Yeast and molds	−0.321	−0.053	0.117
Lactic acid bacteria	0.250	−0.154	0.221
Bacterial endospores	−0.284	−0.205	−0.041
ABTS	−0.176	0.220	0.451
DPPH	0.075	0.233	0.389
FRAP	−0.165	0.294	0.312
δ-tocopherol	−0.002	−0.180	−0.389
γ-tocopherol	−0.010	0.499	−0.064
α-tocopherol	−0.244	0.212	−0.329
Cumulative %	41.07	58.97	71.98

4. Conclusions

In conclusion, FFPs could be employed as feed in *T. molitor* rearing with effects on the chemical composition of the larvae and growth performance. Results point out a high plasticity of mealworm larvae and the potential to tailor the final outcomes. Single FFPs or mixes of multiple FFPs could be employed in relation to the requests and benefit–cost criteria determined by the producers. In general, it seems that feeds with high protein contents determine a better performance in rearing time, at the expense of the lipid content of the larvae.

Author Contributions: Conceptualization, S.M. (Simone Mancini) and F.F.; Data curation, S.M. (Simone Mancini); Formal analysis, S.M. (Simone Mancini), F.F., S.M. (Simona Mattioli) and S.N.; Investigation, S.M. (Simone Mancini), F.F., S.M. (Simona Mattioli) and S.N.; Methodology, S.M. (Simone Mancini), F.F. and S.M. (Simona Mattioli); Supervision, G.P.; Writing–original draft, S.M. (Simone Mancini), F.F. and S.M. (Simona Mattioli); Writing–review & editing, S.M. (Simone Mancini), F.F., B.T., S.M. (Simona Mattioli), A.D.B., T.T. and G.P.

References

1. Van Huis, A.; Van Itterbeeck, J.; Klunder, H.; Mertens, E.; Halloran, A.; Muir, G.; Vantomme, P. *Edible Insects: Future Prospects for Food and Feed Security*; Food and Agriculture Organization of the United Nations (FAO): Rome, Italy, 2013; Volume 171, ISBN 978-92-5-107595-1.

2. Anankware, J.P.; Fening, K.; Obeng-Ofori, D. Insects as food and feed: A review. *Int. J. Agric. Res. Rev.* **2015**, *3*, 143–151.
3. Verbeke, W. Profiling consumers who are ready to adopt insects as a meat substitute in a Western society. *Food Qual. Prefer.* **2015**, *39*, 147–155. [CrossRef]
4. Tan, H.S.G.; van Berg, E.V.; Stieger, M. The influence of product preparation, familiarity and individual traits on the consumer acceptance of insects as food. *Food Qual. Prefer.* **2016**, *52*, 222–231. [CrossRef]
5. Hartmann, C.; Ruby, M.B.; Schmidt, P.; Siegrist, M. Brave, health-conscious, and environmentally friendly: Positive impressions of insect food product consumers. *Food Qual. Prefer.* **2018**, *68*, 64–71. [CrossRef]
6. Mancini, S.; Moruzzo, R.; Riccioli, F.; Paci, G. European consumers' readiness to adopt insects as food. A review. *Food Res. Int.* **2019**, *122*, 661–678. [CrossRef] [PubMed]
7. Oonincx, D.G.A.B.; van Broekhoven, S.; van Huis, A.; van Loon, J.J.A. Feed Conversion, Survival and Development, and Composition of Four Insect Species on Diets Composed of Food By-Products. *PLoS One* **2015**, *10*, e0144601. [CrossRef]
8. Pinotti, L.; Giromini, C.; Ottoboni, M.; Tretola, M.; Marchis, D. Review: Insects and former foodstuffs for upgrading food waste biomasses/streams to feed ingredients for farm animals. *Animal* **2019**, *13*, 1365–1375. [CrossRef] [PubMed]
9. Wilkinson, J.M. Re-defining efficiency of feed use by livestock. *Animal* **2011**, *5*, 1014–1022. [CrossRef]
10. Van Broekhoven, S.; Oonincx, D.G.A.B.; van Huis, A.; van Loon, J.J.A. Growth performance and feed conversion efficiency of three edible mealworm species (Coleoptera: Tenebrionidae) on diets composed of organic by-products. *J. Insect Physiol.* **2015**, *73*, 1–10. [CrossRef]
11. Azagoh, C.; Ducept, F.; Garcia, R.; Rakotozafy, L.; Cuvelier, M.-E.; Keller, S.; Lewandowski, R.; Mezdour, S. Extraction and physicochemical characterization of Tenebrio molitor proteins. *Food Res. Int.* **2016**, *88*, 24–31. [CrossRef]
12. Osimani, A.; Garofalo, C.; Milanović, V.; Taccari, M.; Cardinali, F.; Aquilanti, L.; Pasquini, M.; Mozzon, M.; Raffaelli, N.; Ruschioni, S.; et al. Insight into the proximate composition and microbial diversity of edible insects marketed in the European Union. *Eur. Food Res. Technol.* **2017**, *243*, 1157–1171. [CrossRef]
13. Dreassi, E.; Cito, A.; Zanfini, A.; Materozzi, L.; Botta, M.; Francardi, V. Dietary fatty acids influence the growth and fatty acid composition of the yellow mealworm Tenebrio molitor (Coleoptera: Tenebrionidae). *Lipids* **2017**, *52*, 285–294. [CrossRef] [PubMed]
14. Fasel, N.J.; Mene-Saffrane, L.; Ruczynski, I.; Komar, E.; Christe, P. Diet induced modifications of fatty-acid composition in mealworm larvae (Tenebrio molitor). *J. Food Res.* **2017**, *6*, 22. [CrossRef]
15. Finke, M.D. Complete nutrient content of four species of commercially available feeder insects fed enhanced diets during growth. *Zoo Biol.* **2015**, *34*, 554–564. [CrossRef] [PubMed]
16. Ramos-Elorduy, J.; González, E.A.; Hernández, A.R.; Pino, J.M. Use of Tenebrio molito (Coleoptera: Tenebrionidae) to Recycle Organic Wastes and as Feed for Broiler Chickens. *J. Econ. Entomol.* **2002**, *95*, 214–220. [CrossRef] [PubMed]
17. Oonincx, D.G.A.B.; de Boer, I.J.M. Environmental impact of the production of mealworms as a protein source for humans A Life Cycle Assessment. *PLoS One* **2012**, *7*, e51145. [CrossRef] [PubMed]
18. Merrill, A.L.; Watt, B.K. Energy value of food: Basis and derivation. *USDA. United States Dep. Agric.* **1973**.
19. Association of Official Analytical Chemists (AOAC). *Official Methods of Analysis*, 19th ed.; Association of Official Analytical Chemists: Washington, DC, USA, 2012; ISBN 9783642312410.
20. Janssen, R.H.; Vincken, J.-P.; van den Broek, L.A.M.; Fogliano, V.; Lakemond, C.M.M. Nitrogen-to-protein conversion factors for three edible insects: Tenebrio molitor, Alphitobius diaperinus, and Hermetia illucens. *J. Agric. Food Chem.* **2017**, *65*, 2275–2278. [CrossRef]
21. Stoops, J.; Vandeweyer, D.; Crauwels, S.; Verreth, C.; Boeckx, H.; Van Der Borght, M.; Claes, J.; Lievens, B.; Van Campenhout, L. Minced meat-like products from mealworm larvae (Tenebrio molitor and Alphitobius diaperinus): microbial dynamics during production and storage. *Innov. Food Sci. Emerg. Technol.* **2017**, *41*, 1–9. [CrossRef]
22. Mancini, S.; Preziuso, G.; Dal Bosco, A.; Roscini, V.; Szendro, Z.; Fratini, F.; Paci, G.; Szendrő, Z.; Fratini, F.; Paci, G. Effect of turmeric powder (Curcuma longa L.) and ascorbic acid on physical characteristics and oxidative status of fresh and stored rabbit burgers. *Meat Sci.* **2015**, *110*, 93–100. [CrossRef]
23. Zaspel, B.J.; Csallany, A.S. Determination of alpha-tocopherol in tissues and plasma by high-performance liquid chromatography. *Anal. Biochem.* **1983**, *130*, 146–150. [CrossRef]

24. R Core Team. *R: A Language and Environment for Statistical Computing*; R Foundation for Statistical Computing: Vienna, Austria. Available online: https://www.r-project.org/ (accessed on 24 July 2019).
25. Herrero, M.; Thornton, P.K. Livestock and global change: Emerging issues for sustainable food systems. *Proc. Natl. Acad. Sci.* **2013**, *110*, 20878–20881. [CrossRef] [PubMed]
26. Grabowski, N.T.; Klein, G. Microbiology of processed edible insect products – Results of a preliminary survey. *Int. J. Food Microbiol.* **2017**, *243*, 103–107. [CrossRef] [PubMed]
27. Grabowski, N.T.; Klein, G. Microbiological analysis of raw edible insects. *J. Insects as Food Feed* **2017**, *3*, 7–14. [CrossRef]
28. Wynants, E.; Crauwels, S.; Verreth, C.; Gianotten, N.; Lievens, B.; Claes, J.; Van Campenhout, L. Microbial dynamics during production of lesser mealworms (Alphitobius diaperinus) for human consumption at industrial scale. *Food Microbiol.* **2018**, *70*, 181–191. [CrossRef] [PubMed]
29. Vandeweyer, D.; Crauwels, S.; Lievens, B.; Van Campenhout, L. Microbial counts of mealworm larvae (Tenebrio molitor) and crickets (Acheta domesticus and Gryllodes sigillatus) from different rearing companies and different production batches. *Int. J. Food Microbiol.* **2017**, *242*, 13–18. [CrossRef] [PubMed]
30. Wynants, E.; Crauwels, S.; Lievens, B.; Luca, S.; Claes, J.; Borremans, A.; Bruyninckx, L.; Van Campenhout, L. Effect of post-harvest starvation and rinsing on the microbial numbers and the bacterial community composition of mealworm larvae (Tenebrio molitor). *Innov. Food Sci. Emerg. Technol.* **2017**, *42*, 8–15. [CrossRef]
31. Mancini, S.; Fratini, F.; Tuccinardi, T.; Turchi, B.; Nuvoloni, R.; Paci, G. Effects of different blanching treatments on microbiological profile and quality of the mealworm (Tenebrio molitor). *J. Insects as Food Feed* **2019**, *5*, 225–234. [CrossRef]
32. Wynants, E.; Frooninckx, L.; Van Miert, S.; Geeraerd, A.; Claes, J.; Van Campenhout, L. Risks related to the presence of Salmonella sp. during rearing of mealworms (Tenebrio molitor) for food or feed: Survival in the substrate and transmission to the larvae. *Food Control.* **2019**, *100*, 227–234. [CrossRef]
33. Mancini, S.; Paci, G.; Ciardelli, V.; Turchi, B.; Pedonese, F.; Fratini, F. Listeria monocytogenes contamination of Tenebrio molitor larvae rearing substrate: Preliminary evaluations. *Food Microbiol.* **2019**, *83*, 104–108. [CrossRef]
34. Arnold, K.E.; Ramsay, S.L.; Henderson, L.; Larcombe, S.D. Seasonal variation in diet quality: antioxidants, invertebrates and blue tits Cyanistes caeruleus. *Biol. J. Linn. Soc.* **2010**, *99*, 708–717. [CrossRef]

9

Quality of Eggs and Albumen Technological Properties as Affected by *Hermetia Illucens* Larvae Meal in Hens' Diet and Hen Age

Giulia Secci [1], Fulvia Bovera [2], Giuliana Parisi [1,*] and Giuseppe Moniello [3]

[1] Department of Agriculture, Food, Environment and Forestry, University of Firenze, via delle Cascine 5, 50144 Firenze, Italy; giulia.secci@unifi.it
[2] Department of Veterinary Medicine and Animal Production, University of Napoli Federico II, via Federico Delpino 1, 80137 Napoli, Italy; fulvia.bovera@unina.it
[3] Department of Veterinary Medicine, University of Sassari, via Vienna 2, 07100 Sassari, Italy; moniello@uniss.it
* Correspondence: giuliana.parisi@unifi.it.

Simple Summary: Concurrently with the actual challenges in the poultry industry, we aimed to assess the changes induced by the inclusion of an alternative protein source (partially defatted *Hermetia illucens* larva meal, HI) at two different levels as well as hen age on the overall egg quality and deepened their effect of albumen technological properties. This study could provide useful information for the egg supply chain in order to optimize egg utilization, as a whole or as egg products, thus preventing food losses or undesirable wastes. Specifically, based on the obtained results, the eggs laid by hens fed the highest inclusion of HI would be directed towards egg product industry due to their reduced eggshell percentage and thickness which could increase their fragility. Contrariwise, due to the impaired albumen technological properties, as foaming, found in the egg laid by 27–35 wk-old hens, they could be preferentially sold as a whole.

Abstract: The impact on the egg quality and the albumen technological properties were here evaluated as affected by diet and hen age (A) of 162 Hy-line Brown hens. Three isoproteic and isoenergetic diets were formulated respecting the requirements for Hy-line hens: the control diet (C) based on conventional protein sources, and other two where vegetable proteins were substituted at 25% (HI25) and 50% (HI50) by partially defatted *Hermetia illucens* larva meal (HI). Ten eggs collected from each group at the hen ages of 20, 27, and 35 weeks were evaluated. The eggshell percentage and thickness were significantly reduced in the HI50 eggs (11.93% and 476 μm, respectively) compared to the C (12.34%, 542 μm) and HI25 (12.54%, 516 μm). The aging lowered ($p = 0.05$) the protein and increased ($p < 0.001$) water contents of the eggs. Although the foam capacity of the HI50 albumen was halved than the C group ($p < 0.05$), it was unaffected by the aging. Additionally, this did not impair the volume and the textural properties of a batter (angel cake) in which it was included. On the opposite, the textural characteristics of the cake made by the oldest hens (i.e., 35 wk-old) were compromised. In conclusion, the diet and hen age differently affected egg quality and its technological properties, which could be positive to obtain eggs to destine directly to the market or to the egg industry.

Keywords: black soldier fly; foam capacity; angel cake; texture; week of deposition

1. Introduction

Standing on the data proposed by the Food and Agriculture Organization in 2018 [1], eggs are the second-fastest growing industry in the world, with more than 50% growth forecast in the next 2 decades. Europe production in 2019 amounts to 7.5 million tonnes of eggs thanks to the 400 million laying hens kept throughout the European Union (27 Member States) [2]. Italy is the fourth producer with approximately 817,000 tons of eggs consumed as a whole or destined to the egg industry. Indeed, despite eggs play a very important role in human nutrition as a precious source of proteins, essential amino acids, lipid and several trace elements, the industry is continuously asking for egg products, such as liquid, powdered, concentrated whole egg or its separated main components, yolk and albumen. For instance, the egg allowance directed toward processing chains has been estimated at 32% of the whole Italian production [3]. This interest is linked to the unique functional properties of eggs, such as foaming, gelling and emulsification. The foaming capacity belongs to the egg white protein and it is defined as "the ability to rapidly adsorb on the air-liquid interface during whipping or bubbling and by its ability to form a cohesive viscoelastic film by way of intermolecular interactions" [4]. Thanks to these characteristics, albumen is widely utilized as an ingredient in bakery products, like bread, cakes and meringues, ice creams and several other processed foods.

The increase in egg and egg-derived product demand is however dependent on the laying hen farming, a sector that is affected by some critical aspects, such as feed supplying. As one of the main concerns of the last decade, the unsustainability of feed, especially soybean meal, is quite debated. Indeed, despite being an important source of essential amino acids, soybean production for feed directly competes with human nutrition and is largely associated with water pollution and exploitation of the lands [5].

A big opportunity for companies looking at more sustainable protein sources has been recently identifying in insect meals. Although processed animal proteins (PAPs) are still banned as livestock feeds in Europe, the European Commission voted in 2017 to introduce seven insect species, processed as meal or oil, into fish feeds and pet food [6]. That is why the stakeholders are confident that these protein sources would be allowed in other feeds in the coming years. As the interest in this alternative ingredient has grown, researchers are checking the feasibility of partial or complete substitution of the conventional protein sources with insect meal in poultry feed. Among the other species, *Hermetia illucens* meal (HI) has deserved major attention for laying hen diets. Indeed, in addition to the other advantages related to the rearing process of this species of insect, HI is a valuable protein source (40–44 g/100 g of dry matter, DM) and it provides calcium and phosphorous (50 to 80 g/kg dry matter, 6 to 15 g/kg dry matter, respectively), which are fundamental for laying nutrition [7]. HI has been proved to be a feasible substitute for vegetable protein sources in the hen diet when considering animal growth and performance [8–10] and egg quality [11–13]. However, the authors focused on the egg deposition, quality characteristics as the albumen height and the yolk color and its chemical composition, more than the albumen technological properties, which could be of interest to the egg industry. Thus, this study aimed to test the effect of partial substitution of the conventional protein sources with a partially defatted *Hermetia illucens* larva meal on the overall egg quality and on the technological properties of the albumen from the eggs produced by Hy-line Brown laying hens and collected at three different hen ages.

2. Materials and Methods

2.1. *In Vivo Trial*

All the hens were treated in accordance with the Directive of the European Parliament of the Council on the Protection of Animals Used for Scientific Purpose and in agreement with the Institutional Animal Care and Use Committee of the University of Naples Federico II, D.lgs n. 26 04/03/2014. All the experiments involving hens were approved by the Bioethical Committee of the University of Naples Federico II, under the number of protocol 2017/0017676.

The HI tested in this study was purchased from Hermetia Deutschland GmbH & Co KG (Baruth/Mark, Germany), chemically characterized as reported in Bovera et al. [14] and utilized to formulate the experimental diets, presented as follows. A control diet was formulated based on the conventional protein sources (corn-soybean based diet, C) while the other two were based on two different substitution levels of the conventional protein source with the partially defatted *Hermetia illucens* larva meal (HI, ether extract equal to 8.34% as fed). Specifically, the 25 and 50% of the dietary proteins were substituted by HI in the H25 and HI50 groups, respectively. The diets, whose ingredients and chemical composition are shown in Table 1, were formulated in order to result isoproteic and isoenergetic and to satisfy the Hy-line Brown requirements, as specified in Bovera et al. [14]. A total of 162 sixteen weeks-old Hy-line Brown hens (average live weight 1.41 kg ± 0.13) were housed in a commercial private laying hens farm (total flocks housed: 40,000 hens), located in Sardinia (Italy), for 20 weeks. The hens were equally divided into three feeding groups. Each group was housed in modified cages (18 hens/cage, 800 cm^2/hen, nine replicates of six hens each) adjacent to a nest box for egg deposition. The trial started after four weeks of adaptation, when hens were 20 weeks old. The mashed feeds were administrated *ad libitum* to the hens through a feeding belt, manually loaded. A dark: light cycle of 9:15 h was set. More details concerning the farming conditions as well as the data on feed intake, feed conversion ratio, weight gains and laying percentage can be retrieved in Bovera et al. [14]. The collection of the eggs for the purposes of the present trial occurred at three different hen ages (A), i.e., namely 20-wk-old (A20), 27-wk-old (A27), and 35-wk-old (A35). Specifically, the eggs were collected every day, then the overall weekly production per each dietary treatment was pooled. Ten undamaged eggs were selected from each pool and destined to the present trial, hence stored at −80 °C and moved to the Department of Agriculture, Food, Environment and Forestry (University of Firenze, Italy) where they were subjected to the scheduled analyses.

2.2. Egg Quality and Albumen Foaming Properties

After an over-night thawing in a refrigerated room (4 °C), the eggs were weighed prior to being broken and to separate each component (eggshell, albumen, and yolk) one from each other. The weights of the albumen and yolk were recorded and utilized for the calculation of their relative percentage. Concerning the eggshells, they were weighed after the removal of the inner membrane and the drying of the internal surface with common kitchen paper. After this, the thickness of the eggshell was measured by a manual caliper (Salmoiraghi, Milan, Italy) on three points (equator, round and apex). After this, the whole eggshell was grounded prior to ash quantification [15]. The color of the yolk was measured on three points by a Konica Minolta colorimeter CR-200 (Chiyoda, Japan) and the data were expressed as lightness (L*), redness (a*) and yellowness (b*) indexes [16].

The following analyses were conducted one-by-one on each albumen, without pooling the ones belonging to the same treatments. The chalazae were removed from each albumen prior to gently homogenate it by stirring for 15 min (speed set n. 4 of the magnetic stirrer, Type Age, Velp Scientific, Usmate Velate, Italy) in order to avoid foam formation and water losses, then the pH was determined using a SevenGo pH meter (Mettler-Toledo, Columbus, OH, USA) at room temperature. Ten mL of each albumen were allotted to the evaluation of egg white technological properties, as foam ability and foam stability, following the methodology recently proposed [17]. Briefly, the albumen was whipped for 1 min at 20 °C in a 50 mL plastic cylinder by a high-speed mixer (IKA Werke T25, Staufen, Germany) operating at 9500 rpm and the volume of the foam was recorded 30 s after whipping (V0) and holding 30 min (V30). The data were inserted into the following formula for the determination of the foam capacity (FC, %) and the foam stability (FS, %) [18]:

$$FC = (V0/Valb) \times 100 \tag{1}$$

$$FS = (V30/V0) \times 100 \tag{2}$$

where Valb is the initial volume of the albumen.

The other 20 g of the egg white from each sample were destined to the angel cake preparation, as detailed in the next section. The remaining albumen (around 8 g) was freeze-dried prior to being subject to water and crude protein content quantification [15].

Table 1. Ingredients (g/kg), chemical composition (g/100 g as fed), and metabolizable energy (MJ/kg) of the experimental diets C (control diet based on maize grain and soybean meal), HI25 (*Hermetia illucens* larva meal substituted the 25% of dietary vegetable protein) and HI50 (*Hermetia illucens* larva meal substituted the 50% of dietary vegetable protein).

	Diets		
	C	**HI25**	**HI50**
Ingredients			
Maize grain	605.5	597.5	630.5
Soybean meal	265	200	95
Insect meal	-	73	145
$CaCO_3$ grains	80	80	80
Vegetable oil	10	10	10
MinVit *	10	10	10
Methionine	2.5	2.5	2.5
Monocalcium phosphate	5	5	5
Celite	20	20	20
Salt	2	2	2
Chemical Composition			
Dry matter [1]	91.53	91.39	91.62
Crude protein [1]	16.45	16.32	17.03
Ether extract [1]	3.17	3.61	4.06
Metabolizable Energy [2]	11.85	11.90	11.89
Ash	14.00	14.21	14.13
Ca	4.63	4.75	4.51
Total P	0.68	0.69	0.65
Lysine [2]	0.86	0.97	1.00
Methionine [2]	0.63	0.58	0.61
Methionine + Cysteine	0.84	0.80	0.82
Isoleucine	0.72	0.73	0.75
Threonine	0.63	0.64	0.65
Tryptophan	0.18	0.15	0.16
Valine	0.93	1.07	1.14

[1] Determined according to AOAC (2004). [2] Calculated according to NRC (1994).* Provided per kilogram: vitamin A (retinyl acetate), 20,000 IU; vitamin D3 (cholecalciferol), 6000 IU; vitamin E (dl-α-tocopheryl acetate), 80 IU; vitamin B1(thiamine monophosphate), 3 mg; vitamin B2 (riboflavin), 12 mg; vitamin B6 (pyridoxine hydrochloride), 8 mg; vitamin B12 (cyanocobalamin), 0.04 mg;, vitamin K3 (menadione), 4.8 mg; vitamin H (d biotin), 0.2 mg; vitamin PP (nicotinic acid), 48 mg; folic acid, 2 mg; calcium pantothenate, 20 mg; manganous oxide, 200 mg; ferrous carbonate, 80 mg; cupric sulphate pentahydrate, 20 mg; zinc oxide, 120 mg; basic carbonate monohydrate, 0.4 mg; anhydrous calcium iodate, 2 mg; sodium selenite, 0.4 mg; choline chloride, 800 mg; 4–6-phytase, 1800 FYT; D.L. methionine, 2600 mg; canthaxanthin, 8 mg.

2.3. *Angel Cake Preparation and Characterisation*

Twenty g of egg white, 14 g of white sugar (Italia Zuccheri, Minerbio, Italy), 8 g of wheat flour "00" (Il Molino Chiavazza, Casalgrasso, Italy), and 2 g of rice starch (Pedon S.p.A., Molvena, Italy) were utilized for the production of the angel cake (one for each sample), which is a common batter utilized to test the foaming capacity of the protein ingredients. Albumen and sugar were whipped together for 60 s with a hand mixer equipped with a wire whisk attachment (Kenwood HDP408WH Triblade Mixer, Kenwood, Woking, UK), set at its maximum speed level. All the other dry ingredients were manually incorporated into the foam by beating on high until stiff peaks form (approximately 40 s). An aliquot corresponding to 31.711 ± 1.072 g of batter was spooned into one cavity of an ungreased silicone pan (mold size: 23.5 × 21.2 × 2.5 cm, total 6 cavities, cavity size: 8.0 × 5.5 × 2.5 cm, Shen zhen shi jun yang ke ji you xian gong si, Guangdong, China). The batters were baked for 15 min in a pre-heated static oven (180 °C, Mod. S370EB, Smeg, Guastalla, Italy), then the angel cakes were left to cool at room temperature for 1 h. Some physical parameters such as the weight and the maximum height of each cake were recorded, then baking loss (BL, %) was calculated as proposed [19]:

$$BL = (WCk\text{-}Wb) \times 100/WCb \quad (3)$$

where WCk was the weight of the angel cake, Wb was the weight of the batter and WCb was the water content of the batter obtained following the AOAC method [15]. Both angel cake dough and the baked cakes were analyzed for water and protein contents [15].

The color and texture profile analyses were assessed. Specifically, the cakes were cut vertically in the middle. Once opened, the color values were determined with a Konica Minolta colorimeter (Chiyoda, Japan) [16] on two points of the two halves. Finally, 4 × 4 cm pieces were cut from the two halves of each cake, the crust was removed and then both pieces were analyzed, using a Zwick Roell® texturometer (model KAF-TC 0901279, Zwick GmbH & Co., Ulm, Germany) equipped with a 50 N load cell. The texture profiles analysis was determined through a double compression cycle (crosshead speed: 1 mm s^{-1}) until the 60% deformation of the initial height, as previously done [19]. The hardness (N), cohesiveness (the area of work during the second compression divided by the area of work during the first compression), springiness (the distance of the detected height during the second compression divided by the original compression distance) and chewiness (Hardness x Cohesiveness x Springiness) were calculated from the force–time diagram (Test-Xpert2 software version 3.0, Zwick GmbH & Co., Ulm, Germany) [20].

2.4. Statistical Analysis

The data collected were processed by a 2-way ANOVA using the PROC GLM of SAS/STAT Software, Version 9 [21] by using Diet (D), Hen Age (A) and their interaction (D × A) as fixed factors. The comparison among the means was achieved by conducting Tukey's test [21] and the significance was set at $p < 0.05$. The means are accompanied by the Root Mean Square Error (RMSE) for characterizing the variability of the groups. Each egg within the feeding group and hen age was considered as a biological replicate.

3. Results

The egg, eggshell and yolk weights significantly increased due to the inclusion of HI in the feed for laying hens, irrespective of the insect inclusion level utilized (Table 2). Despite this, while looking at the percentage of the main components of the egg, only the eggshell resulted significantly diminished in the HI50 eggs compared to the C and HI25 groups. Accordingly, the thickness of the eggshell was reduced while increasing HI in the feed.

Table 2. Egg weight (g), egg components weight (g) and percentage (% on egg weight) and eggshell thickness (μm) as affected by the diet (C, HI25, HI50) and hen ages (20, 27, and 35 weeks).

Item	Diet, D			Hen Age, A			D	A	D × A	RMSE
	C	HI25	HI50	A20	A27	A35				
Egg Weight, g	57.56 ^{b}	61.10 ^{a}	60.82 ^{a}	60.79	59.36	59.43	*	ns	ns	4.413
Eggshell, g	7.10 ^{b}	7.65 ^{a}	7.24 ^{a,b}	7.26	7.60	7.14	*	ns	ns	0.643
Yolk, g	12.93 ^{b}	14.30 ^{a}	14.34 ^{a}	13.92	13.46	14.19	***	ns	***	1.298
Albumen, g	37.53	39.15	38.47	39.62	37.43	38.10	ns	ns	ns	4.091
Eggshell, %	12.34 ^{a}	12.54 ^{a}	11.93 ^{b}	11.96 ^{Y}	12.83 ^{X}	12.02 ^{Y}	*	***	ns	0.741
Yolk, %	22.49	23.44	23.59	22.92	22.71	23.89	ns	ns	ns	1.896
Albumen, %	65.18	64.02	63.21	65.12	63.19	64.10	ns	ns	ns	4.016
Eggshell Thickness, μm	542 ^{a}	516 ^{a,b}	476 ^{b}	499	540	494	*	ns	ns	72.249

RMSE: Root Means Square Error. a, b are significant different means among diets. X, Y are significant different means among hen ages. ns: not significant ($p > 0.05$), * $p < 0.05$, *** $p < 0.001$.

Furthermore, the eggshell percentage varied ($p < 0.05$) along with the hen age, showing the highest value at A27. These modifications in the eggshell characteristics cannot be attributed to the different ash content since it resulted unaffected by both D and A and equal to 76.4 ± 2.5 g/100 g of eggshell. Since L*, a*, and b* values resulted significantly affected by the interaction D × A at $p < 0.026$, 0.0001, and 0.049, respectively, we summarised the means obtained for this interaction in Table 3. No difference in the lightness of yolks was found as affected by the diet in the A20 group (Table 3). Conversely, the H50 eggs laid at A27 and A35 showed L* value significantly different from the C ones. The major differences were in the a* value, found significantly higher in the C yolks than in the H25 and H50 ones at A20 and A27. A greater decrease of the a* value was recorded in the C group increasing the time of laid, whereas the H50 raised its maximum a* value in the yolk from the eggs collected at A35. The b* index appeared scarcely affected by the diet, indeed the H50 yolks were significantly less yellow than the other two groups only in the A20 group. Furthermore, the same group significantly increased the b* index from the A27.

Table 3. Color parameter (lightness, L*, redness, a*, and yellowness, b*, indexes) values of the yolk from hens fed three different diets (C, HI25, HI50) and collected at three different hen ages (20, 27, and 35 weeks).

Colour Parameter	Diet	Hen Age (Weeks)		
		A20	A27	A35
L*	C	65.804 ^{ab}	64.686 ^{b,Y}	67.865 ^{a,XY}
	H25	66.665	68.553 ^{X}	68.671 ^{X}
	H50	68.858 ^{a}	68.723 ^{a,X}	65.537 ^{b,Y}
a*	C	6.514 ^{ab,X}	8.338 ^{a,X}	2.874 ^{b,X}
	H25	1.353 ^{ab,Y}	2.395 ^{a,Y}	−1.747 ^{b,Y}
	H50	−0.437 ^{b,Y}	−2.422 ^{b,Z}	5.268 ^{a,X}
b*	C	31.907 ^{X}	32.428	32.558
	H25	33.463 ^{X}	34.551	31.020
	H50	20.458 ^{b,Y}	28.391 ^{a}	33.300 ^{a}

^{a,b}: different superscript letters are significantly different means within the row. ^{X,Y,Z}: different superscript letters are significantly different means within the column.

Table 4 shows the results of the technological properties and chemical composition analyses of the albumen. The FC was decreased by the presence of HI in the diet of laying hens, whereas the FS ad the pH value, as well as the water and protein contents were unaffected. A deep effect of the A can be observed for pH and chemical components of the albumen. For instance, a significant decrease in the pH value ($p < 0.01$) was found after the A27. The albumen of the eggs collected at the A35 showed a significantly ($p = 0.05$) lower protein content than those collected at the A20, while their water content increased with the aging.

Table 4. Technological properties (foam capacity and foam volume stability as %) and chemical composition (g/100 g of albumen) of the albumen from the eggs laid by hens fed three different diets (C, HI25, HI50) and collected at three different hen ages (20, 27, and 35 weeks).

Item	Diet, D			Hen Age, A			D	A	D × A	RMSE
	C	HI25	HI50	A20	A27	A35				
pH	8.12	8.08	8.05	8.19 ^{X}	8.04 ^{Y}	8.03 ^{Y}	Ns	**	ns	0.143
Foam Capacity	43.33 ^{a}	29.44 ^{b}	20.74 ^{b}	34.82 ^{X}	39.44 ^{X}	19.26 ^{Y}	***	***	*	14.036
Foam Stability	59.20	52.55	63.05	52.58	57.34	64.89	Ns	ns	ns	19.331
Water Content	87.30	87.47	87.11	87.39 ^{XY}	85.44 ^{Y}	89.05 ^{X}	Ns	***	ns	3.003
Crude Protein Content	11.20	11.03	11.39	11.11 ^{X}	13.06 ^{X}	9.45 ^{Y}	Ns	***	ns	3.003

RMSE: Root Means Square Error. ^{a,b} are significant different means among diets. ^{X,Y} are significant different means among hen ages. ns: not significant ($p > 0.05$). * $p < 0.05$, ** $p < 0.01$, *** $p < 0.001$.

Data related to the significant interaction between D and A emerged for the FC are reported in Figure 1. Although the H50 albumens had a halved ($p < 0.05$) FC than the C ones, they kept this property constant while increasing the hen age, while both the C and H25 groups reduced their FC between the A20 and the A35.

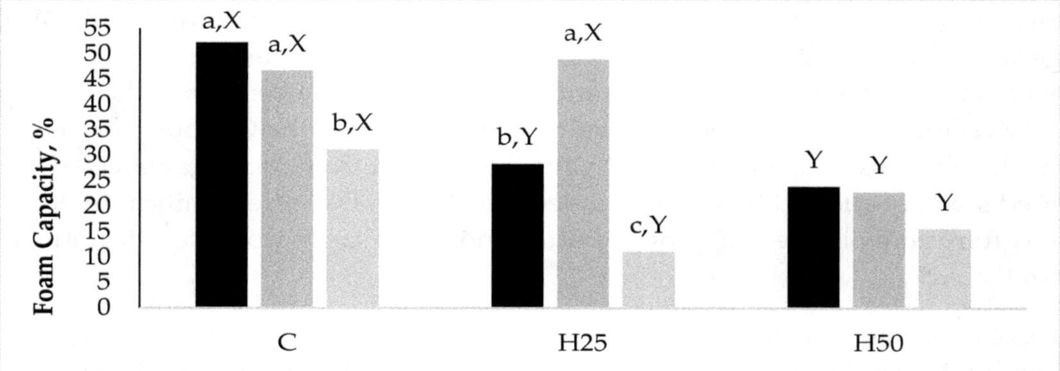

Figure 1. Effect of the interaction Diet x Hen Age on the foam capacity (%) of albumen from eggs laid by hens fed three experimental diets (C, H25, and H50) at three different hen ages (A20, black bars; A27, dark grey bars; A35, grey bars). a, b, c indicate significantly different means among the hen ages, within the diets. X, Y indicate significantly different means among the diets, within the hen ages.

The FC of the egg white was even determined by the analyses of physical parameters conducted on the angel cakes (Table 5). The textural properties of the angel cake were not affected by the dietary inclusion of *Hermetia illucens*, while A20 induced a significant reduction of the chewiness.

Table 5. Physical and chemical characteristics of the angel cakes obtained by the albumen from the eggs laid by hens fed three different diets (C, HI25, HI50) and collected at three different hen ages (20, 27, and 35 weeks).

Item	Diet, D			Hen Age, A			D	A	D × A	RMSE
	C	HI25	HI50	A20	A27	A35				
Height, mm	20.25 ^{a}	15.95 ^{b}	18.65 ^{a}	17.58	19.31	17.97	***	ns	*	2.39
Baking Loss, %	43.28 ^{a}	44.67 ^{a}	35.19 ^{b}	41.28 ^{X}	45.23 ^{X}	36.63 ^{Y}	***	***	*	7.871
Colour Parameters										
L*	87.956	87.653	88.360	88.257	87.690	88.023	ns	ns	ns	1.509
a*	−2.705	−2.742	−2.644	−2.669	−2.650	−2.772	ns	ns	ns	0.630
b*	13.494 ^{a}	13.335 ^{a}	11.936 ^{b}	13.034	13.024	12.707	***	ns	***	0.847
Texture										
Hardness, N	6.39	5.52	5.95	5.90	6.85	5.13	ns	ns	ns	3.088
Cohesiveness	0.71	0.72	0.70	0.71	0.70	0.72	ns	ns	ns	0.027
Springiness	8.92	9.22	8.67	9.94 ^{X}	7.89 ^{Y}	8.98 ^{X}	ns	***	ns	1.565
Chewiness	36.63	34.23	34.29	40.64 ^{X}	36.42 ^{X}	28.01 ^{Y}	ns	***	ns	11.121

RMSE: Root Means Square Error. a, b are significant different means among diets. X, Y are significant different means among hen ages. ns: not significant ($p > 0.05$). * $p < 0.05$, *** $p < 0.001$.

A significant interaction (D × A) emerged for height, b* value and baking loss of the samples (Table 6). The height of the angel cakes produced from the C egg whites was unaffected by the A, whereas the angel cakes obtained by the H25 eggs collected during the A35 significantly lost their height compared to those produced from the H25 eggs of the A20. While looking at the difference among the dietary treatments, the angel cake from the C eggs was higher than the H50 only considering the A20. In addition, the H25 egg whites collected at the A27 and A35 significantly decreased the height of the angel cake compared both to the C and H50 groups. Concerning the baking loss, the age did not affect the water retention ability of the angel cake made with the C albumen, while the highest inclusion of HI in the diet increased the ability of the batter to retain water, especially at the A35. The b* value of the angel cake obtained with the C albumens was unaffected by the age, while the yellowness was significantly higher using the albumen from the H25 eggs collected at the A35 than that collected at the A20. Conversely, the H50 angel cake was slight but significantly discoloured when prepared with the egg whites from the A27 and A35. Regarding the differences among the diets, the angel cakes produced with the H25 albumens collected at the A20 had a lower ($p < 0.001$) b* value than those of the C and H50 groups. In contrast, the H50 cakes prepared with the eggs collected at the A27 and A35 showed a lower ($p < 0.001$) b* index than the C and HI25.

Table 6. Results of the interaction between the treatments (Diet × Hen Age) for angel cake characteristics.

Item	Diet	Hen Age (Weeks)		
		A20	A27	A35
Height, mm	C	19.45 ^{X}	21.41 ^{X}	19.88 ^{X}
	H25	17.11 ^{a,XY}	16.49 ^{ab,Y}	14.26 ^{b,Y}
	H50	16.17 ^{b,Y}	20.03 ^{a,X}	19.77 ^{a,X}
Baking Loss, %	C	44.97	42.26 ^{Y}	42.60 ^{X}
	H25	40.86 ^{b}	54.30 ^{a,X}	38.84 ^{b,X}
	H50	38.01 ^{a}	39.12 ^{a,Y}	28.44 ^{b,Y}
b* Value	C	13.59 ^{X}	13.78 ^{X}	13.113 ^{X}
	H25	12.584 ^{b,Y}	13.49 ^{ab,X}	13.932 ^{a,X}
	H50	12.929 ^{a,XY}	11.803 ^{b,Y}	11.077 ^{b,Y}

^{a,b} are significant different means within the row. ^{X,Y} are significant different means within the column.

Finally, Table 7 depicts that the batters prepared with the egg whites collected at the A27 had a slight but significantly lower water content than those from the egg whites collected at the other two considered ages.

Table 7. Chemical composition (g/100 g fresh sample) of the angel cake as raw and baked batter.

Item	Diet, D			Hen Age, A			D	A	D × A	RMSE
	C	HI25	HI50	A20	A27	A35				
Angel Cake Batter										
Water	40.75	40.68	40.52	40.79 ^{X}	39.74 ^{Y}	41.42 ^{X}	ns	***	ns	1.254
Crude protein	8.11	8.07	8.36	8.38	8.22	7.95	ns	ns	ns	0.744
Angel Cake Baked										
Water	28.61	29.13	27.72	29.68 ^{X}	26.08 ^{Y}	29.70 ^{X}	ns	***	ns	2.736
Crude protein	9.77	9.65	10.16	9.96	10.08	9.54	ns	ns	ns	0.963

RMSE: Root Means Square Error. ^{a,b} are significant different means among diets. ^{X,Y} are significant different means among hen ages. ns: not significant ($p > 0.05$), *** $p < 0.001$.

4. Discussion

The main factors affecting the egg weight are the dietary metabolized energy [22] and the size of the yolk that, in turn, is influenced by the body weight of the hens [23]. Since the administered diets in the present trial were isoenergetic, it seems clear to attribute the major changes in HI egg weights to the increase in the size of their yolks. Nevertheless, this modification cannot be attributable to the different body weight of the hens, resulted unaffected by the dietary intervention as well as the feeding intake and the feed conversion ratio [14]. Thus, the diet could have influenced the egg and yolk weights. Leeson and Summers [23] suggested that egg weight is very sensitive to methionine and total sulphur amino acid levels, which however were balanced in all the three administered diets.

Our hypothesis is that the significantly ($p = 0.032$) augmented length of the jejunum found in the HI25 and HI50 hens, being respectively 4.68 and 4.64 (as percentage of the live weight), compared to the C one (3.45% live weight), [14] could have promoted the absorption of the amino acids, hence resulting in a higher assumption of these nutrients. This possible explanation strongly needs the support of other comprehensive studies, since the effect of HI meal on hen egg weight is still scarcely investigated and conflicting in results [13,24]. The increased egg weight found with the HI meal inclusion in the hens' diet may require an increase in eggshell weight because of a rising in calcium deposition. Notwithstanding, authors underlined that egg-laying birds have a limited amount of calcium available to produce the shell, approximately 2.0–2.5 g Ca^{2+}, irrespective of the egg size weight [25]. Hence, the production of heavier eggs in the HI50 group might explain the significant reduction in shell percentage and even in its thickness. Since the eggshell weight and thickness are physical variables correlated with the egg strength, resistance to physical and pathogenic challenges from laying to the transportation and selling phase [26], our results should be considered when evaluating the suitability to include the HI meal in the laying hens' feed.

The egg internal quality, determined by several parameters (as albumen height and Haugh Unit), is even associated with the yolk quality, whose color is a fundamental characteristic [27]. The pigments contained in feeds mainly derive by plant ingredients, however, Secci et al. [13] underlined that even the HI larva meal may contain carotenoids and tocopherols. These pigments have a strong affinity for non-polar molecules, such as lipid, and they generally absorb wavelengths ranging from 400 to 550 nm, coloring as yellow, orange, or red. Thus, carotenoids can be easily accumulated in the egg yolk, which is rich in fat, producing the major modification of the yolk redness index, as we found.

Although the egg production and quality as affected by *H. illucens* in laying hen diets have been recently examined [11–13,24], the focus areas were the egg deposition and quality characteristics as the albumen height, the yolk color and its chemical composition. Nonetheless, the albumen has relevant importance for the egg industry mainly because of its unique functional properties, such as foaming that promote its extensive use as an ingredient in several processed food. In the present study, the diet did not affect the egg white pH and protein content, thus supporting previous findings [13]. The protein concentration as well as pH and cooking temperature, in addition to the physical-chemical properties of proteins, are well-established factors affecting the foaming ability [28]. Although in the present study no significant difference among the diets was found for pH and protein values, the foam ability of the HI25 and HI50 egg whites was decreased compared to that of the C ones. Since the temperature was controlled during the experiment, we hypothesize that the HI diets could modify the concentration, or the proportion of the single protein fractions contained in the egg white. As previously noted [4], the albumen foaming properties are the result of the interaction among the different proteins, such as globulins, ovalbumin, ovotransferrin, lysozyme, ovomucoid and ovomucin. Each fraction has its own ability to form a voluminous foam (foam capacity), to maintain it (foam stability) and to increase the volume of the batter containing it (i.e., angel cake) which differs from the overall foaming ability of the whole albumen [4]. For instance, globulins have the highest foaming index, around 4.71 cm^3/g min, more than 7 times higher than that of the ovalbumin (0.59), while ovomucin and ovomucoid show no foaming ability [4]. Nevertheless, Mine [4] reviewed that these last two proteins contributed to the final volume of the angel cakes made with them, whereas the ovalbumin produced cakes with a volume comparable to the one obtained by globulins, being 308 and 330 cm^3, respectively [4]. Standing on the mentioned literature, the possible role of the diet on protein matrix composition should be investigated to support or not our results about the technological properties of the egg white.

The albumen chemical composition varied while increasing the hen age, in line with previous findings [29] which proposed that the increase in hen age and weight led to a reduction in the crude protein content of the albumen. Changes in pH and protein content are consistent with the reduction of the foam capacity while using eggs from the 35th week of age. Instead, the maximum height raised by the angel cake varied due to the interaction D $\times$ WDA, thus supporting the previous hypothesis on the induced modification of proteins.

Baking loss is an important technological characteristic of a batter and plays a key role both for the quality of the final baked products and their shelf life. Indeed, the water content can promote microbiological growth during the storage, inducing a loss of the shelf life length. Considering that proteins may affect the water retention during cooking, the high affinity to the water of the white from eggs laid by hens fed the HI deserves other studies.

It has been previously demonstrated that the HI inclusion in the diet affects the yolk color values of the eggs [11–13], probably because of the pigments contained in the insect meal [13]. Standing on our knowledge, this is the first time that a difference in color emerged in a baked product made by egg white derived by hens fed with HI and, for this reason, little comparisons are possible.

The textural attributes of food are strictly related to its quality and consumers' acceptance. In the present study, the HI meal presence in the diet did not affect the textural properties of the angel cake, similarly to other authors who found that different dietary types of protein (i.e., soybean meal, cottonseed protein, double-zero rapeseed meal, individually or in combination with equal crude protein) administered to the Jinghong laying hens did not affect the cooked yolk hardness and springiness [30]. Furthermore, the deep reduction in the chewiness that occurred between the eggs laid at 20 and 35 weeks of hens' age agrees with the finite effect of the substitution of egg white by vegetable protein on angel cake chewiness [19]. Since chewiness can be enhanced due to a reinforcement of the protein entanglement in the networks [19], it is possible that variations in the egg white protein-gluten-starch connections occurred with hen aging, although more studies are necessary to define what kind of interaction is eventually developed.

5. Conclusions

The challenge for the foreseeable future will be to increase the sustainability of poultry production, including the egg supply chain. If one of the steps could be the substitution of the conventional protein sources, i.e., soybean meal, with the alternative ingredients, as insects in the case of their approval by European legislation, another point could be how to optimize the use of poultry products. In this regard, the present study provides the first information on how *infra-vitam* factors, such as diet and hens' age, could affect egg components and albumen technological properties, thus suggesting a differential use of the eggs based on hens' farming. Here, the use of eggs laid by hens fed the HI50 diet could be directed towards to egg product industry instead of selling as whole eggs because of their reduced eggshell percentage and thickness. Contrariwise, the eggs laid by hens from 27th to 35th weeks of age could be sold as whole because of their impaired albumen technological properties and consequently the chewiness of the baked cakes, which could be negatively perceived by the consumers.

Author Contributions: Conceptualization, F.B., G.M., and G.P.; methodology, G.S.; formal analysis, G.S.; data curation, G.P.; writing-original draft preparation, G.S.; writing-review and editing, F.B., G.M., G.S, and G.P.; supervision, G.P. All authors have read and agreed to the published version of the manuscript.

Acknowledgments: This work was supported by the contribution of the Fondo di Ateneo per la Ricerca 2019, University of Sassari (Responsible: G. Moniello) and by the Fondo di Ateneo per la Ricerca 2019, University of Florence (Responsible: Giuliana Parisi).

References

1. FAO. Global Forum on Food Security and Nutrition. Report of Activity n. 154; 2018. Available online: http://www.fao.org/3/CA3569EN/ca3569en.pdf (accessed on 7 October 2019).
2. Committee for the Common Organization of the Agricultural Markets. EU Market Situation for Eggs. Available online: http://circabc.europa.eu (accessed on 5 December 2019).
3. Unaitalia. Annata Avicola. 2017. Available online: http://www.unaitalia.com/mercato/annata-avicola (accessed on 5 December 2019).

4. Mine, Y. Recent advances in the understanding of egg white protein functionality. *Trends Food Sci. Tech.* **1995**, *6*, 225–232. [CrossRef]
5. FAO. *The State of Food and Agriculture*; FAO: Rome, Italy, 2009.
6. European Commission. Commission Regulation (EU) 2017/893 of 24 May 2017. *O.J. L.* **2017**, *L138*, 92–116.
7. Makkar, H.P.; Tran, G.; Heuzé, V.; Ankers, P. State-of-the-art on use of insects as animal feed. *Anim. Feed Sci. Technol.* **2014**, *197*, 1–33. [CrossRef]
8. Cutrignelli, M.I.; Messina, M.; Tulli, F.; Randazzo, B.; Olivotto, I.; Gasco, L.; Loponte, R.; Bovera, F. Evaluation of an insect meal of the Black Soldier Fly (*Hermetia illucens*) as soybean substitute: Intestinal morphometry, enzymatic and microbial activity in laying hens. *Res. Vet. Sci.* **2018**, *117*, 209–215. [CrossRef] [PubMed]
9. Marono, S.; Loponte, R.; Lombardi, P.; Vassalotti, G.; Pero, M.E.; Russo, F.; Gasco, L.; Parisi, G.; Piccolo, G.; Nizza, S.; et al. Productive performance and blood profiles of laying hens fed *Hermetia illucens* larvae meal as total replacement of soybean meal from 24 to 45 weeks of age. *Poult. Sci.* **2017**, *96*, 1783–1790. [CrossRef]
10. Maurer, V.; Holinger, M.; Amsler, Z.; Früh, B.; Wohlfahrt, J.; Stamer, A.; Leiber, F. Replacement of soybean cake by *Hermetia illucens* meal in diets for layers. *J. Insects Food Feed* **2016**, *2*, 83–90. [CrossRef]
11. Kawasaki, K.; Hashimoto, Y.; Hori, A.; Kawasaki, T.; Hirayasu, H.; Iwase, S.; Hashizume, A.; Ido, A.M.; Miura, C.; Miura, T.; et al. Evaluation of black soldier fly (*Hermetia illucens*) larvae and pre-pupae raised on household organic waste, as potential ingredients for poultry feed. *Animals* **2019**, *9*, 98. [CrossRef]
12. Mwaniki, Z.; Neijat, M.; Kiarie, E. Egg production and quality responses of adding up to 7.5% defatted black soldier fly larvae meal in a corn-soybean meal diet fed to Shaver White Leghorns from wk 19 to 27 of age. *Poult. Sci.* **2018**, *97*, 2829–2835. [CrossRef]
13. Secci, G.; Bovera, F.; Nizza, S.; Baronti, N.; Gasco, L.; Conte, G.; Serra, A.; Bonelli, A.; Parisi, G. Quality of eggs from Lohmann Brown Classic laying hens fed black soldier fly meal as substitute for soya bean. *Animal* **2018**, *12*, 2191–2197. [CrossRef]
14. Bovera, F.; Loponte, R.; Pero, M.E.; Cutrignelli, M.I.; Calabrò, S.; Musco, N.; Panettieri, V.; Lombardi, P.; Piccolo, G.; Di Meo, C.; et al. Laying performance, blood profiles, nutrient digestibility and inner organs traits of hens fed an insect meal from *Hermetia illucens* larvae. *Res. Vet. Sci.* **2018**, *120*, 86–93. [CrossRef]
15. AOAC. *Official Methods of Analysis of AOAC International*; Association of Official Analysis Chemists International: Arlington, TX, USA, 2012.
16. CIE (Commission Internationale de l'Éclairage). *Colorimetry. Publication n. 15*; Bureau Central de la CIE: Wien, Austria, 2004.
17. Fernández-Martín, F.; Pérez-Mateos, M.; Dadashi, S.; Gómez-Guillén, C.M.; Sanz, P.D. Impact of magnetic assisted freezing in the physicochemical and functional properties of egg components. Part 1: Egg white. *Innov. Food Sci. Emerg.* **2017**, *44*, 131–138. [CrossRef]
18. Lomakina, K.; Míková, K. A study of the factors affecting the foaming properties of egg white—A review. *Czech J. Food Sci.* **2006**, *24*, 110–118. [CrossRef]
19. Jarpa-Parra, M.; Wong, L.; Wismer, W.; Temelli, F.; Han, J.; Huang, W.; Eckhart, E.; Tian, Z.; Shi, K.; Sun, T.; et al. Quality characteristics of angel food cake and muffin using lentil protein as egg/milk replacer. *Int. J. Food Sci. Technol.* **2017**, *52*, 1604–1613. [CrossRef]
20. Bourne, M.C. Texture profile analysis. *Food Technol.* **1978**, *32*, 62–66.
21. Statistical Analyses System (SAS). *SAS/STAT Software*; Version 9; SAS Institute Inc.: Cary, NC, USA, 2000.
22. Al-Qazzaz, M.F.A.; Ismail, D.; Akit, H.; Idris, L.H. Effect of using insect larvae meal as a complete protein source on quality and productivity characteristics of laying hens. *Rev. Bras. Zootecn.* **2016**, *45*, 518–523. [CrossRef]
23. Leeson, S.; Summers, J.D. Feeding program for laying hens. In *Commercial Poultry Nutrition*; Nottingham University Press: Nottingham, UK, 2009; pp. 193–214.
24. Ruhnke, I.; Normant, C.; Campbell, D.L.M.; Iqbal, Z.; Lee, C.; Hinch, G.N.; Roberts, J. Impact of on-range choice feeding with black soldier fly larvae (*Hermetia illucens*) on flock performance, egg quality, and range use of free-range laying hens. *Anim. Nutr.* **2018**, *4*, 452–460. [CrossRef]
25. Jonchère, V.; Brionne, A.; Gautron, J.; Nys, Y. Identification of uterine ion transporters for mineralisation precursors of the avian eggshell. *BMC Physiol.* **2012**, *12*, 1–17. [CrossRef]
26. Ketta, M.; Tůmová, E. Eggshell structure, measurements, and quality-affecting factors in laying hens: A review. *Czech J. Anim. Sci.* **2016**, *61*, 299–309. [CrossRef]

27. Roberts, J.R. Factors affecting egg internal quality and egg shell quality in laying hens. *J. Poult. Sci.* **2004**, *41*, 161–177. [CrossRef]
28. Alleoni, A.C.C. Albumen protein and functional properties of gelation and foaming. *Sci. Agric.* **2006**, *63*, 291–298. [CrossRef]
29. Anyaegbu, B.C.; Onunkwo, D.N.; Irebuisi, D. Evaluation of hen age, body weight and egg weight on percent egg component and the internal composition of Harco hen eggs. *Sci. Afr. J. Sci. Issues Res. Essays* **2016**, *4*, 943–945.
30. Wang, X.C.; Zhang, H.J.; Wu, S.G.; Yue, H.Y.; Wang, J.; Li, J.; Qi, G.H. Dietary protein sources affect internal quality of raw and cooked shell eggs under refrigerated conditions. *Asian-Australas. J. Anim. Sci.* **2015**, *28*, 1641–1648. [CrossRef] [PubMed]

10

Methane Emissions and Milk Fatty Acid Profiles in Dairy Cows Fed Linseed, Measured at the Group Level in a Naturally Ventilated Housing and Individually in Respiration Chambers

Jernej Poteko [1,2,†], Sabine Schrade [2], Kerstin Zeyer [3], Joachim Mohn [3], Michael Zaehner [2], Johanna O. Zeitz [1,4], Michael Kreuzer [1] and Angela Schwarm [1,5,*]

1 ETH Zürich, Institute of Agricultural Sciences, 8092 Zurich, Switzerland; jernej.poteko@lfl.bayern.de (J.P.); jzeitz@gmx.de (J.O.Z.); michael.kreuzer@usys.ethz.ch (M.K.)
2 Agroscope, Ruminants Research Unit, 8356 Ettenhausen, Switzerland; sabine.schrade@agroscope.admin.ch (S.S.); michael.zaehner@agroscope.admin.ch (M.Z.)
3 Empa, Laboratory for Air Pollution/Environmental Technology, 8600 Duebendorf, Switzerland; kerstin.zeyer@empa.ch (K.Z.); joachim.mohn@empa.ch (J.M.)
4 Laboklin, Klinische Labordiagnostik, 97688 Bad Kissingen, Germany
5 Department of Animal and Aquacultural Sciences, Norwegian University of Life Sciences, P.O. Box 5003, 1432 Ås, Norway
* Correspondence: angela.schwarm@nmbu.no.
† Current address: Bavarian State Research Center for Agriculture, Institute for Agricultural Engineering and Animal Husbandry, 85586 Grub-Poing, Germany.

Simple Summary: Cows emit the greenhouse gas methane (CH_4) as a result of microbial feed digestion. Methane emissions can be reduced by adopting nutritional strategies, such as dietary supplementation of linseed. Additionally, the oil in linseed increases the proportion of favorable fatty acids in milk fat. This study evaluated the effect of linseed on CH_4 emission and milk fatty acid composition measured in a group of cows in a naturally ventilated barn and in individual cows in respiration chambers. The substantially higher proportions of favorable fatty acids in the milk of linseed-fed cows were detected in individual milk samples and in the milk of the herd. Therefore, the analysis of bulk milk could be a suitable control instrument for retailers. Visualizing the course of CH_4 emissions over a whole day showed slightly lower CH_4 values in linseed-supplemented individuals and groups. However, we found no significant reduction of CH_4 as a result of linseed supplementation. Feed supplements in concentrations that are effective in reducing CH_4 must show whether the reduction potential is comparable when determined at the group and individual levels.

Abstract: The present study evaluated the effects of linseed supplementation on CH_4 emission and milk fatty acid composition in dairy cows measured at the group level in an experimental dairy loose housing using a tracer gas technique and individually in tied stalls and respiration chambers. Cows (2×20) were maintained in two separate sections under loose-housing conditions and received a diet supplemented with extruded linseed (L) lipids (29 g·kg^{-1} dry matter) or a control (C) diet containing corn flour. Subsequently, 2×6 cows per dietary group were investigated in a tied-housing system and respiration chambers. Substantially higher proportions of favorable milk fatty acids were recovered in L cows when compared with C cows at the group level, making the analysis of bulk milk a suitable control instrument for retailers. Linseed supplementation resulted in a slightly lower diurnal course of CH_4 emission intensity than the control at the group and individual levels. However, we found no more than a trend for a CH_4 mitigating effect, unlike in other studies supplementing similar linseed lipid levels. Feed supplements in concentrations that lead to a significant reduction in CH_4 emissions must show whether the reduction potential determined at the group and individual levels is comparable.

Keywords: methanogenesis; methane mitigation; lipid; ruminant; cattle; emission measurement

1. Introduction

Ruminants emit methane (CH_4) as a result of enteric fermentation. Ruminal CH_4 emissions can be mitigated by adopting several nutritional strategies [1,2]. These are advantageous over animal breeding measures as they are immediately effective [3]. In particular, dietary supplementation of crushed or extruded oilseeds has been demonstrated to mitigate enteric CH_4 emission [4,5]. In this context, supplementing the diet with >50 g of linseed lipids per kg feed dry matter (DM) particularly decreased CH_4 emission, but diet digestibility was concomitantly impaired [5,6]. At supplementation levels of ≤20 g linseed lipids per kg feed DM, there was no reduction in the CH_4 emissions [7–9]. Thus, supplementing feed with moderate levels ranging from 26–36 g linseed lipids kg^{-1} feed DM appears to be the most promising approach to achieve a clear CH_4 reduction with marginal effects on digestion and intake [6,10,11]. In addition, extruded linseed increases the proportion of n–3 polyunsaturated fatty acids (FAs) in milk fat, which is considered beneficial to human health (e.g., [12]). Therefore, some retailers in France and Switzerland are currently paying higher prices to farmers for milk with a higher proportion of n–3 FAs in milk fat. Label programs may require evidence of CH_4 reduction at the herd level or changes in the FA profile of bulk milk.

Traditionally, CH_4 emissions are measured continuously on individual animals to quantify the reduction potential of dietary interventions. Continuous measurements in respiration chambers determine total emissions, but locomotion is restrained. The sulfur hexafluoride (SF_6) bolus technique allows free locomotion of the cows during continuous measurements [13] but detects only exhaled CH_4. These limitations do not apply to group measurements, which concomitantly allow more natural behavior of the cows [14,15] and total emission assessment. Furthermore, climate and housing-based effects, such as slurry storage and area soiling, on CH_4 emissions are considered. These differences could promote or reduce the effects of diet on CH_4 emissions. However, studies at the group level in relation to dietary interventions are rare, and none of the reports studied the effect of linseed supplementation on total CH_4 emission at the group level. Schmithausen et al. [16] quantified CH_4 emissions from cows that were fed diets with different levels of condensed tannins in a naturally ventilated experimental dairy housing using the carbon dioxide (CO_2) balance method. Several studies compared CH_4 emissions obtained on individual animals in respiration chambers with those determined at the practical scale using either SF_6 permeation tubes (e.g., [17]), the GreenFeed system (e.g., [18]) or other ventilated hoods (e.g., [13]). To our knowledge, there are currently no reported data on CH_4 emissions for individual animals in respiration chambers vs. groups of animals in naturally ventilated dairy housing.

Therefore, this study aimed to evaluate the effect of extruded linseed supplementation on the CH_4 emission and milk FA composition both at the group level on a practical scale in naturally ventilated housing and in respiration chambers individually. The following hypothesis was tested: moderate levels of lipids exert effects on the CH_4 emission and milk FA composition, which are detectable and similar in extent at the individual animal and group levels. In addition, the effect on feeding behavior was investigated.

2. Materials and Methods

The experiment was carried out from December 2015 to March 2016. Group level measurements were carried out in a naturally ventilated experimental loose-housing system for dairy cows at Agroscope (Waldegg, Switzerland; 47°29′22″ N, 8°55′10″ E, 550 m above sea level). Then, cows were measured individually at the AgroVet-Strickhof in a tied-housing barn (Wülflingen, Switzerland; 47°30′58″ N, 8°41′49″ E) and in respiration chambers (Eschikon, Switzerland; 47°26′53″ N, 8°40′43″ E). The cows were transported over a distance of 30 km from the loose- to the tied-housing and over 13 km from the tied-housing to the respiration chambers. The experimental protocol complied with the Swiss legislation on Animal Welfare and was approved (ZH091/15) by the Cantonal Veterinary Office of Zürich.

2.1. Animals and Experimental Design

2.1.1. Measurements at the Group Level in the Loose-Housing System

For group measurements, 40 primiparous and multiparous lactating cows of Brown Swiss and Swiss Fleckvieh breeds were selected and stratified into two groups in a completely randomized manner, balanced as best as possible for lactation number, breed, body weight (BW), days in milk, and milk yield. The control (C) group consisted of 14 Brown Swiss and six Swiss Fleckvieh cows, and the linseed (L) group consisted of 11 Brown Swiss and nine Swiss Fleckvieh cows. Group level measurements were conducted over 25 days, of which 21 days were used to allow the cows to adapt to the housing and diet, and emissions were measured for four consecutive days. The adaptation period was initiated with a proportionate increase of either corn flour (in the C cows) or linseed (in the L cows) supplementation at the group level during the first 5 days at 0.2, 0.4, 0.6, 0.8, and 1.0 of the final amounts. The feed intake of the two groups was assessed at the group level, whereas feeding behavior (eating and rumination time) was determined individually for ten cows per group.

2.1.2. Measurements at the Individual Level in the Tied-Housing System and in the Respiration Chambers

We randomly selected 2 × 6 exclusively multiparous cows from the 2 × 20 cows in the experimental groups for the experiments where the cows were housed in a tied-housing system. All but one cow were of the Brown Swiss breed. Individual measurements were obtained directly after the measurements at the group level. Cows were given the C and L diets (details on diet ingredients and nutrient composition are given in Section 2.2 and Table 1). The mixed ration was prepared daily at Agroscope (loose housing) and transported to the tied stall barn and respiration chambers. A period of 7 days was set to allow the cows to adapt to the tied-housing system. This was followed by a 7–day period of data and sample collection. Sampling was carried out in two blocks of 2 × 3 cows each. Within the sampling period, cows were moved in pairs, with one cow per experimental diet, into respiration chambers for an adaptation period of 6 h and gas exchange measurement period of 2 days. The tied-housing systems in the tied stall barn and in the respiration chambers were equipped with rubber mats and feces collection trays. Bedding material was omitted during sampling to avoid contamination by feces.

2.2. Feeding

The mixed ration consisted of grass silage, corn silage, hay, beet pulp, and concentrate (Table 1). The linseed product used (TradiLin 135; Trinova AG, Wangen, Switzerland) consisted of an extruded mixture of linseed and mill byproducts (0.6:0.4). The latter absorbs the oil released from the linseed during processing. This linseed product was added on top of the mixed ration, providing 112 g TradiLin 135 per kg of DM, equivalent to 67 g linseed and 29 g (as analyzed) of extra lipids. The cows in the C group received corn flour instead of linseed in an iso-energetic manner. The amount of linseed and corn flour was adjusted to the amount of mixed ration, ensuring that the relative intake of linseed and corn flour remained the same. In the loose-housing system, cows were fixed in the feeding barrier during the first 30 min of new feed provision to enable complete consumption of the linseed or corn flour without disturbance by dominant animals. In addition, concentrates that were rich in energy and protein were allocated according to milk yield following recommendations by Agroscope [19] by an automatic feeder (loose housing) or given manually after each feeding event (tied housing) (Table 1). Each animal received 50 g·day^{-1} of sodium chloride and 100 g·day^{-1} of a commercial mineral-vitamin mixture (Minex 976, UFA AG, Sursee, Switzerland) consisting (per kg) of 100 g calcium, 80 g phosphorus, 75 g magnesium, 20 g sodium, 4 g zinc, 3.8 g manganese, 1 g cooper, 1 g sulfur, 100 mg iodine, 40 mg cobalt, 40 mg selenium, 1,000,000 IU vitamin A, 200,000 IU.

Table 1. Ingredients and their nutrient compositions and realized average intakes with the control and the extruded linseed diets [7] in the loose and tied-housing systems (means ± standard deviation).

	Mixed Part of the Diet				Supplementary Concentrates				Linseed
Component	**Grass Silage**	**Corn Silage**	**Hay**	**Beet Pulp**	**Concentrate 1**	**Rich in Protein [2]**	**Rich in Energy [3]**	**Corn Flour**	**Product**
Dry matter ($g \cdot kg^{-1}$)	459	417	899	260	882	863	865	859	933
Nutrients ($g \cdot kg^{-1}$ of dry matter)									
Organic matter	917	970	914	900	917	919	923	989	957
Crude protein	94	62	109	83	563	413	182	79	180
Neutral detergent fiber	463	328	506	435	221	127	161	161	398
Acid detergent fiber	303	193	295	230	135	95	85	52	215
Acid detergent lignin	37.7	33.8	37.1	29.2	50.4	32.5	32.5	22.1	103
Ether extract [4]	20.8	29.6	17.5	3.9	42.0	37.4	62.1	35.7	255
Realized DM intake ($kg \cdot day^{-1}$ per cow)									
Loose housing (group observations) [5]									
Control	6.38 ± 0.13	3.30 ± 0.07	2.32 ± 0.05	0.94 ± 0.02	1.50 ± 0.03	1.55 ± 0.53	1.05 ± 1.72	2.16 ± 0.05	-
Linseed	6.40 ± 0.14	3.31 ± 0.07	2.32 ± 0.05	0.94 ± 0.02	1.50 ± 0.03	1.73	1.14 ± 1.39	-	2.19 ± 0.05
Tied housing (individual observations) [6]									
Control	6.60 ± 0.80	3.41 ± 0.42	2.40 ± 0.29	0.97 ± 0.12	1.55 ± 0.19	1.73	1.14 ± 1.62	2.23 ± 0.27	-
Linseed	6.40 ± 0.82	3.31 ± 0.43	2.33 ± 0.30	0.94 ± 0.12	1.51 ± 0.19	1.73	1.15 ± 1.67	-	2.19 ± 0.28

[1] Composed of soybean meal, dried distillers grains with solubles, rapeseed cake, rapeseed meal, milling byproducts, malt culms, corn gluten meal, and minerals (UFA AG, Herzogenbuchsee, Switzerland). [2] Composed of soybean meal, corn gluten meal, rapeseed cake, dextrose, sugar beet molasses, vitamin–mineral mixture, fruit syrup, vegetable fat, and minerals (Thurtalfutter AG, Frauenfeld, Switzerland). [3] Composed of barley, triticale, maize, rapeseed cake, soybean meal, peas, wheat, oat, sugar beet molasses, vegetable fat, vitamin-mineral mixture, corn gluten meal, fruit syrup, and minerals (Thurtalfutter AG, Frauenfeld, Switzerland). [4] Proportions of the fatty acids as analyzed ($g \cdot 100\ g^{-1}$ of total analyzed fatty acids) C16:0, C18:0, C18:1 n-9, C18:2 n-6, and C18:3 n-3 were 16.4, 1.98, 15.5, 33.0, and 24.8 in the total mixed part of the diet; 10.9, 2.18, 26.0, 54.0, and 4.51 in corn flour; and 7.04, 4.10, 16.8, 17.7, and 52.1 in linseed, respectively. [5] Average of group observations in the loose-housing system (two treatments × 4 days). Dry matter intake (DMI) per cow was calculated by dividing the group DMI by the number of cows. [6] Average of individual observations in the tied-housing system (two treatments × six animals). [7] Contents (g/kg dry matter) of organic matter, crude protein, neutral detergent fiber, acid detergent fiber, acid detergent lignin, and ether extract were 934, 153, 351, 212, 35, and 28 for the complete control diet and 930, 165, 377, 230, 44, and 53 for the complete linseed diet, respectively.

Vitamin D_3, 3 g vitamin E, and 100 mg biotin. New portions of diets provided with ad libitum access were offered at 16:45 after milking at the group level. Specialized equipment automatically moved feed towards the cows 18 times a day. At the individual level, feeding took place at 11:00 (10:00 in the respiration chambers), 13:00, 16:30, 23:00, and 05:30. The feed amounts offered to the two groups and to the individual cows were recorded daily. Leftovers were removed and weighed once daily. All the cows were given permanent access to water. The cows were milked at 05:30 and 16:30 at all experimental sites.

2.3. *Measurement of Methane and Carbon Dioxide Emissions*

2.3.1. Emission Measurement in the Experimental Dairy Loose-Housing System (Group Level)

At the group level, emissions were measured as described in detail by Mohn et al. [20]. Briefly, the housing system provided two spatially separated experimental sections with three rows of cubicles with straw. A milking parlor and analytical devices were located between the two sections. The covered underground slurry storages were separated for each section. Management routines such as milking, feeding, and 12 dung removals per day remained the same for both groups. Each group was fed separately to avoid cross-contamination. Curtains in the facades remained closed, ensuring that wind and temperature conditions in the two sections were equivalent. Adequate ventilation was ensured by spaced boards in the upper part of the longitudinal facades. There were no other natural (e.g., ruminants) or technical emission sources near the housing. The position of the housing axis was orthogonal to prevailing wind directions, minimizing cross-contamination between the experimental sections. The dual tracer ratio method [20,21] was applied to quantify emissions for each experimental section independently and detect potential cross-contamination. It involved constant dosing of the tracer gases SF_6 and trifluoromethyl sulfur pentafluoride (SF_5CF_3), one per section, at floor level using mass flow controllers (Contrec AG, Switzerland) to regulate the total flow and critical steel orifices to achieve homogenous spatial distribution. Representative air sampling in each section was accomplished with critical glass orifices (250 μm in diameter 2.5 m above the ground; Thermo-Instruments, Germany, and Louwers, The Netherlands). Concentrations of the tracer gases and target gases (CH_4 and CO_2) were analyzed in real time by gas chromatography with electron capture detection (GC-ECD, model 7890A, Agilent Technologies AG, Switzerland) and by cavity ring-down spectrometry (CRDS, CH_4, CO_2, model G2301, Picarro Inc., USA). More details on the implemented analytical technique and its performance with respect to suitability for point/areal sources, sensitivity, and uncertainty have been described in Mohn et al. [20]. The applied measurement sequence provided emission data with a temporal resolution of 10 min per section. Milking times were excluded from the analysis, as only part of the group was present in the sections during these periods.

2.3.2. Emission Measurement in the Respiration Chambers (Individual Level)

The two open-circuit respiration chambers used in the present study had a volume of 19.3 m^3 each [22]. The chambers were air-conditioned to 15 °C and 55% relative humidity at an air pressure of −60 Pa (relative to ambient). The airflow was set to 700 $L{\cdot}min^{-1}$ (Promethion FG 1000 flow generator, Sable Systems Europe GmbH, Berlin, Germany). A light:dark cycle of 16:8 was used. The CH_4, oxygen (O_2), and CO_2 concentrations in the spent air were measured with a Promethion GA 4 gas analyzer (Sable Systems, Las Vegas, USA), alternating between in- and outgoing air at 1-min intervals. Calibrations of the gas analyzer were initiated automatically before each 2-day measurement period using pure (99.999%) nitrogen (N_2) and a reference gas (19.8% O_2, 1.0% CO_2, 0.1% CH_4, in N_2). Recovery, averaging at 105%, was assessed before and after the two experimental blocks by burning propane gas. For statistical evaluation, emissions were averaged across the 2 days of measurement. The chambers were entered through the airlock for feeding, milking, and urine collection during the measurements. For feces collection, the back door was opened after 24 and 48 h of measurement.

2.4. *Performance and Feeding Behavior (Individual Measurements in the Loose and Tied-Housing Systems)*

We individually assessed the cows' BWs after evening milking on the days before and after the 4-day measuring period in the loose housing and on the first and the last day of the 7-day sampling period in the tied-housing system. The BW was directly measured with a cattle balance (Modell FX21, Iconix, Rotorua, New Zealand) in the tied housing and estimated from girth size determinations with a measuring tape in the loose- (and tied-) housing system. The close correlation of 0.88 (n = 24, $p < 0.001$) determined between actual weights and estimates with the measuring tape allowed the application of the latter in the loose housing. During all measurement periods, milk yield was recorded individually at each milking using a milk meter (EasyFlow, Fullwood, Ellesmere, United Kingdom) in the loose housing and by weighing the buckets in the tied housing. Chewing and ruminating activity was monitored with nose band pressure sensors of the RumiWatch System equipped with the Converter V.0.7.3.36 (Itin+Hoch GmbH, Liestal, Switzerland) as described by Werner et al. [23]. This was accomplished for ten multiparous cows per group, including those subjected later to individual measurements during the 4-day measuring period at the group level and at the individual level for 5 days during the sampling period (excluding days were the cows were moved). The times of eating and ruminating and the number of chews during eating and ruminating were evaluated. In the loose-housing system, data from one cow in the C treatment group were excluded due to technical problems with the sensor.

2.5. *Sampling*

Grass silage, corn silage, and hay samples were collected at six times, i.e., at the beginning and end of the measurements in the loose- and tied-housing systems (separately in the two animal blocks). Samples of the sugar beet pulp and the three types of concentrates were taken twice during the entire study. The linseed and corn flour were sampled daily and pooled for the measurement period in the loose housing and analyzed separately for the two blocks of animals in the tied-housing system. In the loose housing, samples of leftovers were collected daily per group. In the tied-housing system, the daily feed leftovers were sampled individually and in proportion to the total amount of leftover. These samples were immediately frozen. In the tied-housing system, urine was separated from the feces by attaching urinals fixed by hook-and-loop fastener straps around the vulva of the cows. These straps were glued (ergo 5011, Kisling, Wetzikon, Switzerland) onto shaved skin. The urine was collected in a large and a small container, the latter containing 5 M sulfuric acid to avoid N losses. The total amount of feces and urine was recorded during the 7–day sampling period. Urine and feces subsamples were collected once and twice per day, respectively; samples were a constant fraction of the amounts excreted. The subsamples were frozen and later pooled for each cow. Before analysis, the feed, leftover feed, and fecal samples were dried at 60 °C until a constant mass was achieved and then ground to pass a 1-mm screen of either a cutting mill or a centrifugal mill (extruded linseed, corn flour, and concentrates). In the loose-housing system, milk samples were obtained individually at each milking on sampling days 1 and 4. A constant fraction of the yield at each milking was collected and the samples were pooled per group. In the tied-housing system, samples were collected on each sampling day from each milking, and a constant fraction of the yield at each milking was composited for each cow. Aliquots of individual and pooled milk samples were either frozen without additive or preserved with Bronopol® (D&F Inc., Dublin, CA, USA).

2.6. *Laboratory Analyses*

Feeds, leftovers, and feces were analyzed according to standard procedures [24]. The DM and total ash contents (AOAC method 942.05) were determined with a thermo-gravimetric device (TGA 701, Leco Corporation, St Joseph, MI). For fiber analyses (AOAC method 973.18), a Fibertec System M (1020 Hot Extractor and 1021 Cold Extractor; Tecator, Foss Hillerød, Denmark) was used. Neutral detergent fiber (NDF) analysis was performed with the addition of heat stable α-amylase, but without sodium sulfite [25]. Acid detergent lignin was determined sequentially after acid detergent fiber (ADF) analysis by treatment with sulfuric acid (72%) for 3 h. Fiber data were corrected for ash content. Combustion energy in the feed and feces items to calculate energy intake and energy excretion with feces was quantified by a bomb calorimeter (C7000, IKA-Werke GmbH & Co. KG, Staufen, Germany). Feeds, leftovers, previously frozen milk, non-dried feces, and acidified urine were analyzed for N content (AOAC method 968.06) with a C/N analyzer (TruMac CN, Leco Corporation, St. Joseph, MI). Crude protein in the feed was defined as 6.25 × nitrogen (N). Carbon (C) content in the non-acidified urine was determined on the same C/N analyzer in order to be able to calculate urine energy (as described in Grandl et al. [26]). Ether extract (AOAC method 963.15) of feed items, leftovers, and feces was determined with a Soxhlet extraction system (model Extraktionsapparatur B-811, Büchi, Flawil, Switzerland). Bronopol-preserved milk samples were analyzed for fat, protein, lactose, and milk urea nitrogen (MUN) contents using a Fourier-transform infrared spectrophotometer (FTIR; MilkoScan FT6000, Foss, Hillerød, Denmark) at the Swiss routine milk analysis laboratory (Suisselab AG, Zollikofen, Switzerland). The MUN content was also determined in previously frozen milk by an enzymatic method [27].

For quantification of the proportions of FAs in the lipids of extruded linseed and corn flour, FAs were extracted by a solvent extractor (ASE 200, Dionex Corporation, Sunnyvale, CA) using a hexane:propane-2-ol mixture (3:2 v/v). The FAs were transformed to FA methyl esters (FAME) according to the International Union of Pure and Applied Chemistry (IUPAC) [28] method 2.301. Cleaning was performed as described in Wettstein et al. [29]. For FAME analysis, a gas chromatograph (model HP 6890 equipped with a flame ionization detector, Hewlett Packard, Palo Alto, CA, USA) and a CP7421 column (200 m × 0.25 mm, 0.25 μm; Varian Inc., Darmstadt, Germany) were used. Split injection (1:5) was applied. The internal FA standard was C11:0 (Fluka, Steinheim, Germany), and the external standard for the response factor was sunflower oil. A volume of 1 μL was injected with a constant hydrogen flow of 1.7 mL·min^{-1}. The temperature program was set to 170 °C for 60 min, increased to 230 °C at a rate of 5 °C·min^{-1}, ramped for 32 min, increased to 250 °C at a rate of 5 °C·min^{-1}, and ramped for 15 min. For analysis of the FA in milk, samples were thawed and gently mixed to disperse milk fat. Internal standards (5 mL of *n*-heptane containing triundecanoin, tetradecenoic methylate, and trivaleranoin) were mixed with 0.5 mL of milk. Cold transesterification to FAME was done with sodium methylate [30]. The response factors obtained from C6:0, C13:0, and C19:0 triglyceride standards were used to adjust individual FA data. The same gas chromatograph and column and operational conditions were applied as for FA from feeds, except for the temperature regime which was as follows: initial temperature of 60 °C, ramped for 12 min, increased to 170 °C at the rate of 5 °C·min^{-1}, ramped for 60 min, increased to 250 °C at the rate of 5 °C·min^{-1}, and ramped for 20 min. The FAME was identified using a Supelco 37 component standard (Supelco Inc., Bellefonte, PA, USA). Peaks were identified using chromatograms from Collomb and Bühler [31].

2.7. Calculations and Statistical Analysis

In the loose- and tied-housing systems, the realized dry matter intake (DMI) of each component of the mixed ration was calculated based on the assumptions that the proportions of each component were the same in the offered feed and in the leftovers, and that the extruded linseed and corn flour were consumed completely (as confirmed by observations). One livestock unit (LU), defined as 500 kg BW [32], was used to adjust for differences in BW resulting from using only the multiparous cows in the tied stall system. The energy-corrected milk (ECM) (kg·d^{-1}) was calculated as milk (kg·day^{-1}) × [0.038 × fat (g·kg^{-1}) + 0.024 × protein (g·kg^{-1}) + 0.017 × lactose (g·kg^{-1})]/3.14. [19]. The efficiency traits calculated were milk production efficiency (ECM/BW) and feed conversion efficiency (ECM/DMI). In the loose-housing system, the individual DMI was calculated by dividing the measured total DMI of each group by the number of animals in the group (n = 20). The energy-related variables were calculated with the standard equations used in energy balance studies as listed in detail in Grandl et al. [26], where the two different methods of calculating the proportionate utilization of metabolizable energy (ME) for milk energy formation as used in the present study are described.

In the loose-housing system, the mass flow of the target gases ($\dot{m}_{target}$) was calculated from the ratio between the background-corrected target gas concentration (c_{target}) and tracer gas concentration (c_{tracer}) multiplied with the known mass flow of the tracer gas ($\dot{m}_{tracer}$), as $\dot{m}_{target} = \dot{m}_{tracer} \times c_{target}/c_{tracer}$. For the loose- and tied-housing systems, the CH_4 yield was described as the absolute CH_4 amount in relation to the DMI, digestible organic matter (OM) intake, digestible NDF intake (dNDFI), and gross energy (GE) intake (GEI; CH_4 conversion factor, Y_m). The methane emission intensity was assessed by relating CH_4 to ECM and BW. Methane emissions measured during the 2-day period in the respiration chambers under the tied-housing conditions were related to intake and ECM yield obtained across the entire 7-day sampling periods in the tied stall barn.

Data were subjected to analysis of variance to compare dietary treatments with TIBCO Spotfire+®(version 8.2 for Windows), considering diet as the fixed effect and either day (group level; [33,34]) or animal (individual level) as the experimental units. Spearman correlation coefficients were calculated for some variables. The distributional data properties were visually checked using a normal quantile-quantile plot of residuals. Significance was set at $p < 0.05$, and tendency at $0.05 \leq p < 0.10$. Descriptive statistics were used to evaluate the data and effects of linseed supplementation in the loose- and tied-housing systems.

3. Results

3.1. Individual Performance, Digestibility, and Feeding Behavior (Loose- and Tied-Housing Systems)

The average BW did not differ between the C and L cows in any of the housing systems, but the average BW was approximately 80 kg higher in the subgroup of the groups used for the individual assessment (Table 2). The daily ECM and milk fat yields did not differ between C and L cows but were about 20% lower on average in the cows selected for the tied stall system compared with cows kept in the loose-housing system. Furthermore, there were no significant differences in the DM and OM intakes between the C and L cows. For C cows, the DMI was about 1 kg·day^{-1} higher at the individual level (tied housing) than at the group level (loose housing). The NDF and ADF intakes were higher ($p < 0.05$) in L cows at the group level and the ADF intake was higher at the individual level in L cows than in C cows. The NDF intake was slightly higher at the group level than at the individual level. Milk yield (per BW) and feed conversion efficiencies were not different between the L cows and C cows.

Table 2. Performance of cows from control and linseed treatments measured in the loose- and tied-housing phases (means ± standard deviation).

Housing	Loose (Group)			Tied (Individual)		
Item	Control	Linseed	p–Value	Control	Linseed	p–Value
Body weight (BW, kg) [1]	681 ± 93	682 ± 84	0.971	756 ± 71	761 ± 38	0.883
Energy-corrected milk (ECM, kg·day^{-1}) [1]	33.2 ± 7.9	36.3 ± 10.0	0.294	26.4 ± 5.7	30.5 ± 9.3	0.386
Milk fat (kg·day^{-1}) [1]	1.40 ± 0.34	1.46 ± 0.51	0.661	1.08 ± 0.21	1.16 ± 0.38	0.652
Intake (kg·day^{-1}) [2]						
Dry matter (DMI)	19.3 ± 0.4	19.5 ± 0.4	0.417	20.1 ± 1.6	19.6 ± 2.0	0.665
Organic matter	17.9 ± 0.4	18.0 ± 0.4	0.754	18.7 ± 1.5	18.1 ± 1.9	0.564
Neutral detergent fiber	6.58 ± 0.14	7.14 ± 0.17	0.002	6.91 ± 0.29	7.24 ± 0.36	0.108
Acid detergent fiber	4.01 ± 0.09	4.40 ± 0.11	0.001	4.15 ± 0.16	4.39 ± 0.20	0.045
Efficiency						
ECM per BW (kg·100 kg^{-1}) [1]	4.93 ± 0.12	5.37 ± 0.14	0.312	3.55 ± 0.97	4.02 ± 1.25	0.488
ECM per BW (g·kg$^{-0.75}$) [1]	251 ± 59	273 ± 71	0.292	185 ± 48	211 ± 65	0.458
ECM [1] per DMI [2] (kg·kg^{-1})	1.72 ± 0.41	1.86 ± 0.51	0.111	1.31 ± 0.23	1.53 ± 0.32	0.204

[1] Individual observations in the loose- (2 treatments × 20 animals) and tied- (2 treatments × 6 animals) housing system. [2] Group observations (two treatments × 4 days) in the loose housing and individual observations (2 treatments × 6 animals) in the tied-housing system.

The L cows spent more time eating (min d^{-1}, $p < 0.05$; min kg^{-1} DMI, $p < 0.10$) in the loose housing but the same time eating in the tied-housing system as the C cows (Table 3). The rumination time was longer for L cows than C cows (min·d^{-1}, $p < 0.10$; min·kg^{-1} DMI and chews per kg^{-1} DMI, $p < 0.01$) in the tied-housing system but similarly long for L cows in the loose-housing system. The daily eating time and number of eating chews (per unit DMI) were on average lower by about 4% and the rumination time and number of ruminating chews (per unit DMI) were lower by about 18% in the tied-stall system than in the loose-housing system. Considering the eating behavior exclusively from the 2 days when cows were in the respiration chambers, the cows spent an average of 5% less time eating and ruminating, and made 9% less chews than in the entire 7-day sampling period in the tied-housing system (data not shown), and there was no difference between the two diet treatments ($p > 0.10$). The digestibility of OM, NDF, and ADF did not differ between the L and C cows at the individual level. Similarly, diet did not affect the absolute and relative N balance traits and N utilization (see Appendix A, Table A1). Only the proportion of urinary N in the total manure N tended to be greater ($p < 0.10$) for the L cows than for the C cows. Energy intake (GE, digestible energy, and ME) did not differ between diets (see Appendix A, Table A2). There was no effect of linseed on energy losses, except for the energy lost as CH_4, which tended ($p < 0.10$) to be lower by 9% in L cows than in C cows. Concomitantly, the intake, retention, turnover, and utilization of energy did not differ between the two dietary treatment groups.

Table 3. Feeding behavior, digestibility, and gaseous emissions of cows from control and linseed treatments determined in the loose- and tied-housing phases (means ± standard deviation).

Housing	Loose (Group)			Tied (Individual)		
Item	**Control**	**Linseed**	***p*–Value**	**Control**	**Linseed**	***p*–Value**
Eating [1]						
min	406 ± 27	455 ± 15	0.019	422 ± 96	445 ± 36	0.354
$min \cdot kg^{-1}$ DMI [2]	21.1 ± 1.64	23.3 ± 0.95	0.051	20.4 ± 4.9	22.1 ± 3.0	0.219
$Chews \cdot kg^{-1}$ DMI	1439 ± 142	1598 ± 58	0.072	1386 ± 386	1536 ± 273	0.182
Rumination [1]						
min	466 ± 38	497 ± 9	0.170	411 ± 56	445 ± 34	0.078
$min \cdot kg^{-1}$ DMI	24.2 ± 2.2	25.4 ± 0.9	0.275	19.8 ± 2.2	22.0 ± 1.7	0.008
$Chews \cdot kg^{-1}$ DMI	1546 ± 146	1664 ± 66	0.162	1197 ± 172	1386 ± 188	0.006
Digestibility (%)						
Organic matter (OM)	-	-	-	76.1 ± 2.9	75.9 ± 3.1	0.935
Neutral detergent fiber (NDF)	-	-	-	61.5 ± 6.2	63.2 ± 4.1	0.580
Acid detergent fiber	-	-	-	63.2 ± 6.5	63.9 ± 4.2	0.829
Carbon dioxide ($kg \cdot LU^{-1}$) [3,4]	9.18 ± 0.35	9.20 ± 0.55	0.952	9.07 ± 1.02	8.54 ± 0.60	0.295
Methane yield [4]						
$g \cdot kg^{-1}$ DMI	22.3 ± 1.6	21.4 ± 2.1	0.496	24.6 ± 3.1	22.9 ± 2.1	0.285
$g \cdot kg^{-1}$ digestible OM intake	-	-	-	34.7 ± 3.6	32.7 ± 3.6	0.361
$g \cdot kg^{-1}$ digestible NDF intake	-	-	-	116.3 ± 10.0	97.9 ± 8.4	0.006
$kJ \cdot MJ^{-1}$ gross energy intake (Y_m)	69.4 ± 5.0	64.1 ± 6.2	0.232	76.4 ± 9.7	68.6 ± 6.4	0.131
Methane emission intensity [4]						
$g \cdot kg^{-1}$ ECM [5]	13.6 ± 1.2	12.6 ± 1.2	0.266	19.2 ± 3.8	15.7 ± 4.4	0.165
$g \cdot LU^{-1}$; [4]	321 ± 5.1	301 ± 6.7	0.292	327 ± 44	294 ± 30	0.155

[1] Individual observations in the loose (two treatments × 20 animals) and tied (two treatments × six animals) housing system. [2] Dry matter intake. [3] Livestock unit equivalent to 500 kg body weight. [4] Group observations (two treatments × 4 days) in the loose housing and individual observations (two treatments × six animals) in the tied-housing system. [5] Energy-corrected milk.

3.2. *Methane and Carbon Dioxide Emissions (Group and Individual Level)*

The carbon dioxide emissions (expressed per LU) did not differ between dietary treatments (L vs. C cows) but were on average 5% lower at the individual level than at the group level. In the loose (group) and tied (individual) housings, the CH_4 emissions followed the typical diurnal pattern (Figure 1), with characteristic peaks after feeding (and milking) with both diets. The average CH_4 production per animal ($g \cdot day^{-1}$) was 4% lower among L cows than among C cows at the group level (414 ± 5.7 vs. 431 ± 9.0) and tended ($p < 0.10$) to be lower by 9% at the individual level (447 ± 34 vs. 491 ± 40; data not shown in table). The trends for the lower CH_4 production by L cows than the C cows were not only observed for temporal averages but also could be seen as a distinct shift in the diurnal pattern of CH_4 emission (Figure 1). There were no significant differences in the CH_4 traits related to intakes of DM, digestible OM or GE between the two dietary treatments (Table 3). However, the CH_4 yield per unit of dNDFI was lower ($p < 0.01$) in the L cows than in the C cows. Methane emission intensities did not differ between the diets at both the group and individual levels. The Y_m and DMI were 1.1-fold higher at the individual level than the group level. The methane emission intensity per unit of ECM was 1.3-fold higher at the individual level than at the group level, whereas the CH_4 production per LU was similar.

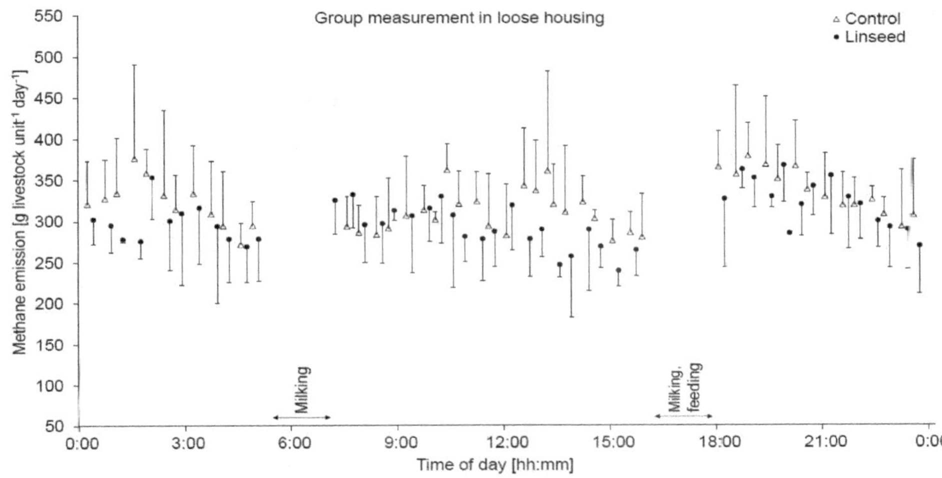

(a)

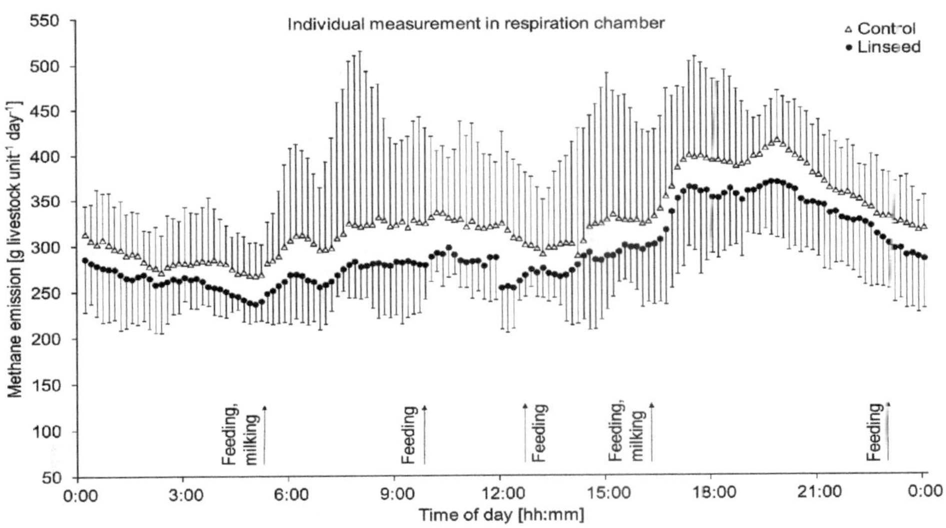

(b)

Figure 1. Diurnal pattern of methane emission per livestock unit (i.e., adjusted to 500 kg of body weight) of lactating dairy cows fed either a control or a linseed-supplemented diet measured (**a**) at the group level (mean of 4-day measurements in a group of 20 animals per dietary treatment) and (**b**) at the individual animal level (mean of 2-day measurements of six animals per diet). The whiskers in the graph extend to minimum (linseed) and maximum (control). The arrows indicate feeding and milking times.

3.3. Milk Composition (Group and Individual Level)

The milk fat and protein contents were lower ($p < 0.01$) in L cows than in the C cows both at the group (bulk milk) and individual levels (Table 4). At the group level, L cows exhibited a higher ($p < 0.01$) milk lactose content than C cows. The milk fat content of the cows selected for the tied stall system was on average slightly higher than that of all cows in the loose-housing system, but the protein and lactose contents were similar at both the individual and group levels. Enzymatic analysis revealed that MUN was significantly higher in the L cows ($p < 0.05$) than in the C cows at both the individual and group levels. The MUN analyzed by FTIR did not reveal these differences. Independent of the type of analysis, the MUN was slightly greater in the cows kept in the tied stall system than in all cows in the loose-housing system.

Table 4. Milk composition of cows from the control and linseed treatment groups measured in the loose[1]- and tied[2]-housing phases (means ± standard deviation).

Housing	Loose (Group) [1]			Tied (Individual) [2]		
Item	**Control**	**Linseed**	***p*–Value**	**Control**	**Linseed**	***p*–Value**
Fat ($g{\cdot}kg^{-1}$ milk)	42.9 ± 2.5	34.3 ± 2.6	0.003	45.6 ± 6.2	36.5 ± 3.1	0.009
Protein ($g{\cdot}kg^{-1}$ milk)	37.7 ± 0.4	33.7 ± 0.3	<0.001	38.8 ± 3.5	33.1 ± 1.8	0.006
Lactose ($g{\cdot}kg^{-1}$ milk)	47.6 ± 0.4	49.1 ± 0.2	<0.001	48.1 ± 0.9	48.7 ± 0.9	0.248
MUN ($mg{\cdot}100^{-1}$ mL milk)						
Enzymatic	10.9 ± 0.5	13.5 ± 2.0	0.041	12.1 ± 2.0	15.6 ± 2.4	0.022
FTIR[3]	10.8 ± 1.0	10.5 ± 1.7	0.777	12.0 ± 2.1	12.8 ± 2.0	0.521
Fatty acids (FA; g 100 g^{-1} of total FA)						
C18:0	7.24 ± 0.07	8.89 ± 0.19	<0.001	8.99 ± 1.64	11.02 ± 1.19	0.034
C18:1 *trans*-6, *trans*-8	0.25 ± 0.01	0.56 ± 0.05	<0.001	0.34 ± 0.05	0.67 ± 0.11	<0.001
C18:1 *trans*-9	0.19 ± 0.01	0.44 ± 0.03	<0.001	0.18 ± 0.02	0.44 ± 0.06	<0.001
C18:1 *trans*-10	0.34 ± 0.00	0.75 ± 0.02	<0.001	0.42 ± 0.06	1.05 ± 0.55	0.018
C18:1 *trans*-11	0.67 ± 0.01	4.39 ± 0.27	<0.001	0.80 ± 0.29	4.37 ± 1.37	<0.001
C18:1 *trans*-12	0.045 ± 0.004	0.105 ± 0.004	<0.001	0.043 ± 0.013	0.097 ± 0.044	0.016
C18:1 *trans*-13, *trans*-14, cis-6, cis-8	0.31 ± 0.00	0.72 ± 0.04	<0.001	0.36 ± 0.04	0.83 ± 0.06	<0.001
C18:1 cis-9	16.6 ± 0.6	19.4 ± 0.4	<0.001	17.4 ± 2.1	24.0 ± 3.3	0.002
C18:1 *cis*-10	0.105 ± 0.025	0.263 ± 0.146	0.076	0.078 ± 0.017	0.137 ± 0.037	0.005
C18:1 *cis*-11	0.67 ± 0.03	1.07 ± 0.08	<0.001	0.74 ± 0.13	1.28 ± 0.12	<0.001
C18:1 *cis*-12	0.208 ± 0.010	0.514 ± 0.028	<0.001	0.259 ± 0.035	0.490 ± 0.072	<0.001
C18:1 *cis*-13	0.077 ± 0.002	0.185 ± 0.019	<0.001	0.092 ± 0.025	0.212 ± 0.019	<0.001
C18:1 *cis*-14, *trans*-16	0.17 ± 0.01	0.33 ± 0.04	<0.001	0.23 ± 0.03	0.54 ± 0.16	<0.001
C18:2 n-6 *trans*	0.11 ± 0.01	0.54 ± 0.07	<0.001	0.12 ± 0.03	0.51 ± 0.16	<0.001
C18:2 *cis*-9, *trans*-13, *trans*-8, *cis*-12	0.22 ± 0.02	0.81 ± 0.08	<0.001	0.26 ± 0.04	1.21 ± 0.30	<0.001
C18:2 *cis*-9, *trans*-12	0.068 ± 0.013	0.178 ± 0.010	<0.001	0.079 ± 0.010	0.257 ± 0.077	<0.001
C18:2 *trans*-11, *cis*-15, *trans*-9, *cis*-12	0.080 ± 0.013	1.282 ± 0.109	<0.001	0.177 ± 0.075	1.482 ± 0.349	<0.001
C18:2 n-6 *cis*	1.95 ± 0.05	1.93 ± 0.05	0.599	2.05 ± 0.23	1.92 ± 0.14	0.278
C18:2 *cis*-9, *cis*-15	0.11 ± 0.01	0.18 ± 0.01	<0.001	0.14 ± 0.02	0.20 ± 0.20	<0.001
C18:2 *cis*-9, *trans*-11	0.21 ± 0.02	0.32 ± 0.07	0.022	0.33 ± 0.07	0.36 ± 0.09	0.523
C18:2 *cis*-9, *cis*-11	0.054 ± 0.003	0.241 ± 0.016	<0.001	0.064 ± 0.014	0.250 ± 0.047	<0.001
C18:2 *trans*-9, *trans*-11	0.015 ± 0.001	0.148 ± 0.009	<0.001	0.015 ± 0.007	0.158 ± 0.046	<0.001
C18:3 n-6	0.062 ± 0.004	0.087 ± 0.012	0.008	0.065 ± 0.017	0.048 ± 0.022	0.181
C18:3 n-3	0.45 ± 0.01	1.10 ± 0.05	<0.001	0.49 ± 0.09	1.23 ± 0.06	<0.001
Σ C18 FA	30.2 ± 0.6	44.4 ± 0.8	<0.001	33.7 ± 4.0	52.8 ± 4.9	<0.001
Σ SFA	72.2 ± 0.7	60.4 ± 0.7	<0.001	70.8 ± 3.3	54.5 ± 4.7	<0.001
Σ MUFA	24.0 ± 0.6	32.4 ± 0.7	<0.001	25.0 ± 2.7	37.6 ± 4.1	<0.001
Σ PUFA	3.80 ± 0.11	7.24 ± 0.24	<0.001	4.26 ± 0.56	7.97 ± 0.79	<0.001
Σ n-6 FA	2.42 ± 0.06	2.79 ± 0.08	<0.001	2.55 ± 0.30	2.68 ± 0.14	0.368
Σ n-3 FA	0.57 ± 0.01	1.24 ± 0.04	<0.001	0.60 ± 0.11	1.35 ± 0.06	<0.001
n-6:n-3 FA	4.28 ± 0.08	2.25 ± 0.13	<0.001	4.29 ± 0.34	1.99 ± 0.07	<0.001

[1] Group observations (two treatments × 4 days) in the loose-housing system. [2] Individual observations (two treatments × 6 animals) in the tied-housing system. [3] Fourier-transform infrared spectroscopy.

The milk FA composition, particularly the proportions of individual C18 FA, was substantially influenced by linseed supplementation to the diet, and this was found at both the group and individual levels. The total C18 FA proportion was higher in the L cows than in the C cows ($p < 0.001$). In particular, the C18:3 n-3 (α-linolenic acid, ALA) proportion was higher in the L cows when compared with the C cows ($p < 0.001$). The total n-3 FA proportion was higher ($p < 0.001$) and the n-6 to n-3 FA ratio was lower ($p < 0.001$) in the milk fat of L cows than that of the C cows. The milk fat of L cows had higher (mostly $p < 0.001$) proportions of the major biohydrogenation intermediates (C18:1 *trans*-11 and C18:2 *cis*-9, trans-11) and terminal products (C18:0 and C18:1 cis-9) than that of C cows. In the milk of the L than the C cows, the proportion of saturated fatty acids (SFAs) was lower and those of monounsaturated fatty acids (MUFAs) and polyunsaturated fatty acids (PUFAs) were higher ($p < 0.001$). In general, the effects of linseed feeding on the FA profile were recovered at both the group and individual levels.

4. Discussion

The main purpose of the present study was to investigate the extent to which a known methane-mitigating feed supplement can reduce methane emissions at both the individual level in respiration chambers and at the group level in naturally ventilated dairy housing.

4.1. Effects of Linseed on the Intake Pattern and Milk Yield in the Loose- and Tied-Housing Systems

In the present study, extruded linseed did not affect feed intake. The cows investigated at the individual level ate more, but this was because individual cows were the exclusively heavier multiparous cows, whereas at the group level, primiparous cows were also included. The addition of extruded linseed products in an iso-energetic manner resulted in a higher dietary content and intake of fiber. This extra fiber might explain the observed longer eating times and higher number of chews per DMI in L cows in the loose housing. In the tied-housing system, linseed also increased chews per DMI; in addition, the rumination time increased, indicating that linseed affected feeding behavior in both the loose- and tied-housing systems. Overall, rumination times were shorter in the tied-housing system than in the loose-housing system. Tied cows, which were also subjected to various experimental procedures, might have expressed less natural behavior than loose-housed cows [14,15]. The numerical increase observed in ECM in the linseed-fed cows in the loose- and tied-housing systems is in contrast with the significant increase reported by Martin et al. [6]. Despite a higher DMI, the overall ECM yield of the cows in the tied-housing system was lower than that of the cows in the loose-housing system, which was probably the result of progressing lactation, repeated transport and change in the environment (tied-housing and respiration chambers).

4.2. Effects of Linseed on Digestibility as well as Nitrogen and Energy Utilization (Individual Level)

At the moderate level of linseed supplementation studied, we expected only marginal effects on digestibility and utilization. We assumed that the high fiber content or the part of the lipids that remained unprotected in the linseed product could adversely affect ruminal nutrient degradation and microbial protein synthesis. In line with previous reports by Martin et al. [6] and Focant et al. [11], using 36 and 26 g linseed lipids·kg^{-1} feed DM, respectively, extruded linseed supplemented at 29 g lipids·kg^{-1} feed DM in the present study did not reduce the digestibility of fiber and organic matter. In another study, it was found that even feeding 20 to 40 g pure linseed oil·kg^{-1} DM did not affect digestibility in dairy cows [12]. Consistent with this, energy and protein retention were not affected either. Thus far, the energy and N balance of cattle fed extruded linseed have been reported for growing cattle, supplemented with only 16 g of additional linseed lipids per kg diet [7], reporting also no effect on energy and N utilization. The observed tendency towards a greater proportion of urinary N in the total manure N for the L cows than the C cows is in accordance with findings by Focant et al. [11] who reported higher urinary N proportion excreted by lactating Holstein cows fed similar linseed levels (26 g linseed lipids·kg^{-1} feed DM). The tendency to lose less energy via CH_4 in L cows than in C cows did not lead to greater energy retention, but in the latter variable, the variation between cows was very high. The results of the present study indicate that the linseed product was indeed iso-energetic to corn flour in the present experiment.

4.3. Effects of Linseed on Methane Emission at Group and Individual Level

In our study, we supplemented a grass silage based diet with a moderate level of 29 g linseed lipids per kg diet at a forage-to-concentrate ratio of 63:37. This lipid supplementation did not cause more than a trend for a CH_4 mitigation effect, although clear effects could have been expected based on the results reported in earlier studies. For example, Martin et al. [6]), Engelke et al. [10], and Focant et al. [11] supplemented extruded linseed at a similar moderate level (26–36 g lipids·kg^{-1} diet DM) and found CH_4 suppression by approximately 10–20%. This difference in response to extruded linseed could have resulted from the different forage type (corn vs. grass silage) and the forage-to-concentrate ratio used in the experiments. Accordingly, the effect of moderate levels of linseed on CH_4 were weaker in the studies of Engelke et al. [10] and Martin et al. [6], when exchanging corn silage with grass silage and when reducing the concentrate proportion from 50% to 40%, respectively. This is consistent with the findings of Machmüller et al. [35], who found that using a lower forage-to-concentrate proportion led to lower methane-mitigating effects of medium-chain FA. The ruminal pH is higher in forage-based diets, which may weaken or even prevent the CH_4-suppressing effect of linseed lipids [36]. In our study, linseed reduced the CH_4 yield per unit of dNDFI, which could have been simply the consequence of the higher NDF intake with the linseed diet, assuming that this extra fiber from the linseed hulls and bran was not easily digestible. However, the NDF digestibility seemed to increase in linseed-fed cows, possibly due to the longer rumination time, in response to higher levels of dietary fiber [37]. When calculating correlations between the variables involved (Table 5), no relationship was found between daily rumination time and daily CH_4 production as in Watt et al. [38]. It seems that rumination has no independent direct effect on CH_4 production, although they are both influenced by the same factors, such as intake and fiber level [37–39]. Accordingly, the dry matter intake of cows was positively correlated with rumination time and ECM in the tied-housing phase. Conversely, no relationship was found between these parameters for the cows in the loose-housing phase, most likely because individual feed intake had to be calculated from group level measurements which leveled out individual differences among cows. It is noteworthy that the multiparous cows in the loose-housing phase spent more time ruminating when eating longer, but the rumination time of the same cows in the tied housing was not related to eating time. As mentioned in Section 4.1, tied cows, which were also subjected to various experimental procedures, might have expressed less natural behavior than loose-housed cows [14,15]. In addition, the observations might be explained by a compensatory relationship between eating and ruminating time [39].

Table 5. Correlation coefficients between performance, ruminating behavior, and methane emission of cows determined in the loose- and tied-housing phases (entire sampling period).

Trait Housing System	ECM [1] ($kg \cdot day^{-1}$)	Methane [2] ($g \cdot day^{-1}$)	Rumination [1] ($min \cdot day^{-1}$)	Eating [1] ($min \cdot day^{-1}$)
DMI ($kg \cdot day^{-1}$) [2]				
Loose	0.494	−0.538	−0.211	−0.107
Tied	0.777 **	0.156	0.610*	−0.156
ECM ($kg \cdot day^{-1}$) [1]				
Loose		−0.387	0.143	0.397
Tied		0.151	0.466	0.156
Methane ($g \cdot day^{-1}$) [2]				
Loose			−0.322	−0.184
Tied			0.198	0.254
Rumination ($min \cdot day^{-1}$) [1]				
Loose				0.828*
Tied				−0.296

** $p < 0.01$; * $p < 0.05$. [1] Individual observations in the loose (two treatments × 20 animals) and tied (two treatments × six animals) housing system. [2] Group observations (two treatments × 4 days) in the loose housing and individual observations (two treatments × six animals) in the tied-housing system.

4.4. Effects of Linseed Supplementation on Milk Composition at the Group and Individual Levels

As is known from other dietary oils, feeding cows with linseed oil in the form of extruded linseed supplementation at sufficiently high levels may decrease the milk fat and protein contents in dairy milk (e.g., Martin et al. [5]). This is caused by adverse effects on fiber digestibility mediated by the lower production of acetate, which is the major precursor of milk fat and microbial protein. Accordingly, in loose- and tied-housing systems, the milk fat and protein contents were suppressed by supplementing the cows' diets with extruded linseed. Feeding cows extruded linseed increased the enzymatically determined MUN contents at both the group and individual levels, which was probably due to the higher crude protein content of the extruded linseed-containing diet. In the study by van Zijderveld et al. [40], a dietary mixture with extruded linseed containing less crude protein was used, which led to a slight decrease in the MUN content (estimated with a pH difference technique). Our data also show that the FTIR analysis did not result in sufficiently accurate MUN values to differentiate between dietary treatments. In this context, it is puzzling that the variation among individual animals was not higher than in the values determined enzymatically. It could be speculated that there were matrix effects that affected the FTIR measurements, such as changes in the milk fat composition of linseed-fed cows.

Linseed oil is particularly rich in the dietetically valuable C18:3 n-3. However, when fed as pure oil, a large proportion of the C18:3 n-3 is biohydrogenated by the ruminal microorganisms. This is why linseed is often provided in crushed or extruded form, as the oil in this form is at least partially rumen-protected but still able to exert some inhibiting effects on ruminal microorganisms [41]. Our results showed that part of the C18:3 n-3 was transferred intact to the milk, as its proportion was substantially elevated by linseed supplementation. However, ruminal biohydrogenation also occurred, as evidenced by elevated proportions of major biohydrogenation products such as C18:1 *trans*-11 (vaccenic acid) and C18:2 *cis*-9, trans-11 (rumenic acid; the most important conjugated linoleic acid), and C18:0 [42]. The two intermediates mentioned, but not the terminal product C18:0, are considered valuable to human health. Linseed has been reported to enhance both the n-3 FA and biohydrogenation intermediates [8,10,11]. The effects of linseed on the FA profile of milk fat in our study were similar at both the individual and group levels (bulk milk), indicating that changes can be clearly recovered in a few bulk milk samples obtained from linseed-fed herds of dairy cattle. Therefore, this method may be implemented as a control instrument by retailers who pay higher prices for milk produced from cows fed linseed.

4.5. Comparability of Methane Emission Measurements in an Experimental Housing System and Individually in Respiration Chambers

The methane-mitigation efficacy of extruded linseed was limited, both at the group and at individual levels. Systematic differences in CH_4 emissions between measurements at the group and individual levels are likely at least partly attributable to indirect effects, as the loose-housed cows were exposed to transport, being tied in stalls, respiration chambers, and further experimental procedures. In addition, the subpopulation of cows that underwent measurements in respiration chambers were exclusively multiparous cows. A system-dependent difference was that CH_4 emissions in the experimental dairy housing at the group level included housing-based sources, such as the floor's soiling. Integrated animal and housing emissions are highly relevant from an environmental point of view [16]; however, they may mask differences between feeding treatments. In particular, Hassanat et al. [43] recently showed that CH_4 emissions from the manure of linseed-fed cows were higher than those from the manure of cows fed a control diet devoid of linseed. In addition, certain differences in linseed intake between different group members cannot be ruled out, although this should have been minimized by fixing the cows for a certain period after the linseed supplement was provided. Simultaneous measurements performed by Schmithausen et al. [16] in a naturally ventilated experimental housing in two separate sections, similar to the setting of the present experiment, also revealed slight numerical reductions in the CH_4 emissions in lactating cows fed a CH_4-mitigating supplement (30 g of condensed tannins·kg^{-1} diet DM; effect reviewed in Beauchemin et al. [4]). This report illustrated that it may be possible to detect the CH_4 reduction potential of a feed supplement in a group-level assessment. However, this is analytically more challenging, the experimental design is more complex and detecting differences in CH_4 emissions at the group level may require a higher level of supplementation than at the individual level. In contrast to CH_4 emissions, CO_2 emissions were comparable at the group and individual levels. Unlike CH_4, the majority of CO_2 originates from nutrient oxidation in the animal's metabolism, which is less variable than ruminal fermentation. The good comparability of CO_2 emissions indicates that both concepts, i.e., respiration chambers and the tracer ratio technique, provide reliable results (if source strengths are comparable), whereas CH_4 emissions at the group and individual levels might differ to some extent.

5. Conclusions

This is the first study to assess the CH_4-mitigating effect of a feeding strategy evaluating measurements at the individual level in respiration chambers and measurements at the group level conducted in a naturally ventilated dairy loose-housing system. The moderate level of extruded linseed chosen only resulted in a slightly lower diurnal course of CH_4 emission intensity than the control. A significant emission reduction was solely observed for CH_4 per unit NDF digested. Additionally, the favorable effect of linseed on milk FA composition at both the group and individual levels suggests that no individual feed allocation and recording is necessary to control the use of linseed in diet. Overall, the results indicate that group-level measurements are more challenging with regard to the experimental conditions in the loose-housing system, but are likely suitable to detect the CH_4 reduction potential of feed supplements provided at higher concentrations. However, appropriate analytical approaches and experimental designs are needed. Due to the higher number of influencing factors, practical measurements are relevant for estimating the effective mitigation potential in a complex barn-manure storage and removal environment.

Author Contributions: Conceptualization, S.S., J.M., J.O.Z., and A.S.; methodology, J.P., S.S., K.Z., J.M., M.Z. and A.S.; formal analysis, J.P., A.S., M.Z.; investigation, J.P., K.Z. and M.Z..; resources, S.S., J.M. and M.K.; writing—original draft preparation, J.P.; writing—review and editing, S.S., J.M, M.Z., J.O.Z., M.K. and A.S.; visualization, J.P.; supervision, S.S., M.K. and A.S.; project administration, S.S.; funding acquisition, S.S., J.M. and A.S. All authors have read and agreed to the published version of the manuscript.

Acknowledgments: We are grateful to the expert advice, assistance, and help of numerous persons involved in the experiment from Agroscope, ETH Zurich and Strickhof (C. Bühler, F. Ferrari, F. Grandl, M. Hunziker, S. Ineichen, P. Jenni, G. Jöhl, M. Keck, M. Keller, C. Kunz, B. Kürsteiner, H. Lüthi-Wettstein, S. Mathis, M. Mergani, S. Meese, D. Näf, H.R. Ott, M. Schlatter, P. Stirnemann, R. Stoz, M. Terranova, S. Wang, and M. Wymann). We gratefully acknowledge the provision of extruded linseed by Trinova AG (Wangen, Switzerland).

Appendix A

Table A1. Nitrogen balance of cows from the control and linseed treatment groups determined individually in the tied-housing phase (six cows per treatment; means ± standard deviation).

Item	Control	Linseed	*p*-Value
N balance ($g \cdot day^{-1}$)			
Intake	526 ± 49	552 ± 59	0.419
Feces	176 ± 26	167 ± 26	0.561
Urine	167 ± 67	214 ± 68	0.250
Milk	125 ± 33	156 ± 53	0.246
Body	58 ± 66	15 ± 59	0.255
N balance ($g \cdot kg^{-1}$)			
Feces	334 ± 29	303 ± 45	0.187
Urine	312 ± 97	383 ± 89	0.211
Milk	235 ± 45	277 ± 65	0.222
Body	119 ± 122	36 ± 105	0.234
N utilization			
Apparent N digestibility (%)	64.6 ± 11.2	68.7 ± 7.2	0.468
N intake ($g \cdot kg^{-1}$ ECM [1])	20.5 ± 4.1	19.1 ± 4.0	0.551
Feces N ($g \cdot kg^{-1}$ ECM)	6.9 ± 1.9	5.8 ± 1.5	0.266
Manure N traits			
Manure N ($g \cdot kg^{-1}$ of N intake)	250 ± 16	232 ± 26	0.174
Urinary N ($g \cdot kg^{-1}$ of manure N)	473 ± 73	554 ± 78	0.094

[1] Energy-corrected milk.

Table A2. Energy balance of cows from control and linseed treatments determined individually in the tied-housing phase (six cows per treatment; means ± standard deviation).

Item	Control	Linseed	*p*-Value
Energy intake (MJ/day)			
Gross energy (GE)	357.8 ± 30.8	361.9 ± 37.6	0.840
Digestible energy	273.5 ± 14.3	276.5 ± 33.8	0.843
Metabolizable energy (ME)	236.5 ± 14.8	241.5 ± 34.3	0.750
Energy loss (MJ/day)			
Feces	84.3 ± 17.1	85.3 ± 13.3	0.906
Urine	9.9 ± 0.9	10.4 ± 3.1	0.717
Methane	27.1 ± 2.2	24.7 ± 1.9	0.063
Heat	119.1 ± 10.7	118.2 ± 6.2	0.858
Total energy loss	240.4 ± 25.5	238.6 ± 16.7	0.887
Energy retention (MJ/day)			
Milk	83.1 ± 17.7	95.7 ± 29.2	0.386
Body	34.3 ± 12.8	27.6 ± 17.1	0.458
Energy turnover ($kJ \cdot MJ^{-1}$ of GE intake)			
Feces	234 ± 26	236 ± 34	0.885
Urine	28 ± 4	29 ± 10	0.750
Methane	76 ± 10	69 ± 6	0.131
Heat	333 ± 17	328 ± 24	0.700
Total energy loss	671 ± 17	663 ± 52	0.712
Milk	231 ± 40	260 ± 54	0.313
Body	98 ± 41	77 ± 47	0.440
Heat ($kJ \cdot MJ^{-1}$ of ME intake)	503 ± 24	495 ± 54	0.743
Energy utilization (%)			
Apparent digestibility	76.6 ± 2.6	76.4 ± 3.4	0.885
Metabolizability (q)	66.2 ± 1.6	66.6 ± 3.7	0.822
k_L [1] (estimated from q)	62.2 ± 0.4	62.3 ± 0.9	0.882
k_L [1] (direct calculation)	57.0 ± 2.6	58.1 ± 2.4	0.493
Supply over estimated requirements			
Net energy for lactation (MJ)	21.8 ± 12.6	12.5 ± 16.6	0.299
% deviation from requirements	18.8 ± 14.1	10.7 ± 13.6	0.335
ME (MJ)	32.5 ± 19.9	17.3 ± 26.0	0.282
% deviation from requirements	17.3 ± 13.5	9.3 ± 13.0	0.327

[1] Proportionate utilization of ME for milk energy formation.

References

1. Hristov, A.N.; Hanigan, M.; Cole, A.; Todd, R.; McAllister, T.A.; Ndegwa, P.M.; Rotz, A. Ammonia emissions from dairy farms and beef feedlots. *Can. J. Anim. Sci.* **2011**, *91*, 1–35. [CrossRef]
2. Knapp, J.R.; Laur, G.L.; Vadas, P.A.; Weiss, W.P.; Tricarico, J.M. Enteric methane in dairy cattle production: Quantifying the opportunities and impact of reducing emissions. *J. Dairy Sci.* **2014**, *97*, 3231–3261. [CrossRef] [PubMed]
3. Cottle, D.J.; Nolan, J.V.; Wiedemann, S.G. Ruminant enteric methane mitigation: A review. *Anim. Prod. Sci.* **2011**, *51*, 491–514. [CrossRef]
4. Beauchemin, K.A.; Kreuzer, M.; O'Mara, F.; McAllister, T.A. Nutritional management for enteric methane abatement: A review. *Aust. J. Exp. Agr.* **2008**, *48*, 21–27. [CrossRef]
5. Martin, C.; Rouel, J.; Jouany, J.P.; Doreau, M.; Chilliard, Y. Methane output and diet digestibility in response to feeding dairy cows crude linseed, extruded linseed, or linseed oil. *J. Anim. Sci.* **2008**, *86*, 2642–2650. [CrossRef] [PubMed]
6. Martin, C.; Ferlay, A.; Mosoni, P.; Rochette, Y.; Chilliard, Y.; Doreau, M. Increasing linseed supply in dairy cow diets based on hay or corn silage: Effect on enteric methane emission, rumen microbial fermentation, and digestion. *J. Dairy Sci.* **2016**, *99*, 3445–3456. [CrossRef]

7. Hammond, K.J.; Humphries, D.J.; Crompton, L.A.; Kirton, P.; Reynolds, C.K. Effects of forage source and extruded linseed supplementation on methane emissions from growing dairy cattle of differing body weights. *J. Dairy Sci.* **2015**, *98*, 8066–8077. [CrossRef]
8. Livingstone, K.M.; Humphries, D.J.; Kirton, P.; Kliem, K.E.; Givens, D.I.; Reynolds, C.K. Effects of forage type and extruded linseed supplementation on methane production and milk fatty acid composition of lactating dairy cows. *J. Dairy Sci.* **2015**, *98*, 4000–4011. [CrossRef]
9. Judy, J.V.; Bachman, G.C.; Brown-Brandl, T.M.; Fernando, S.C.; Hales, K.E.; Harvatine, K.J.; Miller, P.S.; Kononoff, P.J. Increasing the concentration of linolenic acid in diets fed to Jersey cows in late lactation does not affect methane production. *J. Dairy Sci.* **2019**, *102*, 2085–2093. [CrossRef]
10. Engelke, S.W.; Daş, G.; Derno, M.; Tuchscherer, A.; Wimmers, K.; Rychlik, M.; Kienberger, H.; Berg, W.; Kuhla, B.; Metges, C.C. Methane prediction based on individual or groups of milk fatty acids for dairy cows fed rations with or without linseed. *J. Dairy. Sci.* **2019**, *102*, 1788–1802. [CrossRef]
11. Focant, M.; Froidmont, E.; Archambeau, Q.; Dang Van, Q.C.; Larondelle, Y. The effect of oak tannin (*Quercus robur*) and hops (*Humulus lupulus*) on dietary nitrogen efficiency, methane emission, and milk fatty acid composition of dairy cows fed a low-protein diet including linseed. *J. Dairy Sci.* **2019**, *102*, 1144–1159. [CrossRef] [PubMed]
12. Benchaar, C.; Romero-Pérez, G.A.; Chouinard, P.Y.; Hassanat, F.; Eugene, M.; Petit, H.V.; Côrtes, C. Supplementation of increasing amounts of linseed oil to dairy cows fed total mixed rations: Effects on digestion, ruminal fermentation characteristics, protozoal populations, and milk fatty acid composition. *J. Dairy. Sci.* **2012**, *95*, 4578–4590. [CrossRef] [PubMed]
13. Troy, S.H.; Duthie, C.-A.; Ross, D.W.; Hyslop, J.J.; Roehe, R.; Waterhouse, A.; Rooke, J.A. A comparison of methane emissions from beef cattle measured using methane hoods with those measured using respiration chambers. *Anim. Feed. Sci. Technol.* **2016**, *211*, 227–240. [CrossRef]
14. Storm, I.M.; Hellwing, A.L.F.; Nielsen, N.I.; Madsen, J. Methods for measuring and estimating methane emission from ruminants. *Animals* **2012**, *2*, 160–183. [CrossRef] [PubMed]
15. Huhtanen, P.; Cabezas-Garcia, E.H.; Utsumi, S.; Zimmerman, S. Comparison of methods to determine methane emissions from dairy cows in farm conditions. *J. Dairy Sci.* **2015**, *98*, 3394–3409. [CrossRef]
16. Schmithausen, A.J.; Schiefler, I.; Trimborn, M.; Gerlach, K.; Südekum, K.-H.; Pries, M.; Büscher, W. Quantification of methane and ammonia emissions in a naturally ventilated barn by using defined criteria to calculate emission rates. *Animals* **2018**, *8*, 75. [CrossRef]
17. Grainger, C.; Clarke, T.; McGinn, S.M.; Auldist, M.J.; Beauchemin, K.A.; Hannah, M.C.; Waghorn, G.C.; Clark, H.; Eckard, R.J. Methane emissions from dairy cows measured using the sulfur hexafluoride (SF_6) tracer and chamber techniques. *J. Dairy Sci.* **2007**, *90*, 2755–2766. [CrossRef]
18. Hammond, K.J.; Humphries, D.J.; Crompton, L.A.; Green, C.; Reynolds, C.K. Methane emissions from cattle: Estimates from short-term measurements using a GreenFeed system compared with measurements obtained using respiration chambers or sulphur hexafluoride tracer. *Anim. Feed. Sci. Technol.* **2015**, *203*, 41–52. [CrossRef]
19. Agroscope. Feeding Recommendations and Nutrient Tables for Ruminants. 2020. Available online: https://www.db-agroscope.ch/member/fmdb/7Fuumltterungsempfehlungenfuum_113.pdf (accessed on 11 January 2020). (In German).
20. Mohn, J.; Zeyer, K.; Keck, M.; Keller, M.; Zähner, M.; Poteko, J.; Emmenegger, L.; Schrade, S. A dual tracer ratio method for comparative emission measurements in an experimental dairy housing. *Atmos. Environ.* **2018**, *179*, 12–22. [CrossRef]
21. Schrade, S.; Zeyer, K.; Gygax, L.; Emmenegger, L.; Hartung, E.; Keck, M. Ammonia emissions and emission factors of naturally ventilated dairy housing with solid floors and an outdoor exercise area in Switzerland. *Atmos. Environ.* **2012**, *47*, 183–194. [CrossRef]
22. Buehler, K.; Wanner, M. Chapter 6: Metabolic Centre of the University of Zurich and ETH Zurich (under construction). In *Technical Manual on Respiration Chamber Designs*; Leahy, S., Parlane, K., Eds.; Ministry of Agriculture and Forestry: Wellington, New Zealand, 2018; pp. 89–106. Available online: https://globalresearchalliance.org/wp-content/uploads/2018/02/LRG-Manual-Facility-BestPract-Sept-2018.pdf (accessed on 23 April 2020).
23. Werner, J.; Leso, L.; Umstatter, C.; Niederhauser, J.; Kennedy, E.; Geoghegan, A.; Shalloo, L.; Schick, M.; O'Brien, B. Evaluation of the RumiWatchSystem for measuring grazing behaviour of cows. *J. Neurosci. Meth.* **2018**, *300*, 138–146. [CrossRef] [PubMed]

24. AOAC. *Official Methods of Analysis*; Association of Official Analytical Chemists: Arlington, VA, USA, 1995.
25. Van Soest, P.V.; Robertson, J.B.; Lewis, B.A. Methods for dietary fiber, neutral detergent fiber, and nonstarch polysaccharides in relation to animal nutrition. *J. Dairy Sci.* **1991**, *74*, 3583–3597. [CrossRef]
26. Grandl, F.; Zeitz, J.O.; Clauss, M.; Furger, M.; Kreuzer, M.; Schwarm, A. Evidence for increasing digestive and metabolic efficiency of energy utilization with age of dairy cattle as determined in two feeding regimes. *Animal* **2018**, *12*, 515–527. [CrossRef] [PubMed]
27. ISO (International Organization for Standardization). *ISO 14637: Milk—Determination of Urea Content-Enzymatic Method Using Difference in pH (Reference Method)*; ISO: Geneva, Switzerland, 2004.
28. IUPAC (International Union of Pure and Applied Chemistry). *Standard Methods for the Analysis of Oils, Fats and Derivates*, 7th ed.; Blackwell Scientific Publications: Oxford, UK, 1987.
29. Wettstein, H.-R.; Scheeder, M.R.L.; Sutter, F.; Kreuzer, M. Effect of lecithins partly replacing rumen-protected fat on fatty acid digestion and composition of cow milk. *Eur. J. Lipid. Sci. Tech.* **2001**, *103*, 12–22. [CrossRef]
30. Suter, B.; Grob, K.; Pacciarelli, B. Determination of fat content and fatty acid composition through 1-min transesterification in the food sample; principles. *Z. Lebensmitteluntersuchungs- Und -Forschung A* **1997**, *204*, 252–258. [CrossRef]
31. Collomb, M.; Bühler, T. Analysis of the fatty acid composition of the milk fat. I. Optimisation and validation of a general method adapted to the research. *Mitt. Lebensm. Hyg.* **2000**, *91*, 306–332.
32. KTBL (Kuratorium für Technik und Bauwesen in der Landwirtschaft), 2019. Grossvieheinheitenrechner 2.1. Available online: http://daten.ktbl.de/gvrechner/gvHome.do;jsessionid=C389DE28C48F29CD7A3EEA53225A478E#start (accessed on 23 April 2020).
33. Kinsman, R.; Sauer, F.D.; Jackson, H.A.; Wolynetz, M.S. Methane and Carbon Dioxide Emissions from Dairy Cows in Full Lactation Monitored over a Six-Month Period. *J. Dairy Sci.* **1995**, *78*, 2760–2766. [CrossRef]
34. Sauer, F.D.; Fellner, V.; Kinsman, R.; Kramer, J.K.G.; Jackson, H.A.; Lee, A.J.; Chen, S. Methane output and lactation response in Holstein cattle with monensin or unsaturated fat added to the diet. *J. Anim. Sci.* **1998**, *76*, 906–914. [CrossRef]
35. Machmüller, A.; Soliva, C.R.; Kreuzer, M. Methane-suppressing effect of myristic acid in sheep as affected by dietary calcium and forage proportion. *Br. J. Nutr.* **2003**, *90*, 529–540. [CrossRef]
36. Zhou, X.; Zeitz, J.O.; Meile, L.; Kreuzer, M.; Schwarm, A. Influence of pH and the degree of protonation on the inhibitory effect of fatty acids in the ruminal methanogen *Methanobrevibacter ruminantium* strain M1. *J. Appl. Microbiol.* **2015**, *119*, 1482–1493. [CrossRef]
37. Zetouni, L.; Difford, G.F.; Lassen, J.; Byskov, M.V.; Norberg, E.; Løvendahl, P. Is rumination time an indicator of methane production in dairy cows? *J. Dairy Sci.* **2018**, *101*, 11074–11085. [CrossRef] [PubMed]
38. Watt, L.J.; Clark, C.E.F.; Krebs, G.L.; Petzel, C.E.; Nielsen, S.; Utsumi, S.A. Differential rumination, intake, and enteric methane production of dairy cows in a pasture-based automatic milking system. *J. Dairy Sci.* **2015**, *98*, 7248–7263. [CrossRef] [PubMed]
39. Beauchemin, K.A. Current perspectives on eating and rumination activity in dairy cows. *J. Dairy Sci.* **2018**, *101*, 4762–4784. [CrossRef]
40. Van Zijderveld, S.M.; Dijkstra, J.; Perdok, H.B.; Newbold, J.R.; Gerrits, W.J.J. Dietary inclusion of diallyl disulfide, yucca powder, calcium fumarate, an extruded linseed product, or medium-chain fatty acids does not affect methane production in lactating dairy cows. *J. Dairy Sci.* **2011**, *94*, 3094–3104. [CrossRef] [PubMed]
41. Doreau, M.M.; Ferlay, A. Linseed: A valuable feedstuff for ruminants. *Oilseeds. Fats. Crops. Lipids.* **2015**, *22*, 1–9. [CrossRef]
42. Chilliard, Y.; Glasser, F.; Ferlay, A.; Bernard, L.; Rouel, J.; Doreau, M. Diet, rumen biohydrogenation and nutritional quality of cow and goat milk fat. *Eur. J. Lipid. Sci. Tech.* **2007**, *109*, 828–855. [CrossRef]
43. Hassanat, F.; Benchaar, C. Methane emissions of manure from dairy cows fed red clover or corn silage-based diets supplemented with linseed oil. *J. Dairy Sci.* **2019**, *102*, 11766–11776. [CrossRef]

11

Effect of Age and Weaning on Growth Performance, Rumen Fermentation, and Serum Parameters in Lambs Fed Starter with Limited Ewe–Lamb Interaction

Shiqin Wang [1,2,†], Tao Ma [2,†], Guohong Zhao [2], Naifeng Zhang [2], Yan Tu [2], Fadi Li [1], Kai Cui [2], Yanliang Bi [2], Hongbiao Ding [2] and Qiyu Diao [1,2,*]

1 State Key Laboratory of Grassland Agro-Ecosystems, Key Laboratory of Grassland Livestock Industry Innovation, Ministry of Agriculture and Rural Affairs, College of Pastoral Agriculture Science and Technology, Lanzhou University, Lanzhou 730020, China; wshq1988@163.com (S.W.); lifd@lzu.edu.cn (F.L.)

2 Feed Research Institute, Chinese Academy of Agricultural Sciences/Key Laboratory of Feed Biotechnology of the Ministry of Agriculture and Rural Affairs, Beijing 100081, China; matao@caas.cn (T.M.); zhaoguoh@foxmail.com (G.Z.); zhangnaifeng@caas.cn (N.Z.); tuyan@caas.cn (Y.T.); cuikai@caas.cn (K.C.); vetbi2008@163.com (Y.B.); dinghongbiao@caas.cn (H.D.)

* Correspondence: diaoqiyu@caas.cn.

† Shiqin Wang and Tao Ma contribute equally to the article.

Simple Summary: Early weaning is a common practice in the modern lamb industry, which shortens the breeding cycle of ewes and improves the flock productivity by increasing the frequency of lambing. Recently, it was suggested that weaning age (21 or 35 days of age) and milk replacer feeding level had limited effect on apparent digestibility of nutrients in lambs. However, artificial rearing may increase the cost due to the use of milk replacer, labor, and feeding facilities. In this study, we compared the difference in growth performance of lambs weaned at 21 days (early weaning) and 49 days (conventional weaning) of age. All lambs were initially reared with ewes and supplemented with starter at 7 days of age. The results showed that diarrhea rate was increased when lambs were weaned at 21 days of age while average daily gain was decreased. Early weaning reduced average daily gain without compromising lambs' overall immunity. In conclusion, early weaning (21 days of age) may have a negative impact on lambs' performance based on a short-term study.

Abstract: Sixty neonatal Hu lambs were weaned at either 21 (n = 30) (early weaning, EW) or 49 days (n = 30) of age (control, CON). The starter intake and body weight (BW) of lambs was recorded weekly from birth to 63 days of age. Diarrhea rate of lambs was measured from birth to 35 days. Six randomly selected lambs from each treatment were slaughtered at 26, 35, and 63 days of age, respectively. Ruminal pH, NH_3-N, and volatile fatty acid (VFA) concentration, as well as serum parameters including immunity, antioxidant status, and inflammatory parameters from randomly selected lambs from each treatment were measured. There was no difference in BW at birth and day 21 between the two groups of lambs ($p > 0.05$). However, BW of the lambs in the EW group was significantly lower than those in the CON group ($p < 0.01$) from 28 to 49 days of age. Average daily gain (ADG) of the lambs in the EW group was significantly lower than those in the CON group ($p < 0.01$) at three weeks after early weaning. Starter intake of the lambs in the EW group was obviously higher than that in the CON group ($p < 0.01$) from day 28 to 49. In addition, the diarrhea rate was significantly higher than that in the CON group from day 5 to 14 after weaning ($p < 0.01$). The EW group had heavier carcasses ($p < 0.01$) and rumen relative to whole stomach weights ($p < 0.01$). Rumen pH was increased by age ($p < 0.01$) and was not affected by early weaning ($p > 0.05$). Early weaning decreased abomasum relative to whole stomach weight ($p < 0.01$) and increased total VFA concentrations ($p < 0.01$) at day 26. There was no difference in lambs' immunity and stress indicators

($p > 0.05$). The results indicated that lambs weaned at 21 days of age had decreased ADG and higher diarrhea rate, although the overall immunity was not compromised. Long-term study is needed to further validate the feasibility of early weaning strategy in lambs.

Keywords: early weaning; weaning stress; lamb; growth performance; diarrhea

---

1. Introduction

Weaning time is critical to the management of ewes and lambs. Although there is no 'best' age to wean, lambs can be weaned at as early as three-weeks old [1], and early weaning is a common practice in modern lamb industry. Early weaning shortens the breeding cycle of ewes, which improves the flock productivity by increasing the frequency of lambing [2,3]. More importantly, lambs utilize diet more efficiently than ewes because lambs directly convert feed to gain while ewes convert the feed to milk and then to lamb gain. Therefore, an increase in economic profits can be expected if early weaning is successful. However, early weaning can be an important stressor for lambs due to separation from dams [4], as it is suggested that suckling is a major factor in the strength of the ewe–lamb contact [5]. In addition, the transition from liquid (milk/milk replacer) to solid diets (starter or grass) during weaning requires the development of a functional rumen [6]. Consequently, early weaning may cause a drop in feed consumption and decrease in growth rate of lambs [7] if the rumen structure and function has not been fully developed before weaning [8].

To minimize the negative effect of early weaning as well as ensure economic benefits, great efforts have been made to investigate the optimal weaning strategy for lambs. Early study suggested that lambs reared under a mixed system, featured by separating lambs from their dams for 15 h while allowing them to suckle for the remaining 9 h daily, had better financial return than those exclusively suckling their dams and weaned at 30 days after birth [9]. On the other hand, it is reported that sufficient starter intake can stimulate the development of a functional rumen as well as the establishment of rumen microbiota [6]. Our previous study suggested that when starter was offered at day 15 of age, lambs separated from their dams at 20 days of age and artificially fed milk replacer had superior growth performance than their ewe-reared counterparts at the end of the feeding trial (90 days of age) [10]. Recently, it was suggested that weaning age (21 or 35 days of age) and milk replacer feeding level had limited effect on apparent digestibility of nutrients, ruminal microbiota, and fermentation of lambs at 50 days [11]. However, artificial rearing may increase the cost due to the involvement of milk replacer, labor, and feeding facilities [12].

Here, we compared the effect of limited ewe–lamb contact (18 h daily) and supplementation of starter at an early age (seven-days old) on growth performance, rumen fermentation, slaughter performance, and serum parameters of lambs weaned at either 21 or 49 days of age in a 63-day feeding trial. We hypothesized that early weaning (21 days of age) may have temporal negative effect on lambs' performance, but the overall performance of lambs may be similar between two groups.

2. Materials and Methods

The study was conducted from October to December, 2018 at Runlin sheep farm (N 36°82′, E 115°83′) in Linqing county, Liaocheng city, Shandong province, China. The experiment protocol was approved by the Animal Ethics and Humane Animal Care of the Chinese Academy of Agricultural Sciences (with protocol FRI-CAAS-20180810).

A total of 60 neonatal Hu lambs (48 male and 12 female) with similar birth weights (3.82 ± 0.46 kg) were used. Lambs were initially reared with ewes in 6 well-ventilated sheep pens (4 m × 5 m) with controlled temperature and humidity, with 10 lambs (8 male and 2 female) and their ewes in each pen. All ewes were fed two times daily according to the farm's feeding management schedule. None of the lambs in the groups had access to the ewes' feed. Ewes were fed a total mixed ration consisting of 35%

corn silage, 28% peanut straw, 7% garlic straw, 3% soybean residue, 15% corn, 6% wheat bran, and 6% soybean meal.

From day 7 of age, all lambs were separated from the ewes and offered pelleted starter for 6 h daily (08:00–10:00, 12:00–14:00, and 16:00–18:00). At other times, the lambs remained with their dams and still had free access to starter. All lambs were fed starter 1 from 7 to 35 and starter 2 from 36 to 63 days of age, respectively. The ingredient and nutritional composition of starters are listed in Table 1. Half of the lambs from three pens (n = 30) were weaned and abruptly separated from their dams at 21 days of age (early weaning, EW) while the other half of the lambs from another three pens (n = 30) were weaned and abruptly separated from their dams at 49 days of age (control, CON). All lambs and ewes had ad libitum access to water. During the experiment, two lambs from the CON group did not have enough milk intake as their dams were sick, and one lamb from the CON group died accidentally. Therefore, those three lambs were removed from further analysis.

Table 1. Ingredients and chemical composition of the starters.

Item	Starter 1 (7–35 Days)	Starter 2 (36–63 Days)
	Ingredients, air dry basis, %	
Alfalfa hay		7.0
Oat grass		5.0
Corn	50.0	45.5
Soybean meal	23.5	20.0
Corn germ meal	12.0	10.0
Wheat bran	10.0	8.0
Vitamin and mineral mixture [1]	1.0	1.0
Limestone	2.0	2.0
$CaHPO_4$	0.5	0.5
Salt	0.5	0.5
$NaHCO_3$	0.5	0.5
	Chemical composition [2], % of DM	
DM, air dry basis	90.2	89.1
CP	21.5	21.5
NDF	15.1	18.9
ADF	6.4	8.2
Ash	8.0	5.7
Ca	0.9	0.8
P	0.5	0.6
ME, MJ/kg	10.9	10.5

[1] Contained per kilogram of supplement: 800,000 IU of vitamin A, 300,000 IU of vitamin D3, 3000 mg of vitamin E, 4 g of Fe, 4 g of Mn, 0.8 g of Cu, 5 g of Zn, 20 mg of Se, 70 mg of I, 40 mg of Co. [2] DM, CP, NDF, Cam and P were measured values, while ME (metabolizable energy) was calculated according to National Research Council 2007 [13].

The body weight (BW) of each lamb was recorded weekly (before morning feeding) from 7 to 63 days of age. The starter was offered to each pen and the intake of starter was measured by each pen daily before morning feeding. Representative samples of starter were collected weekly and were immediately frozen at −20 °C for further analysis. Diarrhea incidence was also monitored on a daily basis from birth to 35 days of age.

Blood sampled from six male lambs per treatment was collected in heparinized tubes by jugular venipuncture before morning feeding at 26, 35m and 63 days of age, respectively. The blood sample was centrifuged at 3500× *g* for 15 min at 4 °C to separate serum, which was pipetted into 2 mL cryotubes and stored at −20 °C until subsequent analysis. After blood sampling, those six lambs from each group were slaughtered at 26, 35, and 63 days of age, respectively. Ruminal digesta sample was collected and pH value was measured immediately. The rumen digesta was then filtered through four layers of cheesecloth and a 10 mL subsample of each strained fluid was collected, stored at −20 °C for analysis of the volatile fatty acids (VFAs) and ammonia nitrogen (NH_3-N). Immediately after slaughter, the rumen, reticulum, omasum, and abomasum of each lamb was dissected and weighted after the digesta was removed. The carcass weight of each lamb was also recorded.

The starter samples were ground to pass through a 1 mm sieve and dried in an oven at 135 °C for 2 h (method 930.15; AOAC, 1990) [14] to measure the dry matter (DM) content. The ash content, nitrogen, neutral detergent fiber (NDF), acid detergent fiber (ADF), calcium, and total phosphorus was measured according to methods described by previous studies [14–17]. Crude protein (CP) was calculated as 6.25 × Nitrogen.

Feces were scored using a 1 to 4 scale classified as firm and well-formed (score 1), soft and pudding-like (score 2), runny and pancake batter-like (score 3), or liquid and splatters (score 4) as described previously [18]. If an animal presented a fecal score ≥ 3 for 3 consecutive days, it was considered diarrheic. Diarrhea rate (%) = number of lambs of diagnosed at least once for diarrhea/the total number of lambs. Diarrhea frequency (%) = the number of diarrhea lambs × diarrhea days/the total number of lambs × experimental days.

Serum concentration of immunoglobulin (IgG, IgM, and IgA), superoxide dismutase (SOD), Glutathione Peroxidase (GSH-Px), catalase (CAT), total antioxidative capacity (T-AOC), and malondialdehyde (MDA) were determined using the Hitachi 7020 autobiochemistry instrument (Hitachi, Tokyo, Japan) with corresponding commercial test kits (Nanjing Jiancheng Bioengineering Institute, Nanjing, Jiangsu, China) following the manufacturer's instructions. Interleukin 1β (IL-1β), interleukin 4 (IL-4), interleukin 6 (IL-6), cortisol, interferon-gamma (IFN-γ), and tumor necrosis factor α (TNF-α) were determined using bovine ELISA kits (Nanjing Jiancheng Bioengineering Institute, Nanjing, Jiangsu, China) following the manufacturer's instructions.

Individual and Total VFA concentrations were determined as described previously [19]. Briefly, 1 mL rumen fluid filtrate was added to 0.2 mL of metaphosphoric acid solution (250 g/L) containing 2 g/L 2-ethyl butyrate, mixed overnight, and analyzed by gas chromatography (SP-3420, Beijing Analytical Instrument Factory, Beijing, China). The concentration of NH_3-N was measured using phenol hypochlorite colorimetric method as described by [20]

Starter intake, BW, and average daily gain (ADG) of lambs from each group were pooled by week. Data for weights of carcasses and stomach compartment, blood variables, as well as rumen fermentation parameters of lambs from each group were pooled at day 21, 35, and 63, respectively. Those data were analyzed according to a two-way ANOVA using PROC GLM of SAS (SAS Inst. Inc., Cary, NC, USA) to examine the effect of treatment, age, and the interaction between treatment and age. The statistical model was as follows:

$$Yij = \mu + Ti + Aj + TAij + Eij \quad (1)$$

where Yij is the dependent variable, μ is the overall mean, Ti is the treatment effect, Aj is the age effect, TAij is the interaction of treatment and age, and Eij is the error term.

One-way ANOVA was used to analyze the effect of weaning time on diarrhea rate and diarrhea frequency. The statistical model was as follows:

$$Yi = \mu + Ti + Ei \quad (2)$$

where Yi is the dependent variable, μ is the overall mean, Ti is the Treatment effect, and Ei is the error term. Multiple comparisons of means among treatments were performed using Tukey's tests. Significance was defined as $p < 0.05$.

3. Results

3.1. Growth Performance and Diarrhea Rate

There was no difference in BW from birth until day 21 of age between the two groups of lambs ($p > 0.05$) (Figure 1a). However, BW of lambs in the EW group was significantly lower than BW in the

CON group ($p < 0.05$) after 21 days of age (Figure 1a). No difference was observed in ADG at 21 days of age between two groups of lambs ($p > 0.05$) (Figure 1b). The ADG of lambs in the EW group was significantly lower than ADG of those in the CON group at 28 ($p < 0.01$), 35 ($p < 0.001$), 42 ($p < 0.001$), and 49 ($p < 0.001$) days of age (Figure 1b).

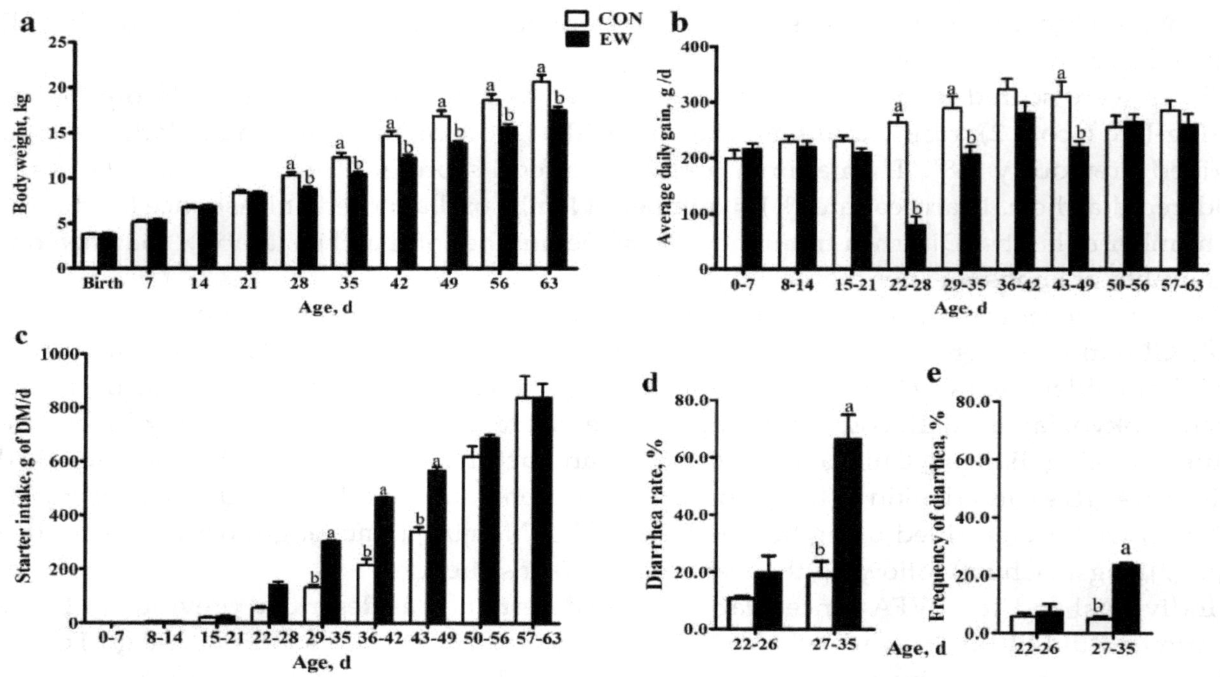

Figure 1. Body weight (**a**), average daily gain (**b**), starter intake (**c**), and diarrhea rate (**d**,**e**) for lambs weaned at day 21 as early weaned group (EW; black columns) or lambs weaned at day 49 as conventional weaned group (CON; white columns). Columns with different letters at a single time point indicate means that differed based on the means separation ($p < 0.05$). Error bars indicate SEM for the treatment × age interaction.

No difference was observed in starter intake from 8 to 28 days of age between the two groups of lambs ($p > 0.05$) (Figure 1c). The starter intake of lambs in EW group was significantly higher than that in CON group ($p < 0.05$) from 29 to 49 days of age (Figure 1c). No difference in starter intake between two groups of lambs was observed from 50 to 56 ($p > 0.05$) and 57 to 63 days of age ($p > 0.05$) (Figure 1c).

No diarrhea incidence was observed for any lamb from birth to 21 days of age. From 22 to 26 days of age, the diarrhea rate as well as frequency was not different between lambs in EW and CON group ($p > 0.05$) (Figure 1d). From day 27 to 35, the diarrhea rate and frequency of lambs in EW group was significantly higher than that in CON group ($p < 0.001$). No diarrhea incidence was observed after 35 days of age for lambs in both groups.

3.2. Slaughter Performance and Development of Visceral Organs

Both slaughter BW ($p = 0.010$) and hot carcass weight ($p = 0.007$) of lambs from EW group was significantly lower than that of lambs from CON group (Table 2). Dressing percentage decreased over time for both groups ($p < 0.001$) and tended to be lower in EW than CON group ($p = 0.065$). No change was found in the weight of whole stomach ($p = 0.705$), but abomasum relative to whole stomach weight ($p = 0.003$) was lower, while rumen ($p = 0.007$) and omasum ($p = 0.035$) relative to whole stomach weight was higher for lambs in EW than those in CON group. Reticulum relative to whole stomach weight was not different between two groups of lambs ($p = 0.821$). Both treatment and age interactively affected dressing percentage ($p = 0.021$) and reticulum relative to whole stomach weight ($p = 0.049$).

Table 2. Carcass weight, weight of rumen, reticulum, omasum, and abomasum relative to whole stomach weight on 26, 35, and 63 days of age [1].

Item	Treatment	Day of Age			SEM [2]	*p*-Value [2]		
		26	35	63		Treatment	Age	Treatment × Age
Slaughter BW, kg	CON	9.8 ^{bc}	12.3 ^{b}	22.2 ^{a}	0.779	0.010	<0.001	0.675
	EW	8.8 ^{c}	10.2 ^{bc}	20.0 ^{a}				
Hot carcass, kg	CON	4.7 ^{b}	5.9 ^{b}	10.2 ^{a}	0.443	0.007	<0.001	0.376
	EW	4.4 ^{b}	4.5 ^{b}	8.7 ^{a}				
Dressing percentage, %	CON	48.3 ^{ab}	47.6 ^{abc}	45.8 ^{bdc}	0.491	0.065	<0.001	0.021
	EW	49.8 ^{a}	44.4 ^{dc}	43.5 ^{d}				
Whole stomach, g	CON	135.5 ^{c}	237.0 ^{b}	566.6 ^{a}	21.400	0.705	<0.001	0.936
	EW	127.7 ^{c}	238.7 ^{b}	552.7 ^{a}				
		relative to whole stomach weight, %						
Rumen	CON	45.1 ^{e}	55.5 ^{dc}	67.9 ^{ab}	1.784	0.007	<0.001	0.300
	EW	51.3 ^{de}	61.0 ^{bc}	68.9 ^{a}				
Reticulum	CON	7.8	9.1	9.9	0.502	0.821	0.258	0.049
	EW	9.1	8.7	8.7				
Omasum	CON	4.5 ^{bdc}	3.7 ^{d}	6.3 ^{ab}	0.434	0.035	<0.001	0.648
	EW	5.7 ^{abc}	4.1 ^{dc}	7.0 ^{a}				
Abomasum	CON	42.6 ^{a}	31.7 ^{b}	15.9 ^{c}	1.852	0.003	<0.001	0.102
	EW	33.9 ^{b}	26.2 ^{b}	15.4 ^{c}				

[1] Lambs weaned at day 21 as early weaned group (EW) or lambs weaned at day 49 as conventional weaned group (CON). Values are presented as means; n = 6 per group, BW= body weight. [2] Treatment = the effect of weaning time; age = the effect of age; treatment × age= the interaction between treatment and age. SEM = standard error of the mean. $^{a–d}$ Means within two rows (six values including age and treatment) with different superscripts differ ($p < 0.05$).

3.3. Rumen Fermentation Parameters

Both rumen pH ($p = 0.007$) and NH_3-N concentration ($p < 0.001$) increased over time and was not affected by treatment ($p = 0.694$ and 0.999, respectively) (Table 3). Total VFA concentration was higher in EW than CON group ($p = 0.027$). Molar proportion of acetate ($p = 0.804$), propionate ($p = 0.154$), butyrate ($p = 0.171$), isobutyrate ($p = 0.991$), isovalerate ($p = 0.159$), or acetate/propionate ratio ($p = 0.253$) was not affected by treatment. Molar proportion of valerate tended to be higher in the EW than that in the CON group ($p = 0.098$). Treatment and age interactively affected the NH_3-N concentration ($p = 0.040$) and molar proportion of isovalerate ($p = 0.041$).

Table 3. Rumen pH, ammonia nitrogen (NH_3-N), and volatile fatty acid (VFA) profiles of lambs at 26, 35, and 63 days of age [1].

Item	Treatment	Day of Age			SEM [2]	*p*-Value [2]		
		26	35	63		Treatment	Age	Treatment × Age
pH	CON	5.7 ^{b}	5.5 ^{b}	6.3 ^{a}	0.081	0.694	0.007	0.684
	EW	5.6 ^{b}	5.6 ^{b}	6.0 ^{a}				
NH_3-N, mg/100 mL	CON	10.6 ^{bc}	15.7 ^{ab}	17.5 ^{ab}	1.048	0.999	<0.001	0.040
	EW	8.1 ^{c}	12.9 ^{bc}	22.8 ^{a}				
Total VFA, mmol/L	CON	66.4 ^{b}	106.5 ^{ab}	92.9 ^{ab}	4.884	0.027	0.049	0.197
	EW	109.7 ^{ab}	122.0 ^{a}	97.6 ^{ab}				
		Molar proportion of VFA, mol/100 mol						
Acetate	CON	50.9	50.1	52.8	1.234	0.804	0.676	0.340
	EW	49.3	55.8	50.2				
Propionate	CON	24.1	26.4	25.0	0.775	0.154	0.744	0.482
	EW	29.4	27.4	26.2				
Butyrate	CON	20.1	16.5	16.7	1.331	0.171	0.379	0.526
	EW	14.8	10.0	17.1				
Valerate	CON	1.9 ^{b}	5.3 ^{a}	3.1 ^{ab}	0.378	0.098	0.015	0.185
	EW	4.9 ^{ab}	5.6 ^{a}	3.5 ^{ab}				
Isobutyrate	CON	1.0 ^{ab}	0.8 ^{ab}	0.9 ^{ab}	0.075	0.991	0.033	0.068
	EW	0.8 ^{ab}	0.6 ^{b}	1.3 ^{a}				
Isovalerate	CON	1.9 ^{a}	0.9 ^{ab}	1.5 ^{ab}	0.135	0.159	0.01	0.041
	EW	0.7 ^{b}	0.7 ^{b}	1.8 ^{a}				
Acetate/Propionate ratio	CON	2.1	2.1	2.3	0.099	0.253	0.71	0.616
	EW	1.7	2.1	1.9				

[1] Lambs weaned at day 21 as early weaned group (EW) or lambs weaned at day 49 as conventional weaned group (CON). Values are presented as means; n = 6 per group. [2] Treatment = the effect of weaning time; age = the effect of age; treatment × age = the interaction between treatment and age. SEM = standard error of the mean. $^{a–c}$ Means within two rows (six values including age and treatment) with different superscripts differ ($p < 0.05$).

3.4. Serum Parameters

The concentration of IgG (p = 0.761), IgA (p = 0.809) or IgM (p = 0.889) was not affected by treatment (Table 4). The concentration of IgG ($p < 0.001$) was lower, while that of IgA ($p < 0.001$) and IgM ($p < 0.001$) was higher at 63 than 26 days of age for both groups. No difference was observed in activity of SOD (p = 0.839), GSH-Px (p = 0.290), CAT (p = 0.145), T-AOC (p = 0.605), or MDA (p = 0.464) between the two groups of lambs (Table 4). Activity of SOD ($p < 0.001$), GSH-Px ($p < 0.001$), CAT ($p < 0.001$), and T-AOC ($p < 0.001$) was higher, while that of MDA ($p < 0.001$) was lower at 35 than those at 26 or 63 days of age for both groups. There was no difference in the concentration of IL-1β (p = 0.659), IL-4 (p = 0.361), IL-6 (p = 0.085), cortisol (p = 0.946), IFN-γ (p = 0.843), or TNF-α (p = 0.927) between the two groups of lambs. The concentration of IL-1β ($p < 0.001$), IL-4 ($p < 0.001$), IL-6 ($p < 0.001$), cortisol ($p < 0.001$), IFN-γ ($p < 0.001$), and TNF-α ($p < 0.001$) were higher, while that of MDA ($p < 0.001$) was lower at 35 than those in 26 or 63 days of age for both groups. There was no interactive effect of treatment and age on any serum parameter ($p > 0.05$).

Table 4. Serum parameters of lambs at 26, 35, and 63 days of age [1].

Item [2]	Treatment	Day of Age			SEM [3]	p-Value [3]		
		26	35	63		Treatment	Age	Treatment × Age
IgG, g/L	CON	24.0^{a}	22.5^{a}	19.0^{b}	0.422	0.761	<0.001	0.454
	EW	22.8^{a}	22.8^{a}	19.4^{b}				
IgA, g/L	CON	0.5^{c}	0.6^{bc}	0.7^{a}	0.014	0.809	<0.001	0.658
	EW	0.6^{c}	0.6^{c}	0.7^{ab}				
IgM, g/L	CON	1.1^{b}	1.1^{b}	1.3^{a}	0.026	0.889	<0.001	0.846
	EW	1.1^{b}	1. 1^{b}	1.3^{a}				
SOD, U/mL	CON	93.5^{b}	111.5^{a}	94.8^{b}	1.548	0.839	<0.001	0.556
	EW	96.7^{b}	110.4^{a}	93.8^{b}				
GSH-Px, μmol/L	CON	754.7^{bc}	818.6^{a}	732.4^{c}	8.645	0.290	<0.001	0.199
	EW	712.7^{c}	811.1^{ab}	743.4^{c}				
CAT, U/mL	CON	7.6^{c}	8.9^{a}	7.8^{bc}	0.118	0.145	<0.001	0.844
	EW	7.5^{c}	8.6^{ab}	7.6^{c}				
T-AOC, U/mL	CON	8.4^{c}	9.8^{a}	8.8^{bc}	0.115	0.605	<0.001	0.134
	EW	8.9^{bc}	9.5^{ab}	8.9^{bc}				
MDA, nmol/mL	CON	5.3^{a}	4.3^{b}	5.3^{a}	0.112	0.464	<0.001	0.948
	EW	5.5^{a}	4.4^{b}	5.4^{a}				
Cortisol, μg/dL	CON	1.5^{b}	2.7^{a}	1.6^{b}	0.107	0.946	<0.001	0.695
	EW	1.6^{b}	2.6^{a}	1.5^{b}				
IL-1β, pg/mL	CON	98.9^{b}	111.4^{a}	99.7^{b}	1.534	0.659	<0.001	0.471
	EW	97.1^{b}	114.0^{a}	96.1^{b}				
IL-4, pg/mL	CON	13.3^{c}	14.3^{ab}	13.2^{c}	0.124	0.361	<0.001	0.850
	EW	13.4^{bc}	14.6^{a}	13.3^{c}				
IL-6, pg/mL	CON	45.9^{c}	54.2^{ab}	45.8^{c}	1.055	0.085	<0.001	0.356
	EW	47.5^{bc}	59.2^{a}	46.2^{c}				
IFN-γ, pg/mL	CON	143.2^{b}	159.0^{a}	141.9^{b}	1.443	0.843	<0.001	0.772
	EW	144.2^{b}	157.1^{a}	141.9^{b}				
TNF-α, pg/mL	CON	54.4^{b}	67.5^{a}	55.7^{b}	1.133	0.927	<0.001	0.549
	EW	56.4^{b}	66.3^{a}	54.5^{b}				

[1] Values are presented as means; n = 6 per group. Lambs weaned at day 21 as early weaned group (EW) or lambs weaned at day 49 as conventional weaned group (CON). [2] IgG = immunoglobulin G, IgM = immunoglobulin M, IgA = immunoglobulin A, SOD = superoxide dismutase, GSH-Px = Glutathione Peroxidase, CAT = catalase, T-AOC = total antioxidative capacity, MDA = malondialdehyde IL-1β = interleukin 1β, IL-4 = interleukin 4, IL-6 = interleukin 6, IFN-γ = interferon-gamma, TNF-α = tumor necrosis factor α. [3] Treatment = the effect of weaning time; age = the effect of age; treatment × age = the interaction between treatment and age. SEM = standard error of the mean. [a–c] Means within two rows (six values including age and treatment) with different superscripts differ ($p < 0.05$).

4. Discussion

It has been generally accepted that intake of solid feed before weaning stimulates rumen development and functionality [21,22]. Early starter feeding (7 days of age) was reported to increase ruminal VFA concentration in lambs compared with late starter feeding (42 days of age) [23]. Therefore, lambs were provided starter at 7 days of age in the current study. However, a sharp decrease in ADG (from 21 to 28 days of age) of lambs weaned at 21 days of age was not prevented. This might be partly due to the short time for ewe-lamb separation (6 h daily), which was insufficient for lambs to get used to the abrupt separation from their dams at 21 days of age. The final BW of lambs weaned at 21 days of age was still lower than that of their counterparts weaned at 49 days of age, which again proved that early-life stress may have a long-lasting negative effect on the performance of lambs. For example, it is suggested that lambs subjected to nutrient restriction at 15 days of age followed by realimentation at 60 days of age still physically lag behind those continuously provided sufficient nutrients, featured by a lower BW at 90 days of age (24.1 vs. 26.1 kg) [24]. On the other hand, we found that the growth performance of lambs weaned and separated from their dams at 49 days of age (>20 kg of BW at 60 days of age) was better than those weaned and separated from their dams at 60 days of age (12.8 kg of BW at 60 days of age) [25]. This discrepancy might be explained by the difference in time of introduction of starter to lambs between current (day 7) and our previous study (day 15), as well as the higher protein and energy level of starter used in the current study than our previous study [25].

Post-weaning diarrhea is an economically important enteric disease due to financial losses [26]. In our present study, diarrhea rate and frequency were increased in the EW group due to early weaning. Similarly, this disease occurs most frequently within the two weeks after weaning and is characterized by a profuse diarrhea, dehydration, significant mortality, and loss of body weight of pigs [27]. Our previous study showed a sudden change in the feeding system due to early weaning resulted in changes of the intestinal morphology and functions [28], that may trigger reductions in digestive and absorptive capacities. Studies suggest that increasing weaning age reduces stress associated with this period and allows animals to have a more mature gastrointestinal tract and become increasingly familiar with solid feed during milk-period with an improvement in growth performance [29,30].

Ruminal pH is crucial for normal development, fermentation, and overall lamb health. Generally, rumen fluid pH is influenced by the rate of fermentation and absorption of VFA, which in turn, are affected by passage rate of digesta and the buffering capacity of the rumen contents [31]. In this experiment, rumen pH of lambs on day 26 or day 35 were lower than day 63 and not affected by weaning time, mainly due to the difference in starter intake between pre-and post-weaning. The increase in pH with age may be explained by the continuous development of rumen and starter intake. Furthermore, when the rumen epithelium is still underdeveloped, early starter feeding in lambs can increase ruminal VFA concentrations and affect the expression levels of genes involved in VFA transport and pH regulation in the ruminal epithelium [23].

Previous studies showed that ruminal VFA concentration increased with increasing solid feed intake and was affected by either dietary (e.g., nutrient fractionation) or animal factors (e.g., absorption and passage rate) [6]. Rumen development is not only driven by the presence of VFA, but also by the nutrients supplied from liquid feed [6]. In our study, weaning and feeding starter to lambs at day 21 promoted the development of the rumen, which is in accordance with the results in previous studies as providing solid feed as early as possible is beneficial to the rapid development of the rumen [32,33]. In the current study, the concentration of Total VFA, molar proportion of acetate and propionate in the EW group increased only at 5 days after weaning. This is because more starter intake in the EW group promoted rumen development, while the CON lambs were mainly fed milk. The consumption of starter in lambs has been shown to increase the ruminal concentrations of VFAs, which have positive effects on the development of ruminal epithelium [23]. This result is similar with Yang et al. [34] who showed no difference in Total VFA concentrations in lambs between pre- and post-weaning.

The effect of passive immunity on mortality, morbidity, and subsequent growth of newborn animals is receiving increasing attention [35]. Among three major immunoglobulin isotypes, namely IgG, IgM, and IgA, IgG usually has the highest concentration in serum and is the main antibody for humoral immunity. A previous study showed that serum IgG concentration was positively associated with calf health, as the morbidity and intensity of disease were the lowest in heifer calves with serum IgG concentration exceeding 10 g/L at 30 to 60 h of life [36]. In our study, the IgG concentration was higher than that of lambs (same age) reported by Hernández-Castellano et al. [37], which may due to the difference in sheep species or different test method of immunodiffusion, as a recent review suggested that the concentration of IgG may differ between methods such as ELISA or radial immunodiffusion [35]. In the present study, no change was found in the concentration of IgG, IgA, or IgM. Similarly, our previous study also showed no difference in serum IgG and IgA between lambs differing in weaning time [25]. Those results are in agreement with Hernández-Castellano et al. [37] who found no difference in the IgG concentration pre- and post-weaning, which may be due to the beginning of the endogenous immunoglobulin production by lambs and increase in the concentration of IgG over time. The authors also found the concentrations of IgG and IgM in blood of lambs decreased after birth with aging until the end of milk feeding period and then increased 30 days after weaning, and this phenomenon is not affected by milk sources (nursing, milk replacer, or cow milk) [37]. In our study, during the first few days of birth, serum immunoglobulin may typically reflect the efficacy of passive immunity transfer through adequate and timely consumption of colostrum. Later on, as feed intake increased, the systematic humoral immune function of lambs developed.

GSH-Px is an antioxidant enzyme that helps to control the levels of hydrogen peroxide and lipid peroxides under normal metabolic conditions. T-AOC indicates the oxidation resistance capacity of the whole body. MDA is one of the final products of polyunsaturated fatty acid peroxidation in cells and is considered as a marker of oxidative stress. The insignificant difference of GSH-Px, T-AOC, and MDA between lambs in different treatments suggested that the antioxidant status was not affected by weaning time. The activity of SOD, GSH-Px, CAT, and T-AOC was the highest, while that of MDA concentration was the lowest at 35 days of age for both groups, suggesting that antioxidant status improved with age. It was reported that weaning induced lower activities of antioxidant enzymes and a greater content of MDA, reflecting an imbalance of redox status in piglets [38]. However, Buchet et al. [39] reported that the observed effects are related with weaning but not with age.

In our study, no changes in the concentrations of cortisol were observed in treatment. Previous study suggested that the plasma concentration of cortisol increased in calves when they were subjected to weaning [40]. On the contrary, it was reported that the change in plasma cortisol concentration was not affected by weaning [41], and a similar result was observed in lambs [42] and piglets [43]. This may be related to the sampling time after weaning, because inflammatory response is a short-term change that only occurs within a few hours after weaning and does not last for a long time [40,44].

Several cytokines biomarkers such as IL-1β, IFN-γ, and TNF-α were used to evaluate stress and inflammatory response before and after weaning [40,42]. Early weaning can stimulate the secretion of proinflammatory cytokines such as IL-1, TNF-α, and stimulate inflammatory response [42]. Yang et al. [45] found that plasma IFN-γ declined with age during weaning period in calves. TNF-α often work synergistically with IFN-γ and IL-1β to improve immune responses [46]. Study in lambs have shown that serum TNF-α concentration was not affected by early weaning or age [25,42], and are responsive to proinflammatory cytokine regulation due to early weaning. In our study, no changes in the concentrations of IL-1β, IL-4, IL-6, IFN-γ, or TNF-α were observed in response to weaning time. Therefore, more sampling time is needed in future study to validate the findings in the current experiment.

5. Conclusions

The results indicate that early starter feeding combined with limited ewe-lamb contact promoted rumen development and increased Total VFA concentration, without compromising lambs' overall immunity weaning at 21 days of age. However, the increase in diarrhea rate and decrease in ADG in lambs weaned at 21 days of age suggested that a long-term effect of weaning on overall performance as well as development in gut microbiota and barrier function needs to be assessed.

Author Contributions: Conceptualization, Q.D. and Y.T.; methodology, S.W. and G.Z.; data curation, S.W.; software, S.W. and Y.B.; formal analysis, S.W.; investigation, S.W.; resources, S.W. and T.M.; writing—original draft, K.C. and H.D.; writing—review and editing, Q.D. and Y.T.; supervision, N.Z. and F.L.; project administration, S.W. and G.Z.; funding acquisition.

Acknowledgments: We thank students in the ruminant nutrition laboratory of Feed Research Institute of Chinese Academy of Agricultural Sciences (CAAS) for their technical assistance. We thank Linqing Runlin Sheep Industry Co. Ltd. (Shangdong Province, China) for their assistance with the animal experiment.

References

1. Jagusch, K.T.; Clark, V.R.; Jay, N.P. Lamb production from animals weaned at 3 to 5 weeks of age on to lucerne. *New Zeal J. Agr. Res.* **1970**, *13*, 808–814. [CrossRef]
2. Cheng, C.S.; Wei, H.K.; Wang, P.; Yu, H.C.; Zhang, X.M.; Jiang, S.W.; Peng, J. Early intervention with faecal microbiota transplantation: An effective means to improve growth performance and the intestinal development of suckling piglets. *Animal* **2019**, *13*, 533–541. [CrossRef] [PubMed]
3. He, X.; Lu, Z.; Ma, B.; Zhang, L.; Li, J.; Jiang, Y.; Zhou, G.; Gao, F. Effects of chronic heat exposure on growth performance, intestinal epithelial histology, appetite-related hormones and genes expression in broilers. *J. Sci. Food Agric.* **2018**, *98*, 4471–4478. [CrossRef] [PubMed]
4. Orgeur, P.; Bernard, S.; Naciri, M.; Nowak, R.; Schaal, B.; Levy, F. Psychobiological consequences of two different weaning methods in sheep. *Reprod. Nutr. Dev.* **1999**, *39*, 231–244. [CrossRef] [PubMed]
5. Hinch, G.N.; Lynch, J.J.; Elwin, R.L.; Green, G.C. Long term associations between Merino ewes and their offspring. *Appl. Anim. Behav. Sci.* **1990**, *27*, 93–103. [CrossRef]
6. Khan, M.A.; Bach, A.; Weary, D.M.; von Keyserlingk, M.A.G. Invited review: Transitioning from milk to solid feed in dairy heifers. *J. Dairy Sci.* **2016**, *99*, 885–902. [CrossRef]
7. Demir, H. Investigations on growing Kivircik lambs weaned at different times. *Vet. Fak. Derg. Istambul* **1995**, *21*, 142–150.
8. Baldwin, R.L.; McLeod, K.R.; Klotz, J.L.; Heitmann, R.N. Rumen development, intestinal growth and hepatic metabolism in the pre- and postweaning ruminant. *J. Dairy Sci.* **2004**, *87*, E55–E65. [CrossRef]
9. McKusick, B.C.; Thomas, D.L.; Berger, Y.M. Effect of weaning system on commercial milk production and lamb growth of East Friesian dairy sheep. *J. Dairy Sci.* **2001**, *84*, 1660–1668. [CrossRef]
10. Chai, J.; Diao, Q.Y.; Wang, H.C.; Tu, Y.; Tao, X.J.; Zhang, N.F. Effects of weaning age on growth, nutrient digestibility and metabolism, and serum parameters in Hu lambs. *Anim. Nutr.* **2015**, *1*, 344–348. [CrossRef]
11. Zhang, Q.; Li, C.; Niu, X.L.; Zhang, Z.F.; Li, F.; Li, F.D. The effects of milk replacer allowance and weaning age on the performance, nutrients digestibility, and ruminal microbiota communities of lambs. *Anim. Feed. Sci. Technol.* **2019**, *257*, 1–9. [CrossRef]
12. Bimczok, D.; Röhl, F.W.; Ganter, M. Evaluation of lamb performance and costs in motherless rearing of German Grey Heath sheep under field conditions using automatic feeding systems. *Small Rumin Res.* **2005**, *60*, 255–265. [CrossRef]
13. National Research Council (NRC). *Nutrient Requirements of Small Ruminants: Sheep, Goat, Cervids, and New World Camelids*; National Academy Press: Washington, WA, USA, 2007.
14. AOAC (Association of Official Analytical Chemists). *Official Methods of Analysis*, 15th ed.; AOAC: Arlington, VA, USA, 1990.
15. Marshall, C.M.; Walker, A.F. Comparison of a short method for Kjeldahl digestion using a trace of selenium as catalyst, with other methods. *J. Sci. Food Agric.* **1978**, *29*, 940–942. [CrossRef]
16. Van Soest, P.J.; Robertson, J.B.; Lewis, B.A. Methods for dietary fiber, neutral detergent fiber, and nonstarch polysaccharides in relation to animal nutrition. *J. Dairy Sci.* **1991**, *74*, 3583–3597. [CrossRef]
17. Goering, H.G.; Van Soest, J.P. *Forage Fiber Analysis. Agricultural Handbook*; UPSDA: Washington, DC, USA, 1970.
18. Kargar, S.; Mousavi, F.; Karimi-Dehkordi, S. Effects of chromium supplementation on weight gain, feeding behaviour, health and metabolic criteria of environmentally heat-loaded Holstein dairy calves from birth to weaning. *Arch. Anim. Nutr.* **2018**, *72*, 443–457. [CrossRef]
19. Zhang, R.; Zhou, M.; Tu, Y.; Zhang, N.F.; Deng, K.D.; Ma, T.; Diao, Q.Y. Effect of oral administration of probiotics on growth performance, apparent nutrient digestibility and stress-related indicators in Holstein calves. *J. Anim. Physiol. Anim. Nutr. (Berl)* **2016**, *100*, 33–38. [CrossRef]

20. Broderick, G.; Kang, J. Automated stimultaneous determination of ammonia and total amino acids in ruminal fluid and in vitro media. *J. Dairy Sci.* **1980**, *63*, 64–75. [CrossRef]
21. Spencer, J.D.; Boyd, R.D.; Cabrera, R.; Allee, G.L. Early weaning to reduce tissue mobilization in lactating sows and milk supplementation to enhance pig weaning weight during extreme heat stress. *J. Anim. Sci.* **2003**, *81*, 2041–2052. [CrossRef]
22. Jesse, B.W. *Energy Metabolism in the Developing Rumen Epithelium*; Elsevier: London, UK, 2005; Volume 3.
23. Liu, T.; Li, F.; Wang, W.M.; Yue, X.P.; Li, F.; Li, C.; Pan, X.Y.; Mo, F.T.; Wang, F.B.; La, Y.F.; et al. Effects of lamb early starter feeding on the expression of genes involved in volatile fatty acid transport and pH regulation in rumen tissue. *Anim. Feed Sci. Technol.* **2016**, *217*, 27–35. [CrossRef]
24. Ma, T.; Wang, B.; Zhang, N.F.; Tu, Y.; Si, B.W.; Cui, K.; Qi, M.L.; Diao, Q.Y. Effect of protein restriction followed by realimentation on growth, nutrient digestibility, ruminal parameters, and transporter gene expression in lambs. *Anim. Feed Sci. Technol.* **2017**, *231*, 19–28. [CrossRef]
25. Chai, J.M.; Ma, T.; Wang, H.C.; Qi, M.L.; Tu, Y.; Diao, Q.Y.; Zhang, N.F. Effect of early weaning age on growth performance, nutrient digestibility, and serum parameters of lambs. *Anim. Prod. Sci.* **2017**, *57*, 110–115. [CrossRef]
26. Amezcua, R.; Friendship, R.M.; Dewey, C.E.; Gyles, C.; Fairbrother, J.M. Presentation of postweaning Escherichia coli diarrhea in southern Ontario, prevalence of hemolytic E-coli serogroups involved, and their antimicrobial resistance patterns. *Can. J. Vet. Res.* **2002**, *66*, 73–78. [PubMed]
27. Fairbrother, J.M.; Nadeau, É.; Gyles, C.L. Escherichia coli in postweaning diarrhea in pigs: An update on bacterial types, pathogenesis, and prevention strategies. *Anim. Health Res. Rev.* **2007**, *6*, 17–39. [CrossRef]
28. Cui, K.; Wang, B.; Zhang, N.F.; Tu, Y.; Ma, T.; Diao, Q.Y. iTRAQ-based quantitative proteomic analysis of alterations in the intestine of Hu sheep under weaning stress. *PLoS ONE* **2018**, *13*, e0200680. [CrossRef] [PubMed]
29. Main, R.G.; Dritz, S.S.; Tokach, M.D.; Goodband, R.D.; Nelssen, J.L. Increasing weaning age improves pig performance in a multisite production system. *J. Anim. Sci.* **2004**, *82*, 1499–1507. [CrossRef]
30. Campbell, J.M.; Crenshaw, J.D.; Polo, J. The biological stress of early weaned piglets. *J. Anim. Sci. Biotechnol.* **2013**, *4*, 19. [CrossRef]
31. Pazoki, A.; Ghorbani, G.R.; Kargar, S.; Sadeghi-Sefidmazgi, A.; Drackley, J.K.; Ghaffari, M.H. Growth performance, nutrient digestibility, ruminal fermentation, and rumen development of calves during transition from liquid to solid feed: Effects of physical form of starter feed and forage provision. *Anim. Feed Sci. Technol.* **2017**, *234*, 173–185. [CrossRef]
32. Liu, J.; Bian, G.; Sun, D.; Zhu, W.; Mao, S. Starter feeding altered ruminal epithelial bacterial communities and some key immune-related genes' expression before weaning in lambs. *J. Anim. Sci.* **2017**, *95*, 910–921. [CrossRef]
33. Liu, J.; Bian, G.; Sun, D.; Zhu, W.; Mao, S. Starter Feeding Supplementation Alters Colonic Mucosal Bacterial Communities and Modulates Mucosal Immune Homeostasis in Newborn Lambs. *Front. Microbiol.* **2017**, *8*, 429. [CrossRef]
34. Yang, B.; He, B.; Wang, S.S.; Liu, J.X.; Wang, J.K. Early supplementation of starter pellets with alfalfa improves the performance of pre- and postweaning Hu lambs. *J. Anim. Sci.* **2015**, *93*, 4984–4994. [CrossRef]
35. McGee, M.; Earley, B. Review: Passive immunity in beef-suckler calves. *Animal* **2019**, *13*, 810–825. [CrossRef] [PubMed]
36. Furman-Fratczak, K.; Rzasa, A.; Stefaniak, T. The influence of colostral immunoglobulin concentration in heifer calves' serum on their health and growth. *J. Dairy Sci.* **2011**, *94*, 5536–5543. [CrossRef] [PubMed]
37. Hernández-Castellano, L.E.; Moreno-Indias, I.; Morales-delaNuez, A.; Sánchez-Macías, D.; Torres, A.; Capote, J.; Argüello, A.; Castro, N. The effect of milk source on body weight and immune status of lambs. *Livest. Sci.* **2015**, *175*, 70–76. [CrossRef]
38. Cao, S.T.; Wang, C.C.; Wu, H.; Zhang, Q.H.; Jiao, L.F.; Hu, C.H. Weaning disrupts intestinal antioxidant status, impairs intestinal barrier and mitochondrial function, and triggers mitophagy in piglets. *J. Anim. Sci.* **2018**, *96*, 1073–1083. [CrossRef] [PubMed]
39. Buchet, A.; Belloc, C.; Leblanc-Maridor, M.; Merlot, E. Effects of age and weaning conditions on blood indicators of oxidative status in pigs. *PLoS ONE* **2017**, *12*, e0178487. [CrossRef]
40. O'Loughlin, A.; McGee, M.; Doyle, S.; Earley, B. Biomarker responses to weaning stress in beef calves. *Res. Vet. Sci.* **2014**, *97*, 458–463. [CrossRef]

41. Hickey, M.C.; Drennan, M.; Earley, B. The effect of abrupt weaning of suckler calves on the plasma concentrations of cortisol, catecholamines, leukocytes, acute-phase proteins and in vitrc interferon-gamma production. *J. Anim. Sci.* **2003**, *81*, 2847–2855. [CrossRef]
42. Zhang, Q.; Li, C.; Niu, X.L.; Zhang, Z.A.; Li, F.; Li, F.D. An intensive milk replacer feeding program benefits immune response and intestinal microbiota of lambs during weaning. *BMC Vet. Res.* **2018**, *14*, 366. [CrossRef]
43. Turpin, D.L.; Langendijk, P.; Chen, T.Y.; Pluske, J.R. Intermittent Suckling in Combination with an Older Weaning Age Improves Growth, Feed Intake and Aspects of Gastrointestinal Tract Carbohydrate Absorption in Pigs after Weaning. *Animals* **2016**, *6*, 66. [CrossRef]
44. Hulbert, L.E.; Moisa, S.J. Stress, immunity, and the management of calves. *J. Dairy Sci.* **2016**, *99*, 3199–3216. [CrossRef]
45. Yang, C.; Zhang, L.; Cao, G.; Feng, J.; Yue, M.; Xu, Y.; Dai, B.; Han, Q.; Guo, X. Effects of dietary supplementation with essential oils and organic acids on the growth performance, immune system, fecal volatile fatty acids, and microflora community in weaned piglets. *J. Anim. Sci.* **2019**, *97*, 133–143. [CrossRef] [PubMed]
46. Schroder, K.; Hertzog, P.J.; Ravasi, T.; Hume, D.A. Interferon-gamma: An overview of signals, mechanisms and functions. *J. Leukoc. Biol.* **2004**, *75*, 163–189. [CrossRef] [PubMed]

12

In Vivo Toxicity and In Vitro Solubility Assessment of Pre-Treated Struvite as a Potential Alternative Phosphorus Source in Animal Feed

Soomin Shim [1,†], Seunggun Won [2,†], Arif Reza [1,3,†], Seungsoo Kim [1], Sungil Ahn [1], Baedong Jung [4], Byungil Yoon [4] and Changsix Ra [1,*]

1 Department of Animal Industry Convergence, College of Animal Life Sciences, Kangwon National University, Chuncheon 24341, Korea; vibrato7@naver.com (S.S.); reza.arif@kangwon.ac.kr (A.R.); seung_su89@nate.com (S.K.); sorzente@naver.com (S.A.)

2 Department of Animal Resources, College of Life and Environmental Science, Daegu University, Gyeongsan 38453, Korea; swon@daegu.ac.kr

3 Department of Environmental Science, College of Agricultural Sciences, IUBAT—International University of Business Agriculture and Technology, Dhaka 1230, Bangladesh

4 Department of Veterinary Science, Kangwon National University, Chuncheon 24341, Korea; bdjung@kangwon.ac.kr (B.J.); byoon@kangwon.ac.kr (B.Y.)

* Correspondence: changsix@kangwon.ac.kr.

† These authors contributed equally to this work.

Simple Summary: Phosphorus (P) is an indispensable element needed for the growth and development of all living organisms. Phosphorus is mined primarily from non-renewable phosphate rock reserves. The use of P as fertilizer and feed additives continues to rise with the increasing global population. We therefore need to find an alternative and renewable, as well as sustainable, source of P to fulfill the demands of agriculture and livestock. Phosphorus recovered as struvite from livestock wastewater could be a sustainable alternative to commercial P sources, while its application has only been limited to arable lands as fertilizer. This study elucidates that struvite can also be used as an alternative P source in animal feed other than fertilizer through proper pre-treatment. Phosphorus recovery from livestock wastewater and its reutilization as an animal feed ingredient would therefore be a good strategy to substitute commercial P sources and ensure societal sustainability.

Abstract: Apart from using as fertilizer for plants, the application of struvite may be expanded to animal feed industries through proper pre-treatment. This study aimed to investigate the safety and efficacy of using pre-treated struvite (microwave irradiated struvite (MS) and incinerated struvite (IS)) in animal feeds. For safety assessment, an in vivo toxicity experiment using thirty female Sprague Dawley rats (average body weight (BW) of 200 ± 10 g) was conducted. The rats were randomly divided into five groups, including a control. Based on the BW, MS and IS were applied daily by oral administration with 1 and 10 mg kg^{-1}-BW (MS1 and MS10; IS1 and IS10) using dimethyl sulfoxide (DMSO) as a vehicle. A series of jar tests were conducted for four hours to check the solubility of the MS and IS at different pH (pH 2, 4, 5, 6 and 7) and compared to a commercial P source (monocalcium phosphate, MCP, control). The toxicity experiment results showed no significant differences among the treatments in BW and organ (liver, kidney, heart, and lung) weight of rats ($p > 0.05$). There were no adverse effects on blood parameters and the histopathological examination showed no inflammation in the organ tissues in MS and IS treated groups compared to the control. In an in vitro solubility test, no significant difference was observed in ortho-phosphate (O-P) solubility from the MCP and MS at pH 2 and 4 ($p > 0.05$), while O-P solubility from MS at pH 5 to 7 was higher than MCP and found to be significantly different ($p < 0.05$). O-P solubility from IS was the lowest among the treatments and significantly different from MCP and MS in all the experiments ($p < 0.05$). The results of this study not only suggest that the struvite pre-treated as MS could be a potential alternative source of P in animal feed but also motivate further studies with more stringent designs to better examine the potential of struvite application in diverse fields.

Keywords: phosphorus; struvite; swine wastewater; pre-treatment; toxicity; solubility; animal feed

---

1. Introduction

Phosphorus (P) is a key element for animal and plant growth as well as an important nutrient for future food security. It has been estimated that the present reserve of phosphate rock is 70,000 million tons as P_2O_5, of which 40 million tons of phosphate rock is mined every year [1,2]. Studies reported that the resource may be exhausted within the next 90–300 years, with an expected annual increase in P consumption of 2% [1,3]. Phosphate rock was recently categorized as one of the 20 critical raw materials in Europe since P utilization has been known as a one-way flow and it is classified as a non-renewable resource. In the world, it is mainly three countries (Morocco, China, and the United States), which take the lead in intensive production for phosphate rock [4]. In addition, due to non-replaceability and high economic importance, there might be a political and social tension built over the phosphate rock reserves and the world could move from an oil-based to a phosphate-based economy, as the aforementioned three countries control more than 85% of the known global phosphorus reserves [5].

The majority (80 to 90%) of the P mined from phosphate rock reserves has been used in the agricultural sector for fertilizer and animal feed production [6]. Whereas, of the total phosphate fertilizer applied in the arable lands, 57% can be lost in soil leaching, erosion, and runoff [7,8]. Moreover, livestock wastewater contains relatively high strengths of P concentration, which leads to eutrophication [9], resulting from excessive P contents inevitably given to achieve high growth performance. Every year 7 million tons of P may be released to the environment through animal manure and excreta [10]. P recovery and recycling from highly polluted livestock wastewater could therefore be a sustainable alternative to commercial P sources.

For P recovery from different wastewater, various physicochemical techniques have been introduced. Among those methods, struvite precipitation has been widely studied with its very straightforward regime and is considered as an eco-friendly and sustainable process in terms of nutrient recovery and recycling from nutrient-rich waste streams.

Along with economic development, swine farming practices in Korea are becoming more intensive. Intensive swine farming practices result in an increase in swine wastewater generation. Swine wastewater treatment practice in Korea is quite different compared to European and North American countries. The Korean government is currently operating more than 200 centralized swine wastewater treatment plants with a treatment capacity of more than 20,000 tonnes per day [11,12]. The bulk recovery of P in the form struvite from the centralized treatment plants might therefore be economical and sustainable. Moreover, the strategy of recovering and recycling P from swine wastewater and its reutilization as an animal feed ingredient can help protect the environment by reducing nutrient contents in discharged effluents as well as initiating a P circular economy.

Currently, the application of struvite has only been limited to arable lands as fertilizer and reported to be non-toxic to plants [13]. Moreover, P bioavailability from struvite has shown similar performance as the chemical fertilizers in crop trials [14]. Furthermore, to our knowledge, none of the studies focused on exploring the feasibility of struvite application in animal feed as an alternative P source. Swine wastewater generally contains high amounts of ortho-phosphate (O-P) and ammonium nitrogen (NH_4-N). NH_4-N is highly toxic for the animals and its accumulation in cells can damage the tissues [15]. It is therefore necessary to pre-treat the struvite recovered from swine wastewater for its further use in animal feed. We have already successfully tested the feasibility of reutilizing P recovered from swine wastewater as a dietary supplement in juvenile far eastern catfish (*Silurus asotus*) by comparing with a commercial P source [16,17]. In this study, we suggest manufacturing options for struvite recovered from swine wastewater to ensure biological safety and present the possibility of struvite application as an alternative P source in animal feed rations, with the results of an in vivo toxicity test on rats and an in vitro solubility test.

2. Materials and Methods

2.1. Struvite Production from Swine Wastewater and Pre-Treatment

Swine wastewater was collected from the storage tank of a gestation pig barn located in Chuncheon, South Korea. After collection, the solid/liquid separation was done using centrifugation at 3000 rpm for 10 min and the liquid part was used for struvite production. The reactor had an effective volume of 20 L, including a compartment of about 2 L as a reaction zone, and the rest was composed of a settling zone and recovery zone. The influent with the flow rate of 0.11 L min^{-1} (160 L d) was introduced from the top and passed through the compartment with holes at the bottom. Crystals formed in the reaction zone were precipitated and the flux flew up to the outlet, which resulted in a hydraulic retention time (HRT) of 3 h in the entire P recovery process. The precipitates were collected from the bottom where the valve was equipped. $MgCl_2$ in the liquid phase was provided into the reactor to meet the appropriate Mg to P ratio (~1.3) in the influent, which was found as an optimal condition in our previous research [18]. An air diffuser was set at the bottom of the compartment in the reaction zone for CO_2 stripping and, thus, a pH of over 8.5 was maintained. An X-ray diffraction (XRD) analysis (X'Pert PRO MPD, PANalytical B.V., Netherlands) was conducted to give the precipitates an identity, where the peaks indicate that the precipitates recovered were identified as struvite. The supernatant was put back into the reactor after collecting precipitates from the bottom of the struvite reactor. The settled materials were spread on the tray and stored in a dry-oven at 20 °C for 2 days, then the moisture content of pre-dried sample was measured at 105 °C following the standard methods [19], which resulted in a moisture content of 35%, but water parts in struvite evaporated during the second drying at 105 °C. To remove all possible toxicity, recovered precipitates were pre-treated in a microwave (80 Hz g^{-1} for 5 min) and a muffle furnace at 550 °C for over 30 min after drying [3,17]. The heavy metal contents in the pre-treated struvites were determined using inductively coupled plasma atomic emission spectroscopy (ICP-AES) (Optima 7300 DV, PerkinElmer, Waltham, MA, USA).

2.2. In Vivo Toxicity Assessment in Rats

An in vivo toxicity experiment using rats was conducted to evaluate the safety of using pretreated struvite (microwave irradiated struvite (MS); incinerated struvite (IS)) as an alternative to commercial P sources in animal feeds. The protocol of the experiment was approved and the rats were cared for according to the guidelines of the Institutional Animal Care and Use Committee of Kangwon National University, Chuncheon, South Korea (ACUC # KW 130620-1). For the toxicity experiment, 30 female Sprague Dawley rats (average body weight (BW) of 200 ± 10 g) were procured from Bio Genomic Inc. (Charles River Technology, Gapyung-Gun, Korea) and were allotted to each of five groups, including a control, in a randomized complete block design, resulting in six rats per treatment group. The sample size was calculated using the resource equation method [20]. All groups of rats were given ad libitum access to water and conventional laboratory diets (5L79, Labdiet, St. Louis, MO, USA) to maintain their normal physiology. The amount of crude protein, crude fat, crude fiber, ash, Ca, and P was 18, 5, 5, 8, 0.85, and 0.70%, respectively, in the supplied lab diet. To observe the toxic effects of test materials, subacute in vivo toxicity tests were conducted for 28 days [21]. Based on the observations regarding the highest dose of ammonia without effect reported by Tuleka et al., pre-treated struvite doses of 1 and 10 mg kg^{-1} BW were designated to evaluate the repeated dose toxicity [22]. MS and IS of 1 and 10 mg kg^{-1} BW were dissolved in 1% dimethyl sulfoxide (DMSO) as a vehicle, which were represented by MS1, MS10, IS1, and IS10, respectively, in 4 treatment groups. Nutrient contents of the treatments are shown in Table 1. The prepared testing materials were given daily by oral administration at once. A control group was only provided with the carrier (1% DMSO). During the experimental period, the cycle of light/dark was maintained at 12 h/12 h, respectively, and the temperature of the experimental cages was maintained at 20–25 °C. The feed intake and body weight were recorded regularly. Rats were monitored for any abnormalities in activities, behavior, and general health status, and fed with respective experimental diets.

Table 1. Nutrient contents in the treatments.

Group	Treatment	Nutrient Content (mg kg^{-1})			
		P	N	Ca	Mg
I	DMSO (Control) [a]	-	-	-	-
II	IS1 [b]	0.26	0.00003	0.10	0.12
III	IS10 [c]	2.60	0.0003	1.03	1.15
IV	MS1 [d]	0.22	0.04	0.09	0.10
V	MS10 [e]	2.21	0.37	0.88	0.98

[a] DMSO = dimethyl sulfoxide; [b] IS1 = rats fed with IS at 1 mg kg^{-1} BW; [c] IS10 = rats fed with IS at 10 mg kg^{-1} BW; [d] MS1 = rats fed with MS at 1 mg kg^{-1} BW; [e] MS10 = rats fed with MS at 10 mg kg^{-1} BW.

At the end of the experimental feeding (28 d), the final BW was measured and all the rats were euthanized and sacrificed. Blood samples of 5 mL were collected from all rats by heart puncture using the sterilized needles and syringes in a disposable vacutainer tube without any anticoagulant and stored at 4 °C for 2 h. After centrifugation (3000 rpm for 10 min at 4 °C), serum samples were separated and stored at −20 °C in Eppendorf tubes. Internal organs (liver, kidney, lung, and heart) were removed and weighed from all rats after they were sacrificed. For histopathological assay of the internal organs, the tissue samples of heart, liver, kidney, and lungs were collected and fixed in 10% neutral buffered formalin, embedded in paraffin wax, and cut into 5 μm-thin sections. For routine examination, all samples were stained with hematoxylin and eosin (H&E) and observed under an optical microscope (×100 or ×400) [23]. The blood tests were conducted at the end of the experimental feeding; i.e., after 24 h from the last dose of testing materials. Assay kits for the analysis of serum biochemicals were obtained from Chemon Inc., South Korea. Serum biochemical parameters including alanine aminotransferase (ALT), aspartate aminotransferase (AST), blood urea nitrogen (BUN), creatinine (CRE), inorganic P (P_i), and calcium (Ca) were examined to assess the effect of the treatments on the internal organs [24,25]. The selected serum biochemical parameters were analyzed using an automated blood biochemical analyzer (Model AU400, Olympus, Tokyo, Japan) and an automated electrolyte analyzer (Model M744 ($Na^+/K^+/Cl^-$) analyzer, Siemens, Deerfield, MA, USA).

2.3. *In Vitro Solubility Test*

A series of jar tests were conducted for four hours at 25 °C to check the solubility of the test materials by simulating the gastrointestinal pH conditions (pH 2, 4, 5, 6, and 7). The required amount of test materials was calculated based on the total P content of the materials. A total of 0.5 g (particle size < 100μm) of each of the test materials were weighed exactly in a 1L beaker. The beakers were filled with de-ionized water. The pH of the solution was monitored constantly and the necessary pH adjustment was done with 1M HCl throughout the experiment. All the experiments were performed in triplicate. Considering the variation of gastrointestinal transit time among animals, immediate release of P at low pH, and the availability for absorption, 2 mL of sample was taken per treatment at 15, 30, 60, 90, 120, 150, 180, 210, and 240 min intervals, diluted with 18 mL de-ionized water, centrifuged at 3000 rpm for 10 min, and O-P content was determined using an auto-analyzer (Quik Chem 8500, LACHAT, Loveland, CO, USA).

2.4. Statistical Analysis

Statistical analysis was conducted using GraphPad Prism (version 5.03, 2009). Data obtained from the in vivo toxicity test were analyzed with the one-way analysis of variance (ANOVA) test, considering treatments as a fixed factor. Multiple comparisons were performed using Tukey's HSD tests. In vitro solubility of O-P was analyzed by comparing the O-P concentrations at the beginning with those at the end of the experiment (240 min) using an ANOVA with O-P concentrations at 9 time points as a repeated measure. A p-value of < 0.05 was designated to determine statistical significance.

3. Results and Discussion

3.1. P Recovery and Manufacturing

Using a lab-scale reactor (160 L d^{-1}), an O-P removal efficiency of 93.1% was observed and the precipitate production rate and yield of 8.7 ± 0.6 kg struvite $m^{-3}{}_{reactor}.d^{-1}$ and 9.8 ± 0.5 g struvite $g^{-1}O\text{-}P_{input}$, respectively, were achieved. The recovered precipitates were identified by XRD analysis, which indicated the recovered material was struvite (Figure 1a). The precipitates contained a P content of 21.6%, which was higher over the theoretical P of 12.7% in struvite since the P in the form of organic compounds might be present in the precipitates.

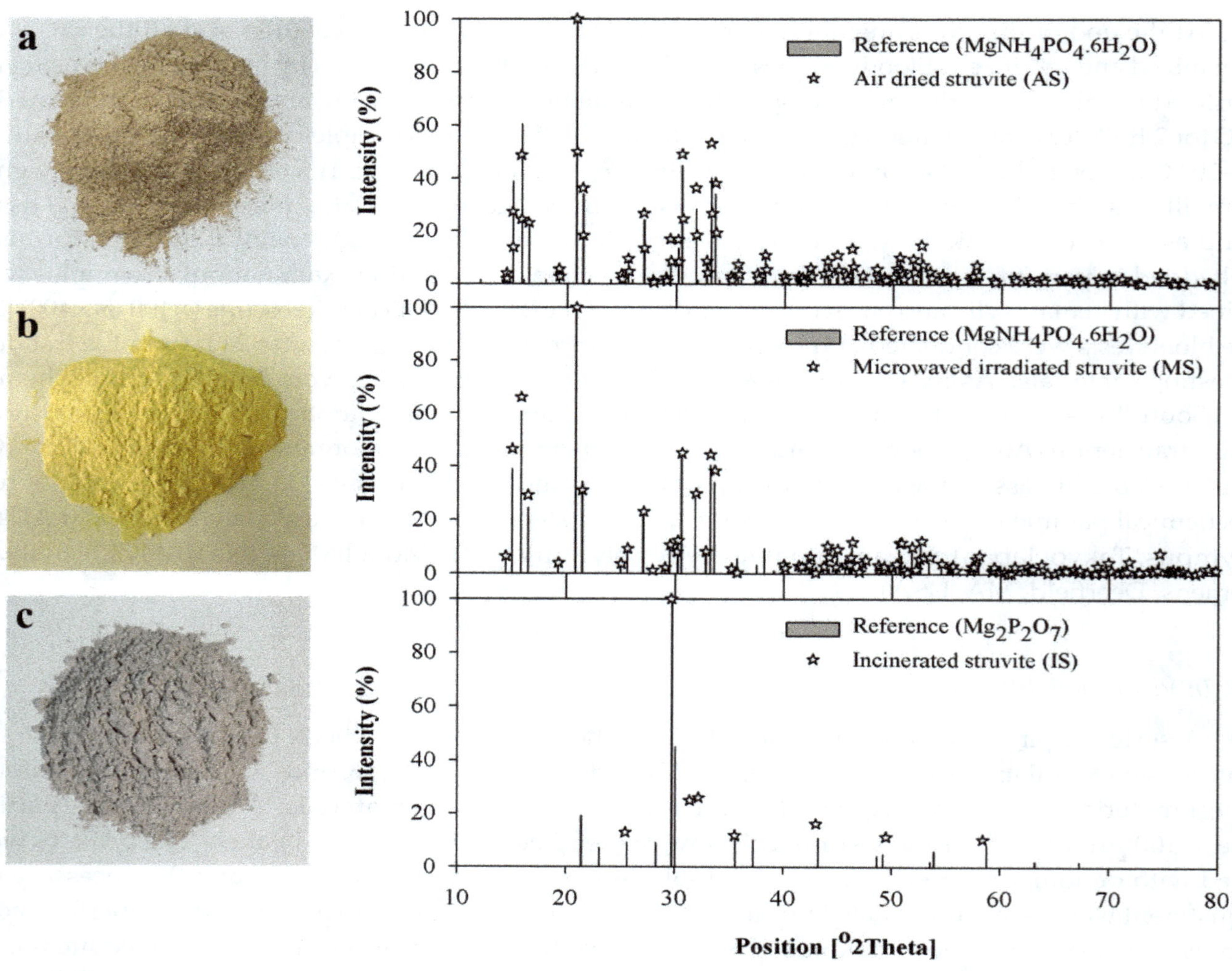

Figure 1. Pictures of recovered and pre-treated struvites with the identification by XRD analyses. (**a**) Air-dried at 20 °C for 2 days, (**b**) microwave irradiated at 80 Hz g^{-1} for 5 min, and (**c**) incinerated at 550 °C for 30 min (* Asterisks in each graph indicate reference materials from XRD analyses).

Different from the struvite recovered from municipal and industrial wastewater, struvite recovered from livestock wastewater is relatively safer for its application to agricultural land [13] because all the ingredients in precipitated struvite from livestock wastewater originate from animal feedstock and include relatively lower amounts of harmful metals. Instead of phosphate rock mined from P reserves, recovered P via struvite precipitation contains lower heavy metal content and needs lesser efforts to purify for target use, such as for fertilizer and chemical industries [26]. During this study, the trace heavy metals in the recovered struvite were rarely found or not detected and satisfied the highest standard limits of feedstock for animal growth as suggested by the Korean Regulatory Authority (Table 2) [27]. In addition to the use in fertilizer sector, struvite as a way of P recycling can therefore be applied in the animal feed industry like phosphate rock. However, it is very important to ensure that the heavy metal contents in the recovered struvite are within the safe limits prior to use as animal feed additive.

Table 2. Chemical composition of recovered and pre-treated struvites and the highest standard limits of heavy metals in feedstock for animals, based on the Korean Regulatory Authority [27].

Parameters	Test Materials			Highest Standard Limits
	AS [c]	MS [b]	IS [a]	
P (g kg^{-1} DM)	216	221	260	
Ca (g kg^{-1} DM)	85	88	103	
N (g kg^{-1} DM)	50	3713	0.03	
Mg (g kg^{-1} DM)	95	98	115	
K (mg kg^{-1} DM)	3567.6		4407	-
Zn (mg kg^{-1} DM)	ND [d]		ND [d]	-
Ni (mg kg^{-1} DM)	ND [d]		ND [d]	-
Cu (mg kg^{-1} DM)	15.7		19.4	-
Cd (mg kg^{-1} DM)	ND [d]	NA [e]	ND [d]	1.0
Pb (mg kg^{-1} DM)	0.0001		0.0001	10.0
As (mg kg^{-1} DM)	0.0012		0.0015	2.0
Cr (mg kg^{-1} DM)	ND [d]		ND [d]	100.0
Hg (mg kg^{-1} DM)	ND [d]		ND [d]	0.4
Se (mg kg^{-1} DM)	ND [d]		ND [d]	2.0

[a] AS = air-dried struvite; [b] MS = microwave irradiated struvite; [c] IS = incinerated struvite; [d] ND = not detected; [e] NA = not analyzed.

The theoretical N content is only around 5.8% in struvite, with NH_4-N being the main N species, and further N content could be decreased through air drying (Table 2). However, N content in air-dried struvite was still higher than the typical N content in animal feeds (1.9 to 4.8%) [28]. Therefore, additional treatment is required to reduce the N content before its application in feed production. To ascertain cost-effective treatments for the reduction of the N content and the removal of organic compounds, microwave irradiation and incineration treatments were selected according to our previous study [3]. Both treatments were conducted to remove mainly NH_4-N from struvite. Microwave irradiation treatment did not show much influence on its morphology and it was still determined as struvite with XRD analysis (Figure 1b), but incineration under 550 °C denatured struvite to magnesium pyrophosphate ($Mg_2P_2O_7$) by removing ammonium and hydrates, which was confirmed by XRD analysis as well (Figure 1c). Ammonium and hydrate contained in struvite started to be released as a gas phase at 55 °C and completed at 250 °C due to the thermal decomposition [29]. The N concentration of 5.0% in air-dried struvite was decreased 3.713 and 0.003% through microwave irradiation and incineration treatments, respectively (Table 2). Moreover, after pre-treatment of recovered struvite with the decrease in moisture content, an increase in the P, Ca, and Mg concentrations in the pre-treated materials were observed (Table 2). Hence, if a specific amount of struvite is added to the animal feed, considering the dietary P requirement, the concentration of other minerals present in the struvite might be higher than the requirement. Further toxicological studies are therefore needed to confirm the safety and efficacy of pre-treated struvites for their use in animal feed, although pre-treatments reduce the N content in struvite.

3.2. Toxicity Test in Rats

Prior to applying struvite recovered swine wastewater as a P additive in the animal feed ration, it is necessary to conduct the toxicity test to ensure biological safety. We therefore pre-treated struvite recovered from swine wastewater as MS and IS. After 28 d of feeding MS and IS, along with control feed, to the rats, the average total BW gain of nearly 121.4 ± 20.3 g was observed in the treated groups, including the control group. However, no statistical differences were found in BW, average daily feed intake, and the internal organs (liver, kidney, heart, and lung) among all the experimental groups fed with different doses of MS and IS, including the control ($p > 0.05$) (Figure 2a).

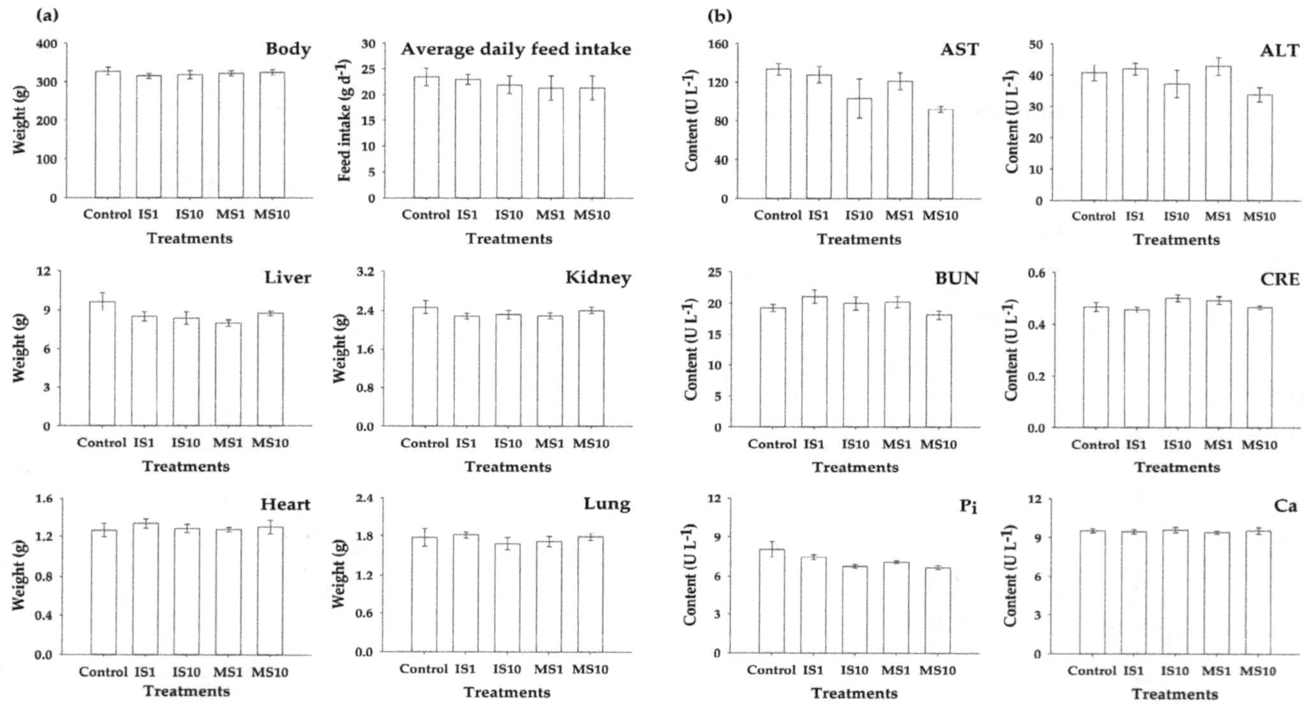

Figure 2. Comparison of (**a**) body and internal organ weights and (**b**) blood metabolites of rats treated with control, IS, and MS groups (Control (only 1% DMSO)); IS1 and IS10, rats fed with IS at 1 and 10 mg kg^{-1} BW; MS1 and MS10, rats fed with MS at 1 and 10mg kg^{-1} BW; Error bar: Standard error of means in each group (n = 6)).

In the administration with only the vehicle (control), the blood metabolites levels in rats were found to be normal (Figure 2b). No significant alteration in AST and ALT levels between IS1, IS10, MS1, MS10 (1 and 10 mg kg^{-1} BW), and the control groups were observed. The BUN, CRE, Pi, and Ca levels (mg dL^{-1}) of blood were found within the normal range and no statistical differences ($p > 0.05$) were observed in MS and IS treated groups compared to the control.

The histopathological analyses of the treatment groups applied 10 mg kg^{-1} BW of MS and IS displayed the normal shape, size, and color of the internal organs. Tissues from each organ (liver, kidney, lung, and heart) were found to be normal and no mark of inflammation, necrosis, edema, and/or any other specific pathological symptoms were noticed at such a higher dose of struvite supplementation (Figure 3).

Struvite contains an ammonium ion, which is thought to be toxic for cellular components for animals. Ammonia and ammonium ions are toxic to animal cells and especially influence the cell membrane [15,30] through interrupting the enzymatic reactions and intracellular pH changes, which may lead to the disturbance of proton gradients and the inhibition of endocytosis and exocytosis [31].

A higher dose of ammonium compound (300 ppm ammonium perfluorooctanoate) in rats caused an elevated liver weight, an increase in the incidence of diffuse hepatocellular hypertrophy, portal mononuclear cell infiltration, mild hepatocellular vacuolation [32], and some reproductive abnormalities in male and female rats were also found with a higher dose of ammonium compounds [33]. The present study did not find any abnormalities or lesions in the internal structures of the lungs, liver, kidney, and heart. It might be stated that, although struvite contains ammonium, it would not be harmful to animals after microwave irradiation treatment or incineration, thus ensuring animal welfare.

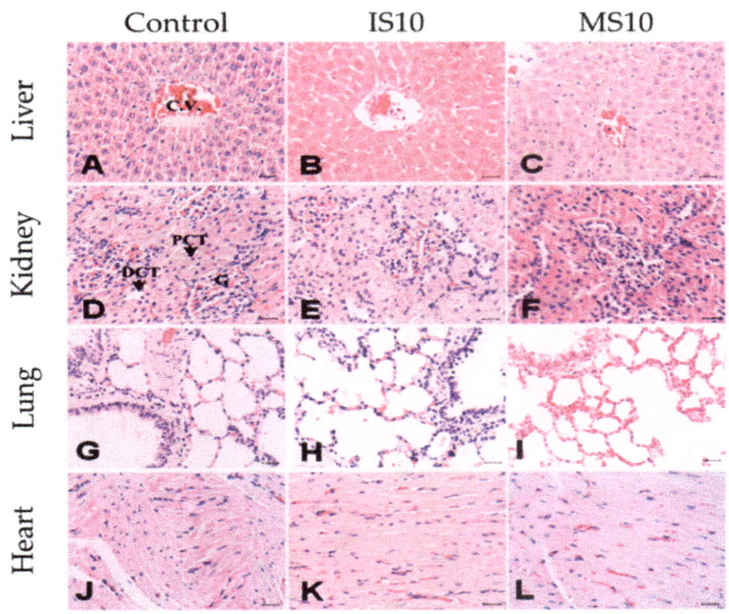

Figure 3. Comparison of histopathological examination of the internal organs of rats treated with control, IS, and MS groups (Control (only 1% DMSO); IS10, rats fed with IS at 10 mg kg^{-1} BW; MS10, rats fed with MS at 10 mg kg^{-1} BW.

The homeostatic condition of P, Ca, and Mg in the body is important as they control regular cellular activities such as energy metabolism and cell signaling [34]. Numerous cellular and tissue injuries including impaired bone mineralization, increased cell death, impaired cell signaling, vascular calcification, pre-mature aging, renal dysfunction, increases tumorigenesis, impaired fertility, skeletal development, etc., are associated with excessive retention of the above-mentioned macronutrients in the body [35–40]. In this study, no abnormalities in internal organ weights and serum biochemicals were observed. Therefore, the results obtained from growth performance, blood biochemical report, and histopathological studies of the internal organs indicated that the administration of MS and IS as a P additive retrieved from swine wastewater had no toxic effects on rats.

3.3. *In Vitro Solubility Test*

Variations in O-P concentration of the test materials at different pH levels (pH 2, 4, and 5) during the in vitro test are shown in Figure 4. All the experiments started with a constant concentration of the test materials. At pH 2, O-P concentrations of MCP, MS, and IS varied from 110.3 to 194.5 mg L^{-1}, 164.4 to 172.9 mg L^{-1}, and 131.4 to 138.2 mg L^{-1}, respectively. At the end of the experiment, 98.5%, 83.4%, and 69.1% of the total P in MCP, MS, and IS, respectively, was dissolved at pH 2 (Table 3). At pH 4, O-P concentrations of MCP, MS, and IS ranged from 111.3 to 158.8 mg L^{-1}, 143.8 to 158.1 mg L^{-1}, and 124.5 to 143.4 mg L^{-1}, respectively, and it was found that at the end of the experiment 79.7%, 77.2%, and 70.4% of the total P in MCP, MS, and IS, respectively, was dissolved. At pH 5, O-P concentrations of MCP, MS, and IM were varied from 109.7 to 119 mg L^{-1}, 146.1 to 159.2 mg L^{-1}, and 119.9 to 135.4 mg L^{-1}, respectively, and, at the end of the experiment, showed solubilities of 57.8%, 78.6%, and 67.6%, respectively. At pH 6, O-P concentrations of MCP, MS, and IS were ranged from 116.6

to 123.1 mg L^{-1}, 142.1 to 154.3 mg L^{-1}, and 104.5 to 128.6 mg L^{-1}, respectively and, at the end of the experiment, it was found that 61.6%, 76.2%, and 64.3% of the total P in MCP, MS, and IS, respectively, was dissolved. At pH 7, O-P concentrations of MCP, MS, and IS were varied from 95.4 to 117.1 mg L^{-1}, 135.9 to 152.5 mg L^{-1}, and 90.1 to 112.9 mg L^{-1}, respectively, and, at the end of the experiment, showed solubility of 58.5%, 76.2%, and 56.4%, respectively. Although the solubility of O-P from MCP was higher than MS at pH 2 and 4, no significant difference was observed ($p > 0.05$). However, the solubility of O-P from MS increased with the increase in pH and was found to be significantly different from MCP ($p < 0.05$). The in vitro solubility of O-P from IS was the lowest among the treatments and significantly different from MCP and MS in all the experiments ($p < 0.05$).

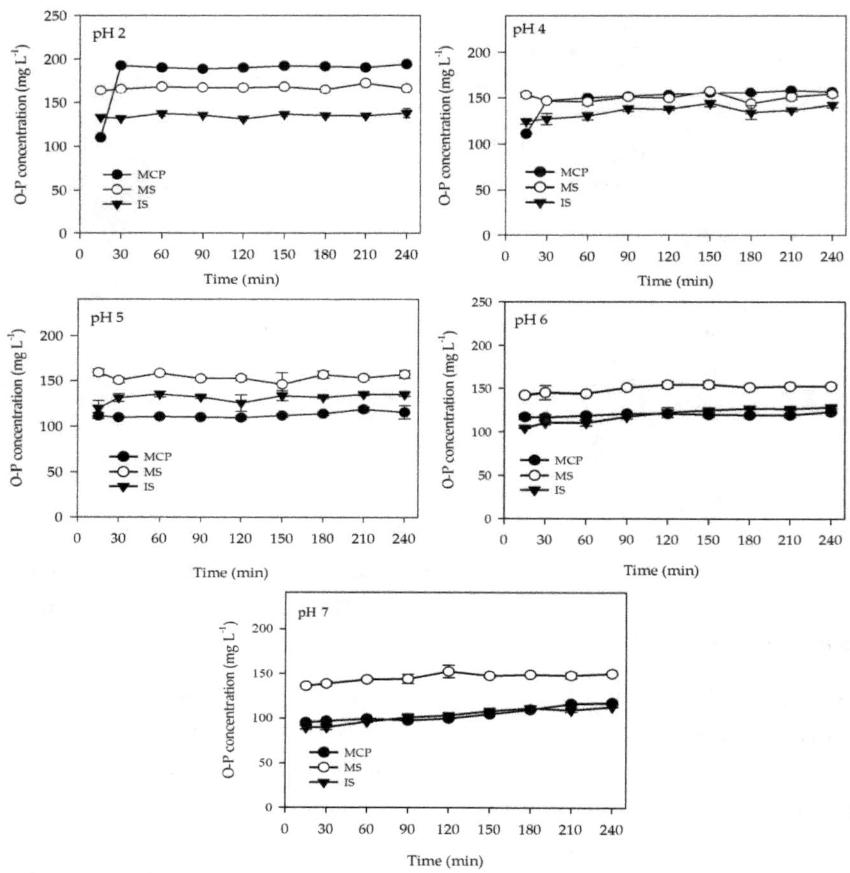

Figure 4. Variations in O-P concentration in an in vitro solubility test of each test materials over time.

Table 3. P solubility from test materials at pH 2, 4, 5, 6, and 7 after four hours of incubation.

P Sources	P Solubility (%)				
	pH 2	pH 4	pH 5	pH 6	pH 7
MCP *	98.5 ± 0.7 [a]	79.7 ± 0.1 [a]	57.8 ± 3.7 [a]	61.6 ± 0.5 [a]	58.5 ± 0.5 [a]
MS	83.4 ± 0.6 [a]	77.2 ± 1.3 [a]	78.6 ± 2.2 [b]	76.2 ± 1.0 [b]	74.8 ± 0.4 [b]
IS	69.1 ± 2.6 [b]	70.4 ± 0.3 [b]	67.6 ± 1.0 [c]	64.3 ± 0.6 [c]	56.4 ± 0.2 [c]

* MCP = Monocalcium phosphate; [a,b,c] Different superscript in the same column indicates statistical differences among the different treatment groups.

Animals are generally fed for higher performance, either in terms of reproduction or growth. Phosphorus plays a vital role in maintaining normal muscle growth, egg formation, and animal reproduction [41]. It also plays an important role in regulating acid-base and osmotic balance, energy and amino acid metabolism, and protein synthesis. Due to limited resources and high cost, nutritionists are trying to formulate diets considering the safety margins as well as dietary P requirements [42].

P solubility is one of the deterministic factors that influences the amount of supplement required for the animals. Factors including transit and retention time, pH, and particle size may control the P solubility in the gastrointestinal tract [43]. Moreover, P from inorganic sources are more soluble in the acid medium than the P from plant materials [44]. The amount of P available for absorption is controlled by its sustained solubility in the gastrointestinal tract in animals [45]. Primarily, P is absorbed in the small intestine within the pH range of 2 to 7 [46]. Studies reported that around 35% of the P is absorbed in the duodenum (pH 2 to 6.4), 5% in the jejunum (pH 6), and 40% in the ileum (pH 7) [47,48]. Struvite, an alternative P source has been extensively used as fertilizers in agriculture [13]; however, it is rarely tested as the P source in animal diets. As struvite is readily soluble under acidic conditions, the in vitro P solubility study was therefore conducted using pre-treated struvite as MS and IS by simulating the conditions in the gastrointestinal tract and compared with MCP. Among the test materials, P solubility from MS and MCP was not significantly different at a low pH. Moreover, due to an increase in P solubility from MS at comparatively higher pH (pH 5, 6, and 7), it can be said that, between the pre-treated struvites, MS could be better absorbed in the gastrointestinal tract. The solubility experiment results provide preliminary support of using MS as a potential alternative P source in animal diets. However, in vitro solubility testing has some limitations in extrapolating conditions in animals and further in vivo experiments are required for a better understanding of P metabolism.

4. Conclusions

In this study, neither any histopathological conditions in the internal organs nor any growth inhibition was found in rats fed with pre-treated struvites (MS and IS). The solubility test showed no significant difference in P solubility from MS and MCP at pH 2 and 4, while P solubility from MS increased at pH 5 to 7 and was found to be significantly different compared to MCP and IS. Considering resource recovery and recycling, efficacy as well as societal sustainability, struvite pre-treated as MS has the potential to be a sustainable alternative source of supplemental P in animal diets. Moreover, the results of this study motivate further studies with more stringent designs to better examine the potential of struvite application in diverse fields. Although commercialization of struvite may not be straightforward and is presumably dependent on circumstances such as legislation and national policies and economic feasibility in each country, P recycling is inevitable for both environmental protection and resource security in the world.

Author Contributions: Conceptualization, S.W., B.J., B.Y., and C.R.; methodology, S.W., B.J., B.Y., and C.R.; software, A.R.; formal analysis, S.S., S.K., and S.A.; investigation, S.S., S.K., and S.A.; resources, C.R.; data curation, S.S. and A.R.; writing—original draft preparation, S.S., S.W., and A.R.; writing—review and editing, A.R., S.W., B.J., B.Y., and C.R.; supervision, S.W. and C.R.; project administration, C.R.; funding acquisition, C.R.

References

1. Phosphate Rock Statistics and Information. Available online: http://https://www.usgs.gov/centers/nmic/phosphate-rock-statistics-and-information (accessed on 25 July 2019).
2. Steen, I. Phosphorus availability in the 21st Century: Management of a non-renewable resource. *Phosphorus Potassium* **1998**, *217*, 25–31.
3. Reza, A.; Shim, S.; Kim, S.; Ahmed, N.; Won, S.; Ra, C. Nutrient Leaching Loss of Pre-Treated Struvite and Its Application in Sudan Grass Cultivation as an Eco-Friendly and Sustainable Fertilizer Source. *Sustainability* **2019**, *11*, 4204. [CrossRef]
4. 20 Critical Raw Materials—Major Challenge for EU Industry. Available online: https://ec.europa.eu/growth/content/20-critical-raw-materials-major-challenge-eu-industry-0_mt (accessed on 20 July 2019).
5. Gilbert, N. Environment: The disappearing nutrient. *Nature* **2009**, *461*, 716–718. [CrossRef] [PubMed]
6. Van Kauwenbergh, S.J.; Stewart, M.; Mikkelsen, R. World reserves of phosphate rock—A dynamic and unfolding story. *Better Crop.* **2013**, *97*, 18–20.

7. Elser, J.; Bennett, E. Phosphorus cycle: A broken biogeochemical cycle. *Nature* **2011**, *478*, 29–31. [CrossRef] [PubMed]
8. Reza, A.; Eum, J.; Jung, S.; Choi, Y.; Jang, C.; Kim, K.; Owen, J.S.; Kim, B. Phosphorus Budget for a Forested-Agricultural Watershed in Korea. *Water* **2019**, *11*, 4. [CrossRef]
9. Khan, F.A.; Ansari, A.A. Eutrophication: An ecological vision. *Bot. Rev.* **2005**, *71*, 449–482. [CrossRef]
10. Cordell, D.; Drangert, J.O.; White, S. The story of phosphorus: Global food security and food for thought. *Glob. Environ. Chang.* **2009**, *19*, 292–305. [CrossRef]
11. Rural Development Administration (RDA). *Centralized Livestock Wastewater Treatment Plan 2019*; Rural Development Administration: Jeonju, Korea, 2019.
12. Ministry of Environment (MoE). *Establishment and Operation of Centralized Livestock Wastewater Treatment Plant*; Ministry of Environment: Sejong, Korea, 2018.
13. Ahmed, N.; Shim, S.; Won, S.; Ra, C. Struvite recovered from various types of wastewaters: Characteristics, soil leaching behaviour, and plant growth. *Land Degrad. Dev.* **2018**, *29*. [CrossRef]
14. Horta, C. Bioavailability of phosphorus from composts and struvite in acid soils. *Rev. Bras. Eng. Agric. Ambient.* **2017**, *21*, 459–464. [CrossRef]
15. Martinelle, K.; Haggstrom, L. Mechanisms of ammonia and ammonium ion toxicity in animal cells: Transport across cell membranes. *J. Biotechnol.* **1993**, *30*, 339–350. [CrossRef]
16. Yoon, T.H.; Lee, D.H.; Won, S.G.; Ra, C.S.; Kim, J.D. Effects of dietary supplementation of magnesium hydrogen phosphate (MgHPO4) as an alternative phosphorus source on growth and feed utilization of juvenile far eastern catfish (*Silurus asotus*). *Asian Austral. J. Anim.* **2014**, *27*, 1141–1149. [CrossRef]
17. Yoon, T.H.; Lee, D.H.; Won, S.G.; Ra, C.S.; Kim, J.D. Optimal incorporation level of dietary alternative phosphate (MgHPO4) and requirement for phosphorus in juvenile far eastern catfish (*Silurus asotus*). *Asian Austral. J. Anim.* **2015**, *28*, 111–119. [CrossRef] [PubMed]
18. Rahman, M.M.; Liu, Y.; Kwag, J.H.; Ra, C. Recovery of struvite from animal wastewater and its nutrient leaching loss in soil. *J. Hazard. Mater.* **2011**, *186*, 2026–2030. [CrossRef] [PubMed]
19. Eaton, A.D. *Standard Methods for the Examination of Water and Wastewater*, 21st ed.; American Public Health Association, American Water Works Association, Water Environment Federation: Washington, DC, USA, 2005; ISBN 978-08-7553-047-5.
20. Charan, J.; Kantharia, N.D. How to calculate sample size in animal studies? *J. Pharmacol. Pharmacother.* **2013**, *4*, 303. [CrossRef] [PubMed]
21. Park, E.; Lee, M.Y.; Seo, C.S.; Yoo, S.R.; Jeon, W.Y.; Shin, H.K. Acute and subacute toxicity of an ethanolic extract of Melandrii Herba in Crl: CD sprague dawley rats and cytotoxicity of the extract in vitro. *BMC Complement. Altern. Med.* **2016**, *16*, 370.
22. Tualeka, A.R.; Faradisha, J.; Maharja, R. Determination of No-Observed-Adverse-Effect Level Ammonia in White Mice Through CD4 Expression. *Dose-Response* **2018**, *16*, 1–9. [CrossRef] [PubMed]
23. Karahan, N.; Işler, M.; Koyu, A.; Karahan, A.G.; BaşyığıtKiliç, G.; Cırış, I.M.; Sütçü, R.; Onaran, I.; Cam, H.; Keskın, M. Effects of probiotics on methionine choline deficient diet-induced steatohepatitis in rats. *Turk. J. Gastroenterol.* **2012**, *23*, 110–121. [CrossRef] [PubMed]
24. Pierre, P.J.; Sequeira, M.K.; Corcoran, C.A.; Blevins, M.W.; Gee, M.; Laudenslager, M.L.; Bennett, A.J. Hematological and serum biochemical indices in healthy bonnet macaques (Macaca radiata). *J. Med. Primatol.* **2011**, *40*, 287–293. [CrossRef]
25. Ahmed, Z.; Ahmed, U.; Walayat, S.; Ren, J.; Martin, D.K.; Moole, H.; Koppe, S.; Yong, S.; Dhillon, S. Liver function tests in identifying patients with liver disease. *Clin. Exp. Gastroenterol.* **2018**, *11*, 301–307. [CrossRef]
26. Liu, Y.H.; Kumar, S.; Kwag, J.H.; Ra, C.S. Magnesium ammonium phosphate formation, recovery and its application as valuable resources: A review. *J. Chem. Technol. Biotechnol.* **2013**, *88*, 181–189. [CrossRef]
27. Control of Livestock and Fish Feed Act. Available online: http://www.law.go.kr/LSW/eng/engLsSc.do?menuId=2&section=lawNm&query=feed&x=0&y=0#liBgcolor0 (accessed on 3 October 2019).
28. Protein Nutrition Requirements of Farmed Livestock and Dietary Supply. Available online: http://www.fao.org/3/y5019e/y5019e06.htm (accessed on 2 October 2019).
29. Bhuiyan, M.I.H.; Mavinic, D.S.; Koch, F.A. Thermal decomposition of struvite and its phase transition. *Chemosphere* **2008**, *70*, 1347–1356. [CrossRef] [PubMed]

30. Ozturk, S.S.; Riley, M.R.; Palsson, B.O. Effects of ammonia and lactate on hybridoma growth, metabolism, and antibody production. *Biotechnol. Bioeng.* **1992**, *39*, 418–431. [CrossRef] [PubMed]
31. Docherty, P.A.; Snider, M.D. Effects of hypertonic and sodium-free medium on transport of a membrane glycoprotein along the secretory pathway in cultured mammalian cells. *J. Cell Physiol.* **1991**, *146*, 34–42. [CrossRef]
32. Butenhoff, J.L.; Kennedy, J.L., Jr.; Chang, S.C.; Olsen, G.W. Chronic dietary toxicity and carcinogenicity study with ammonium perfluorooctanoate in Sprague–Dawley rats. *Toxicol.* **2012**, *298*, 1–13. [CrossRef]
33. Koizumi, M.H.; Fujii, S.; Ono, A.; Hirose, A.; Imai, T.; Ogawa, K.; Ema, M.; Nishikawa, A. Evaluation of the reproductive and developmental toxicity of aluminium ammonium sulfate in a two-generation study in rats. *Food Chem. Toxicol.* **2011**, *49*, 1948–1959.
34. Razzaque, M.S. Phosphate toxicity: New insights into an old problem. *Clin. Sci.* **2011**, *120*, 91–97. [CrossRef]
35. Ohnishi, M.; Nakatani, T.; Lanske, B.; Razzaque, M.S. In vivo genetic evidence for suppressing vascular and soft-tissue calcification through the reduction of serum phosphate levels, even in the presence of high serum calcium and 1,25-dihydroxyvitamin D levels. *Circ.-Cardiovasc. Genet.* **2009**, *2*, 583–590. [CrossRef]
36. Kuro-o, M. Klotho as a regulator of fibroblast growth factor signaling and phosphate/calcium metabolism. *Curr. Opin. Nephrol. Hypertens.* **2006**, *15*, 437–441. [CrossRef]
37. Kuro-o, M. A potential link between phosphate and aging: Lessons from Klotho-deficient mice. *Mech. Ageing. Dev.* **2010**, *131*, 270–275. [CrossRef]
38. Jin, H.; Xu, C.X.; Lim, H.T.; Park, S.J.; Shin, J.Y.; Chung, Y.S.; Park, S.C.; Chang, S.H.; Youn, H.J.; Lee, K.H.; et al. High dietary inorganic phosphate increases lung tumorigenesis and alters Akt signaling. *Am. J. Respir. Crit. Care Med.* **2009**, *179*, 59–68. [CrossRef]
39. Amato, D.; Maravilla, A.; Montoya, C.; Gaja, O.; Revilla, C.; Guerra, R.; Paniagua, R. Acute effects of soft drink intake on calcium and phosphate metabolism in immature and adult rats. *Rev. Invest. Clin.* **1998**, *50*, 185–189. [PubMed]
40. Toba, Y.; Kajita, Y.; Masuyama, R.; Takada, Y.; Suzuki, K.; Aoe, S. Dietary magnesium supplementation affects bone metabolism and dynamic strength of bone in ovariectomized rats. *J. Nutr.* **2000**, *130*, 216–220. [CrossRef]
41. Zhang, B.; Wang, C.; Wei, Z.H.; Sun, H.Z.; Xu, G.Z.; Liu, J.X.; Liu, H.Y. The effects of dietary phosphorus on the growth performance and phosphorus excretion of dairy heifers. *Asian Australas. J. Anim. Sci.* **2016**, *29*, 960–964. [CrossRef] [PubMed]
42. Li, X.; Zhang, D.; Yang, T.Y.; Bryden, W.L. Phosphorus Bioavailability: A Key Aspect for Conserving this Critical Animal Feed Resource with Reference to Broiler Nutrition. *Agriculture* **2016**, *6*, 25. [CrossRef]
43. Walk, C.L.; Bedford, M.R.; McElroy, A.P. Influence of diet, phytase, and incubation time on calcium and phosphorus solubility in the gastric and small intestinal phase of an in vitro digestion assay. *J. Anim. Sci.* **2012**, *90*, 3120–3125. [CrossRef] [PubMed]
44. Lineva, A.; Kirchner, R.; Kienzle, E.; Kamphues, J.; Dobenecker, B. A pilot study on in vitro solubility of phosphorus from mineral sources, feed ingredients and compound feed for pigs, poultry, dogs and cats. *J. Anim. Physio. Anim. Nutr.* **2019**, *103*, 317–323. [CrossRef]
45. Breves, G.; Schröder, B. Comparative aspects of gastrointestinal phosphorus metabolism. *Nutr. Res. Rev.* **1991**, *4*, 125–140. [CrossRef]
46. Kiela, P.R.; Ghishan, F.K. Recent advances in the renal–skeletal–gut axis that controls phosphate homeostasis. *Lab. Invest.* **2009**, *89*, 7–14. [CrossRef]
47. Phosphorus Binders: Relative Potency of Available Agents: Phosphorus Absorption. Available online: https://www.medscape.org/viewarticle/506489_2 (accessed on 4 October 2019).
48. Wagner, A.L. Phytase Impacts Various Non-Starch Polysaccharidase Activities on Distillers Dried Grains with Solubles. Doctoral dissertation, Virginia Tech, Blacksburg, VA, USA, 2008.

Improvement of the Water Quality in Rainbow Trout Farming by Means of the Feeding Type and Management over 10 Years (2009–2019)

Elisa Fiordelmondo [1,*], Gian Enrico Magi [1], Francesca Mariotti [1], Rigers Bakiu [2] and Alessandra Roncarati [1]

[1] School of Bioscience and Veterinary Medicine, University of Camerino, Circonvallazione 93-95, 62034 Matelica, Italy; gianenrico.magi@unicam.it (G.E.M.); francesca.mariotti@unicam.it (F.M.); alessandra.roncarati@unicam.it (A.R.)

[2] Department of Aquaculture, University of Tirana, Place Mother Tereza 183, 695581 Tirana, Albania; bakiurigers@gmail.com

* Correspondence: elisa.fiordelmondo@unicam.it.

Simple Summary: The European Union Water Framework Directive has set the objective to develop a good status for water bodies and showed the importance of promoting the sustainable use of water resources in inland anthropic activities, such as freshwater aquaculture, performed in strict accordance with natural waters. In aquafeed, two elements have mostly contributed to improving aspects of change in fish feeding among rainbow trout—the use of the feed extrusion technique and the adoption of restrictive environmental rules. In this context, the main physico-chemical parameters of water quality were investigated in 2019 and the values obtained were then compared to the parameters obtained 10 years before (2009) in order to show if there were differences in the water quality of an outlet farm. Considering this, the present study aimed to evaluate the suitability of changes adopted in rearing and feeding techniques to improve growth performance, sustainability and the water quality environment. This study was made possible by considering data sampled on a yearly basis in the outlet and compared with the inlet water.

Abstract: Background: In Europe, rainbow trout is one of the main fresh water fish farmed in a constantly developing environment that requires innovative studies to improve farm management, fish welfare and environmental sustainability. The aim of this paper is to investigate the trend of water quality parameters over 10 years, after a feeding strategy change from pellet to extruded feed. Methods: The study was conducted on a farm in central Italy, based on parallel raceways. The cycle started from young rainbow trout (90 ± 2 g) that were grown until they reached market size. A water sample of 500 cm^3 was collected monthly from 2009 to 2019 from the lagoon basin in order to investigate the trends of the total suspended solids (TSS), biochemical oxygen demand (BOD_5), chemical oxygen demand (COD), nitrites (NO_2-N), nitrates (NO_3-N), total ammonia nitrogen (TAN), total phosphorus (TP) and pH. Results: All of the studied parameters (TSS, BOD_5, COD, NO_2-N, NO_3-N, TAN and TP) showed a significant improvement from 2009 to 2019. The pH parameter did not display notable variation during the studied period. The feed conversion ratio (FCR) was also investigated and exhibited a significant improvement from 1.4 to 1.1. Conclusion: Based on the decrease of all the investigated parameters, it is possible to say that extrusion is currently an excellent processing feed technique in aquaculture with a good level of respect for the environment.

Keywords: rainbow trout farming; water quality; feed extrusion technology; total suspended solids

1. Introduction

In the last two decades, new strategies for improving feeding techniques for the main fish species have been developing in order to reach global sustainability, in particular in an environmentally responsible manner. At the start of the new millennium, the European Union adopted the Water Framework Directive (2000/60/EC), which introduced a transnational vision of the management of the freshwater environment able to protect, manage and improve the quality of water resources across the EU. In 2018, the European Environment Agency [1] showed the efforts carried out by countries to monitor and assess the general health status of water bodies and stressed the importance of also promoting the sustainable use of available water resources in inland anthropic activities of the primary sector, such as agriculture and aquaculture, performed in strict accordance with natural waters. Many papers [2–5] have focused on the environmental impact of trout farms when uneaten food, fish catabolites (metabolites, ammonia gill excretion and carbon dioxide) and chemical treatments are not controlled and are discharged into the natural receiving waters. In the case of aquaculture specialized in the growing phase of rainbow trout (*Oncorhynchus mykiss*), the rearing technique is based on the use of flow-through systems which consist of raceways or concrete tanks with water constantly flowing down basins and the removal of waste at the outlet by gravity and the water current [6].

In aquafeed, two elements have mostly contributed to improving aspects of change concerning fish feeding, in particular, in rainbow trout: The use of the feed extrusion technique and the adoption of restrictive environmental rules. First, regarding rainbow trout feeding, the best technique is based on the fact that the feed distribution must be carried out until reaching a level close to satiety [7]. In fact, the feed conversion index and the protein efficiency index improve when rainbow trout are fed with a rationing level equal to 70% of the "ad libitum technique" [7]. This technique drastically reduces the amount of food not eaten by trout.

Feed quality also directly affects water quality, because a proportion of the feed intake by fish is returned to the environment as metabolites or soluble by-products of metabolism [2]. In this context, two main levels of measures (legislative and productive) have been adopted. Restrictive environmental rules of EU countries have imposed settling areas to eliminate solid waste, whilst new feeding technologies have also been considered. The evaluation of the impact of rainbow trout farming on the receiving water quality needs to take into consideration the water quality monitoring over a long period, based on a decade [8] or two years, in different flow-through farms [4]. Concerning feeding technologies, nowadays, modern systems use extruded feed instead of pellet feed. In particular, the metabolite status of fish depends on the degree of gelatinization of the feeding starch. In the modern aquatic system, the removal of metabolites can be decreased using formulations with ingredients able to help bind metabolite matter, allowing these particles to be more easily and thoroughly removed from the water [9–11].

Furthermore, small feed portions not consumed are decreased by a smart diet formulation and processing; nowadays, extruded feed represents the best solution, instead of the pellet one, which was previously the common feeding strategy [9–11]. The digestibility of the extruded feed increases up to 96% relative to the raw wheat starch, while, in the case of pelleted feed, this raw material is digestible to approximately 54% [12]. Moreover, the catabolic residues emitted by rainbow trout, fed with extruded feed with gelatinized starch, are easy to remove from the water, thanks to their sedimentation [11]. Therefore, extruded feed also enables high environmental sustainability due to the increase of digestibility and stability in water, with a consequent reduction of suspended solids and nutrients. With regard to the rearing environment and its traits, it is appropriate to consider how it influences the conditions of well-being. Correct water exchange is essential: Water dilutes and removes the catabolites of fish, as well as feed residues, thus reducing the exposure of the farmed subjects to dangerous nitrogen compounds which are also able to negatively influence the state of well-being.

To check that the water quality is suitable for the farming of rainbow trout, a multitude of physical and chemical parameters must be considered [13]. In the case of suspended solids, their concentration is particularly important, because they directly influence the water turbidity, which can prevent the vision of the fish and finally compromise their life cycle. According to Boyd [3] and Becke [14], water turbidity values higher than 400 mg/L can cause thickening and deformation of the gill filaments, with trout consequently suffering. At the same time, dissolved organic substances, including the sedimentable (undigested portion of the ration and food residues) and non-sedimentable (product of endogenous metabolism) solids in suspension, must be taken into consideration.

For these reasons, checking the water quality is of primary importance for the welfare of trout. Inappropriate rearing conditions, such as inadequate space, excessive densities and poor feeding, can have strong negative repercussions for farmed fish species. Damaged, eroded or hemorrhagic fins are not only correlated with pathological events but also with inadequate environmental factors, connected to stress-related aspects such as a fish stocking density that is too high with a non-optimal water quality [15].

Considering the use of diets administered to salmonids, many papers have focused on the emission of catabolites in the external environment. According to European Environment Agency [16], 15–25% of the total food energy is lost in ammonia and urea through the gills and is released into the environment.

Most of the papers on this topic have focused their attention on relationships between feeding strategies, animal welfare and environmental sustainability and in the current literature, the correlation between water quality and productive performances of farmed rainbow trout is still a key point of discussion.

Based on these considerations, a study focusing on the trend of the most important water quality parameters in rainbow trout farming of central Apennine in terms of long-term activity (2009–2019) was carried out. Before the decade focused on in this study, the first historical monitoring of water quality took place in a trial carried out in 2004, when the owners started to re-think the farming technique adopted until then, based on high stocking densities (40 kg/m^3), in order to produce a higher fish welfare status [17]. Considering this, the present study aimed to evaluate the suitability of changes adopted in rearing and feeding techniques to improve growth performance, sustainability and the water quality environment. The raceway water quality was monitored in terms of the Total Suspended Solids (TSS), Biochemical Oxygen Demand (BOD_5), Chemical Oxygen Demand (COD), Total Ammonia Nitrogen (TAN), Nitrites (NO_2-N), Nitrates (NO_3-N), pH and Total Phosphorus (TP). These parameters were investigated over a decade, starting in 2009 and were compared to the respective annual values of the 10 years after, until the year 2019, in order to show differences in the water quality and feed conversion rate.

2. Materials and Methods

2.1. Trout Farm Description and Water Quality Sampling

A study on the water quality trend was performed on a rainbow trout farm, located in the Apennine area of central Italy, based on raceways in parallel (120 m^3 each).

During the time of the study, the genetic line of rainbow trout did not change and it was directly controlled by the farm's owners, who had a broodstock farm in a different area from which fishes for fattening came and were selected.

On the rainbow trout farm in which the study was conducted, the feeding technique was the same during the entire experimental period (2009–2019) and the feeding rate was the same for fish at the same life cycle stage. The feed was distributed with a semi-moving wagon up to the level close to satiety and at the same time during the day, twice a day.

The water supply system used on the farm allowed at least one complete daily water change. The inlet water came to the farm system with a constant velocity of 0.25 m/s and flowed through four parallel raceways.

A water quality sample was obtained monthly during the last decade (2009–2019) in correspondence with the lagoon basin (outlet water) below the raceways and receiving downstream.

All samples were collected in the early morning before feeding using a polypropylene bottle with a screw cap [18].

In order to monitor the water quality, different physicochemical parameters were investigated. The dissolved oxygen, temperature and pH were measured using portable electronic devices (YSI mod. 55 and 60). At the same time, five samples of 500 cm^3 water were collected for laboratory-based determination of the following parameters: total suspended solids (TSS), biochemical oxygen demand (BOD_5), chemical oxygen demand (COD), total ammonia nitrogen (TAN), nitrites (NO_2-N), nitrates (NO_3-N) and total phosphorus (TP). Nitrogen (N) compounds and TP were determined using a spectrophotometer (Hach mod-2005, Hach Company, Loveland, USA), following the American Water Works Association and Water Pollution Control Federation of American Public Health Association (APHA) standard methods [19]. TSS was recorded following official methods [20]. BOD_5 and COD were determined according to IRSA-CNR [20] and ISPRA [18] methods, respectively.

In order to quantify the results of the waste water of the rainbow trout farm, the quality of the inlet water was also assessed by analyzing the values of TSS, COD, BOD_5, NO_2-N, NO_3-N and TAN, using the same laboratory methodology as described for the outlet water. Each sample of inlet water was collected using a sampler that conducted an average sampling process over three hours. During each year of the study, inlet water was collected seven times, in the months of January, April, May, June, August, October and December. The inlet water showed constant values for every parameter from 2009 to 2019, with the following averages: TSS, 5 mg/L; COD, 5 mg/L; BOD_5, 5 mg/L; NO_2-N, 0.09 mg/L; NO_3-N, 1.4 mg/L; TAN, 1.5 mg/L. Since these values were always suitable for farming rainbow trout, additional investigations and more frequent water samples were not necessary, allowing us to focus our attention on the analysis of the outlet water.

2.2. *Experimental Design and Rearing of the Rainbow Trout*

In order to summarize the methodology that was followed, Figure 1 shows the experimental design step by step. With reference to the first year (2009) and the final year (2019) of the considered decade, fish growth and water quality assessments were also evaluated by considering the balance of nitrogen and phosphorus released in waters. In a fish plant, the productive cycle starts from young rainbow trout, with an average body weight of 90 ± 2 g, which are grown at a stocking density of 20 kg/m^3 until reaching the market size (350 g). In particular, at the beginning of the life cycle, the fish density is low, about 8 kg/m^3 of water, in order to limit the stress of young trout and prevent gill diseases or bacterial infections. This density then increases with the increasing size of the trout until the end of the cycle, when it reaches 20 kg/m^3 of water.

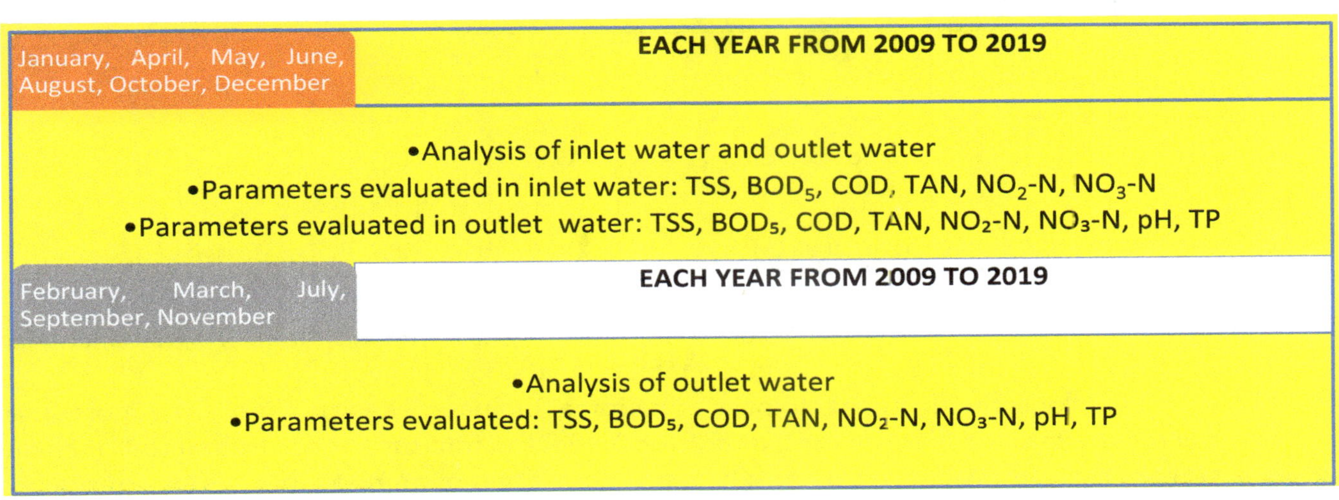

Figure 1. *Cont.*

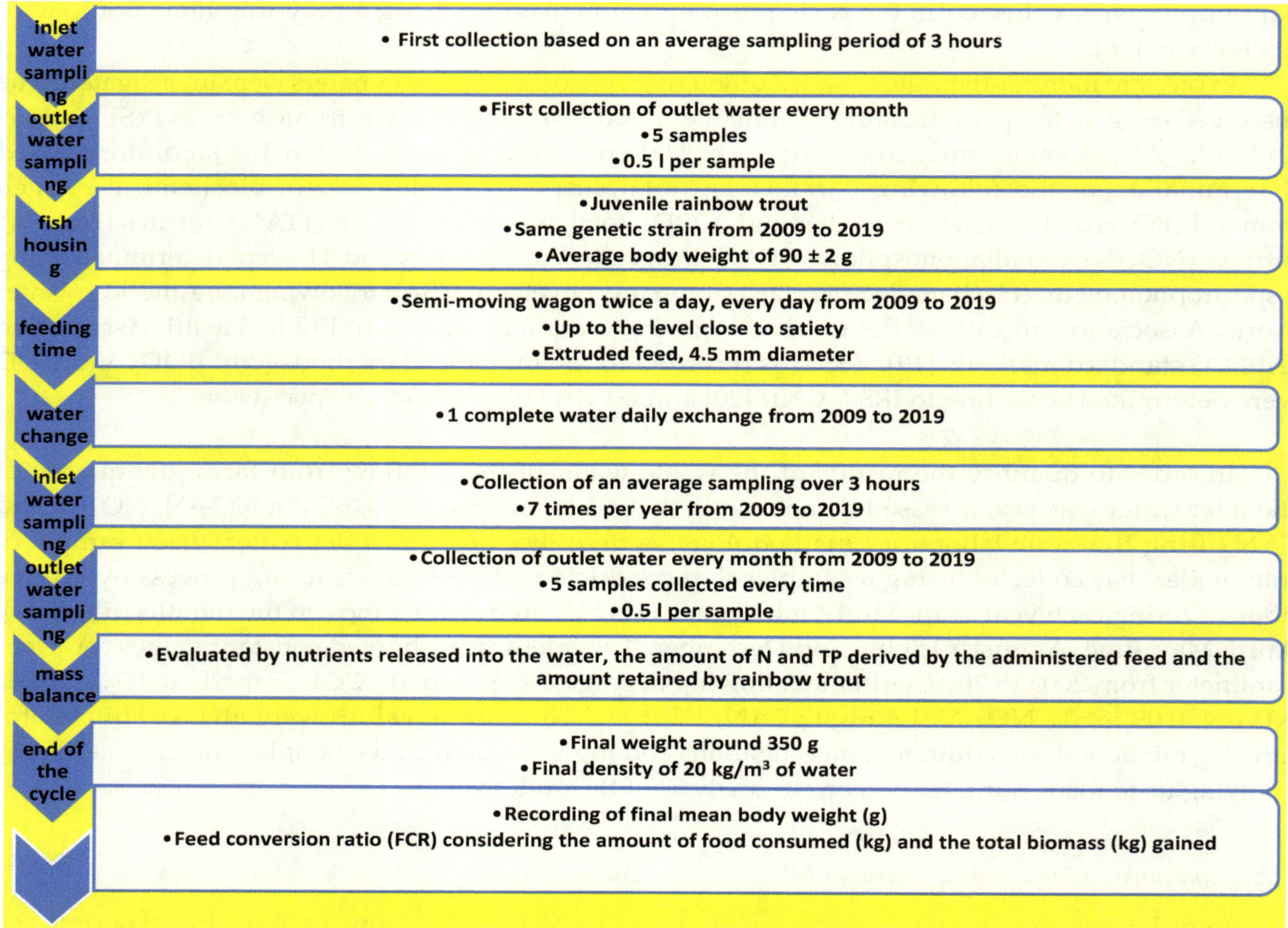

Figure 1. Diagram showing the experimental design adopted during the 10 years of the study.

Fish received extruded feed (closed formula) that was 4.5 mm in diameter and was manufactured by the same fish farmer who had vegetable feedstuffs available at his own land farm. The proximate composition of the two feeding types (pellet, extruded) employed to feed rainbow trout during the decade is reported in Table 1.

Table 1. Proximate composition of the feeds employed for the rainbow trout growing over the decade of water quality monitoring (2009–2019).

Feed	Pellet	Extruded
Chemical composition (%)		
Moisture	6.8	5.5
Crude protein	45.7	44.8
Crude lipid	16.0	21.0
Ash	6.7	8.4
Gross energy (MJ kg)	21.16	18.38

The final mean body weight (g) was recorded and the feed conversion ratio (FCR) was calculated from the amount of food consumed (kg) and the total biomass (kg) gained:

FCR = kg food consumed/kg final biomass − (kg initial biomass + kg sampled fish) + mortalities.

In order to evaluate the mass balance of nutrients released into the water, the amount of N and TP derived from the administered feed and the amount retained by rainbow trout were considered, as shown in Table 2, by comparing the budget of these compounds, expressed as the seasonal mean of the first year (2009) and the last year (2019) of study and applying the coefficients indicated by Bureau et al. [21] and applied by other authors [4].

Table 2. Total ammonia nitrogen (TAN) and phosphorus (TP) budget in the lagoon basin outlet water (mean ± standard deviation).

Year	TAN (mg/L)	TP (mg/L)
2009	0.55 ± 0.03	0.011 ± 0.001
2019	0.46 ± 0.02	0.009 ± 0.002

2.3. Statistical Analysis

All collected data on the outlet water quality were analyzed to determine whether there were significant differences over the 10 considered years. For this aim, the months of the year were divided to define the four seasons: winter included the months of December, January, February and March; spring was represented by April, May and June; summer included July, August and September; and autumn included October and November.

After all seasonal data were collected, the mean of each parameter was calculated season by season for every year and data were finally organized as graphics.

The seasonal mean of the investigated parameters (TSS, BOD_5, COD, NO_2-N, NO_3-N, TAN, TP and pH), recorded per year, was subjected to one-way analysis of variance (ANOVA) using the General Model Procedure of SPSS 25 (IBM Corp., New York, NY, USA) [22], in order to assess if data means were statistically different within the season of the different years. Significance was considered if $p < 0.05$ and the means were compared using the Student-Newman-Keuls (SNK) test.

3. Results

The results of the water quality parameters measured in the outlet lagoon basin in the 10 year-study are reported in Figures 2–9, comparing the means of the same season of the 10 years.

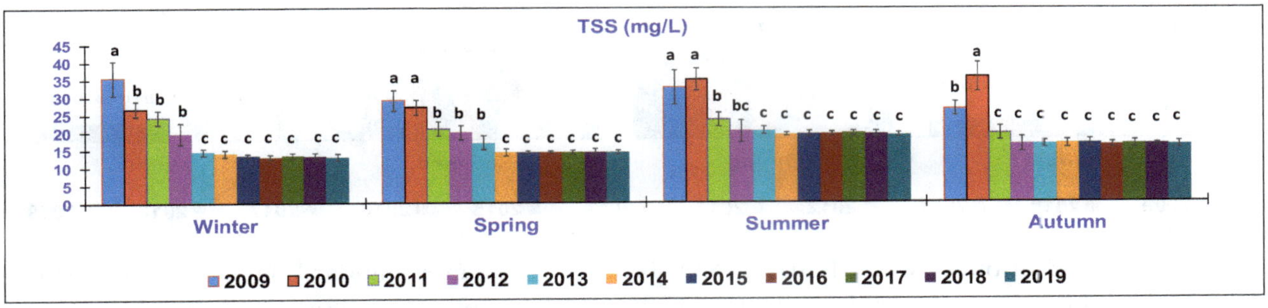

Figure 2. Trend of outlet water Total Suspended Solids (TSS) (mean ± standard deviation) seasonally determined during the 10 year-study. Different letters (a, b, c) per season show significant differences ($p < 0.05$) among the 10 years of sampling.

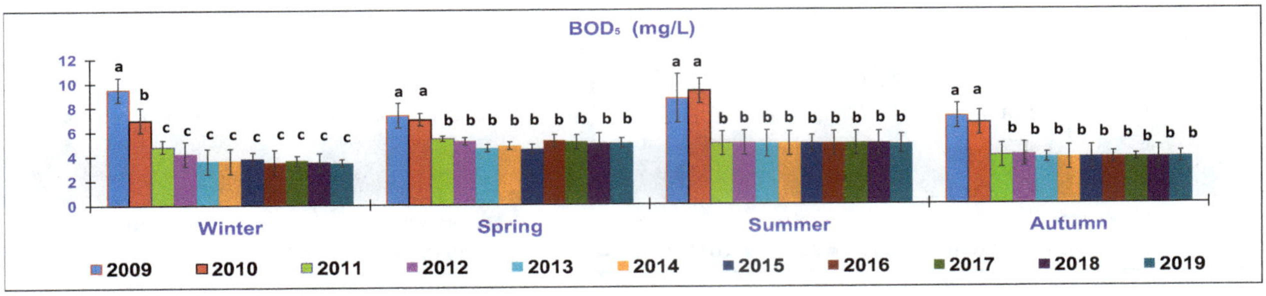

Figure 3. Trend of outlet water Biochemical Oxygen Demand (BOD_5) (mg/L) (mean ± standard deviation) seasonally determined during the 10 year-study. Different letters (a, b, c) per season show significant differences ($p < 0.05$) among the 10 years of sampling.

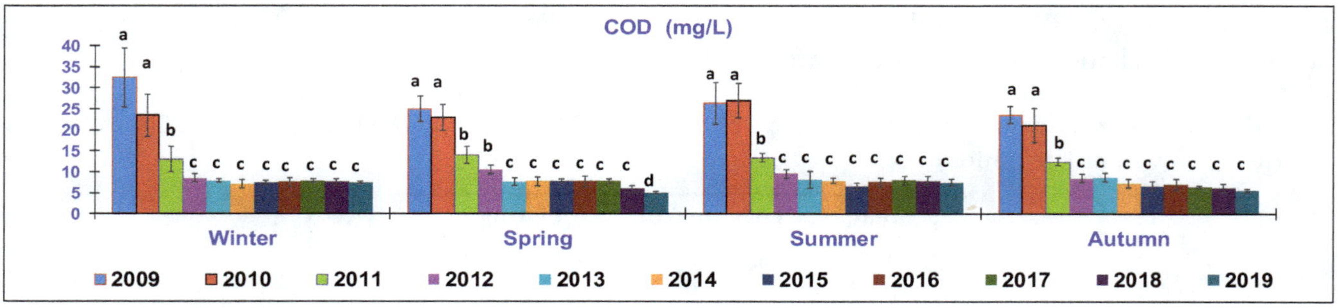

Figure 4. Trend of outlet water Chemical Oxygen Demand (COD) (mg/L) (mean ± standard deviation) seasonally determined during the 10 year-study. Different letters (a, b, c, d) per season show significant differences ($p < 0.05$) among the 10 years of sampling.

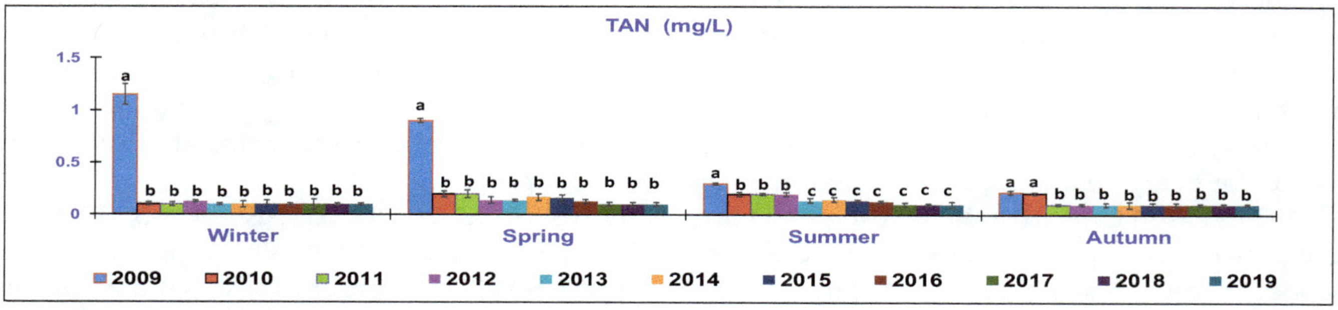

Figure 5. Trend of outlet water Total Ammonia Nitrogen (TAN) (mg/L) (mean ± standard deviation) seasonally determined during the 10 year-study. Different letters (a, b, c) per season show significant differences ($p < 0.05$) among the 10 years of sampling.

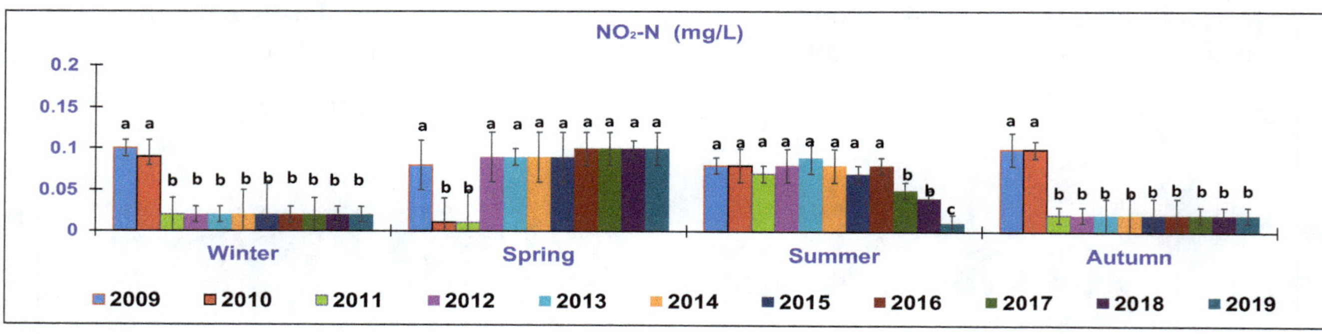

Figure 6. Trend of outlet water Nitrites (NO_2-N) (mg/L) (mean ± standard deviation) seasonally determined during the 10 year-study. Different letters (a, b, c) per season show significant differences ($p < 0.05$) among the 10 years of sampling.

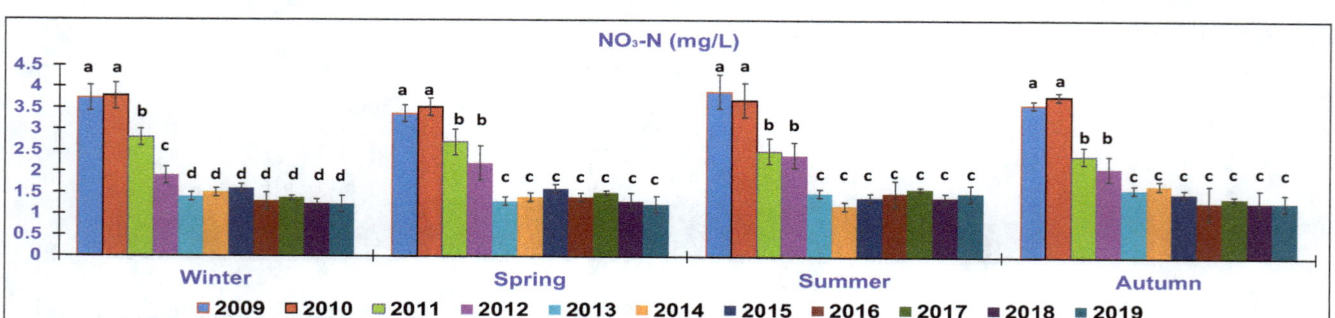

Figure 7. Trend of outlet water Nitrates (NO_3-N) (mg/L) (mean ± standard deviation) seasonally determined during the 10 year-study. Different letters (a, b, c, d) per season show significant differences ($p < 0.05$) among the 10 years of sampling.

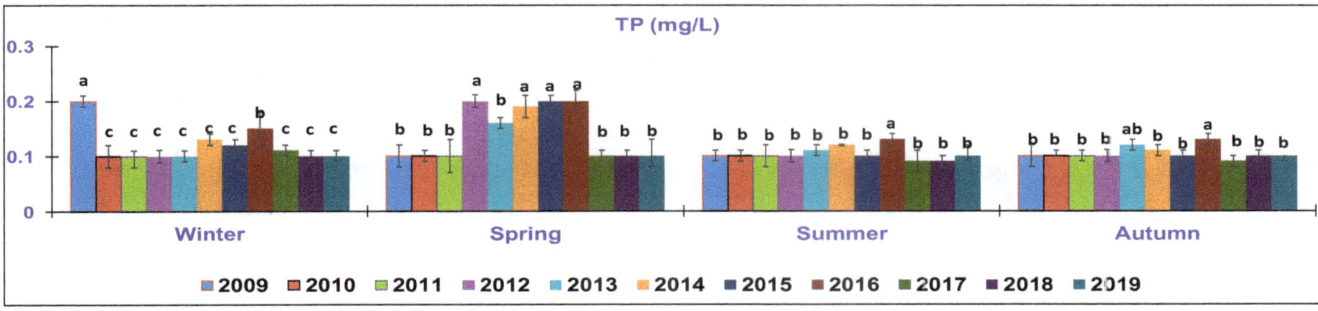

Figure 8. Trend of outlet water Total Phosphorus (TP) (mg/L) (mean ± standard deviation) seasonally determined during the 10 year-study. Different letters (a, b, c) per season show significant differences ($p < 0.05$) among the 10 years of sampling.

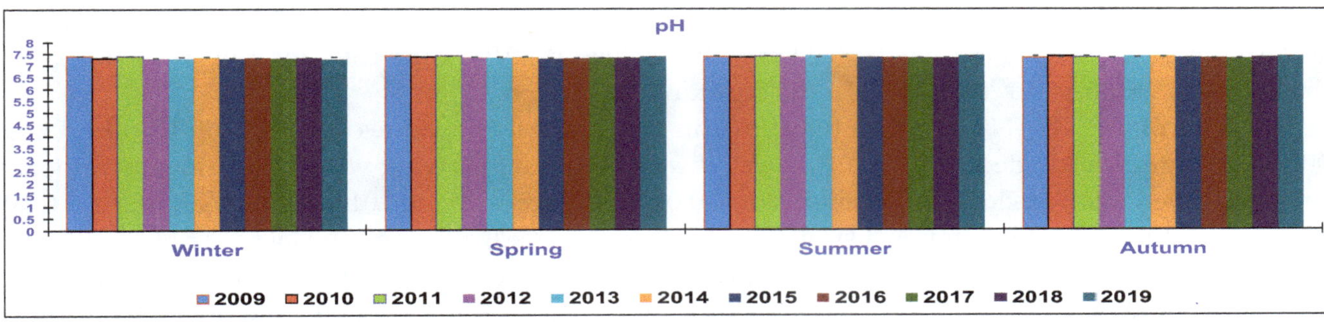

Figure 9. Trend of outlet water pH (mean ± standard deviation) seasonally determined during the 10 year-study.

The TSS (Figure 2) content showed a significant decrease when passing from the first seasons of sampling to the last years. In winter, TSS exhibited a significantly decreased mean value from 35.5 ± 5 mg/L in 2009 to 13 ± 3 mg/L in 2019, with an intermediate reduction notably shown in 2010–2012 (26.75–19.8 mg/L); in 2013 (14.5 ± 2 mg/L), TSS reached the lowest values until 2019. In spring, the same trend was observed, when the means decreased from 29 ± 3 and 27.3 ± 2.6 mg/L (2009–2010, respectively) to around 14.3 mg/L, which was maintained from 2014 to 2019. In the summer season, the recorded means were 37.20 ± 5 mg/L in 2009 and 35.4 ± 4 mg/L in 2010, without significant differences, whereas a marked reduction was observed from the year after (2011, 23.4 ± 3 mg/L), gradually diminishing from 2012 and remaining at around 16 mg/L until 2019. In Autumn, the highest TSS content was recorded in 2010 (35.5 ± 3.9 mg/L), being significantly different from the analyses of the first year of monitoring (2009, 26.5 ± 2 mg/L) but from 2011, the mean value was significantly lower and was maintained at around 16.00 ± mg/L until the end of the study. In particular, considering the seasonal average of TSS detected in Autumn, there was a reduction of almost 55% of this parameter.

Considering the trend of BOD_5 (Figure 3), there was a marked improvement in 2019 compared to the 10 years before. In winter, BOD_5 recorded a significant decrease, passing from the mean value of 9.5 ± 0.9 mg/L in 2009 to that of 7 ± 0.8 mg/L in 2010 and further decreased to around 3.7 ± 2 mg/L in the following years, until 2019. In spring, the highest levels were determined in 2009 (7.33 ± 1 mg/L) and 2010 (7.06 ± 0.7 mg/L); after this, the means significantly stayed at around 5.00 ± 0.7 mg/L until 2019. In summer, the highest levels of BOD_5 were observed in 2009–2010 (8.7–9.3 mg/L), which were significantly different from all of the following means, being around 4.9 ± 1 mg/L until the end of the monitoring period. In autumn, a similar trend was also observed, with a notable difference between the first two years of sampling (7.18–6.72 mg/L in 2009–2010) and the other following years, being around 3.7 ± 0.7 mg/L until the end of the study.

In terms of COD determination (Figure 4), all four seasons showed a similar trend, resulting in a sharp reduction through the years. In winter, the highest levels were recorded in the first two years (2009, 32.5 ± 5.1 mg/L; 2010, 23.5 ± 4.4 mg/L), significantly decreasing in 2011 (13 ± 3.1 mg/L) and reaching the lowest mean in 2012, when the average of around 7.5 mg/L was maintained until 2019.

In spring, the highest values occurred over the first two years (25 ± 1.3 mg/L) and then decreased in 2012–2013 (25–23.1 mg/L), until reaching the significantly lowest level in the last year (5 ± 0.2 mg/L). In summer, the first two years exhibited the highest levels (26.3–27 ± 1 mg/L). The mean value was decreased in 2011 (13.5 ± 0.3 mg/L) but was significantly higher than that in the following years, being around 7.5 ± 0.6 mg/L until the end of the study. In autumn, the mean of this parameter was 23.52 ± 0.5 mg/L in 2009 and 21.14 ± 0.7 in 2010; then, it decreased to 12.5 ± 0.3 mg/L in 2011 and reached 6.8–5.51 mg/L in the last years, without notable differences.

Regarding the TAN parameter (Figure 5), a clear reduction was observed in the range of the years taken into consideration. In the winter season, the data showed a significant decrease from the first year of study, from 1.15 ± 0.3 mg/L in 2009 to 0.1 ± 0.02 mg/L in 2019. In spring, TAN significantly decreased from 0.90 ± 0.2 mg/L in 2009 to 0.10 ± 0.02 mg/L in 2019. During summer time, a significant reduction was observed, that is, the value was 0.30 ± 0.01 mg/L in 2009 and significantly decreased from 2010–2012 (0.21–0.2 mg/L) and then further decreased to 0.1 ± 0.02 mg/L in 2019. In autumn, the TAN value decreased from 0.21–0.22 mg/L in 2009 and 2010 in the following years (2011–2019), when a mean value of 0.1 ± 0.02 mg/L was constantly recorded.

In terms of the NO_2-N parameter (Figure 6), in winter, the mean values (0.1–0.09 mg/L), detected in 2009–2010, significantly decreased to 0.02 mg/L throughout the following years. In spring, an opposite trend was noted, with the lowest concentrations (0.01 mg/L) occurring in 2010–2011 in comparison to all other years (0.1 mg/L). In summer, a similar trend was observed, although the last year of monitoring showed the lowest mean (0.01 ± 0.002 mg/L) recorded among the years. On the contrary, in autumn, the TAN level detected in the two first years (0.1 mg/L) significantly decreased to 0.02 ± 0.001 mg/L in 2019.

In terms of the NO_3-N parameter (Figure 7), a similar trend was observed in all the seasons, with a notable reduction emerging in 2013 and being maintained throughout the last years. In winter, the content dropped from 3.73–3.78 mg/L in 2009–2010 to only 1.22 ± 0.3 mg/L in 2019. In spring, the value significantly changed from 3.37 ± 0.5 mg/L in 2009 to 1.23 ± 0.3 mg/L in 2019. In summer, the value ranged from 3.90 ± 0.5 mg/L in 2009 to 1.50 ± 0.3 mg/L in 2019. In autumn, the concentration ranged from 3.60 ± 0.4 mg/L in 2009 to 1.30 ± 0.2 mg/L in 2019.

Considering the TP parameter (Figure 8), in winter, the values significantly decreased from 0.2 ± 0.01 mg/L in 2009 to 0.1 ± 0.002 mg/L in 2019. In spring, the highest values (0.19–0.2 mg/L) were recorded in the central part of the study (2014–2016); then, a reduction was observed, that is, 0.1 mg/L. In summer and winter, although there was a significantly higher level in 2016 (0.13 ± 00.2 mg/L), TP remained at a low value (0.1 mg/L) until 2019.

The pH parameter (Figure 9) did not show significant variation, remaining approximate neutral during the entire studied period and the obtained values were always in accordance with the range considered appropriate for rainbow trout [8,17].

At the end of the study, the FCR was 1.1 and significantly improved in comparison with the FCR obtained in 2009 (1.4). The survival rate was maintained at around 97% from 2012 onwards compared to the survival rate of the years before (94%).

In Table 2, the nutrient budget of outlet water is reported. In 2009, TAN was approximately 0.55 mg/L, whereas in 2019, the budget had an average value of 0.46 mg/L. The same trend was observed for the TP budget, which ranged from 0.011 mg/L (2009) to 0.009 mg/L over the 10 years.

4. Discussion

The aim of the current study was to investigate the impact of a new type of feed on water quality. Based on a comparison of the values of the analyzed parameters from 2009 to 2019, a notable decrease was observed in the TSS, BOD_5, COD, TAN, NO_2-N, NO_3-N, pH and TP.

As other studies have shown [8,23,24], BOD_5 is a very important parameter for assessing the water quality because it indicates the consumption of oxygen in the processes of indirect oxidation by the metabolic systems of aerobic microorganisms present in the water. Aside from the TSS level [24],

BOD_5 determination is essential for detecting the oxygen demand of the aerobic microbial flora required for the decomposition of organic substances present at a certain temperature in a defined time range. Therefore, it is an indicator that increases as the amount of organic substance to be mineralized increases.

The low values of BOD_5 detected during 2019, together with the decreased COD levels, indicate that aerobic microorganisms prevailed over anaerobic ones and therefore, the self-purifying capacity of the watercourse can be considered good. The reduction of the COD parameter that exceeded 76.5% is proof that the technical and management interventions implemented at the breeding rainbow trout farm effectively resulted in an improvement in the environmental protection. This result is in agreement with data presented by Galezan [24], while Tahar [8] reported lower BOD_5 values.

Based on analysis of the main water quality parameters of the trout farm over the last decade, a significant improvement in the rearing environment was shown. The significant reduction registered can be attributed to the different type of feed adopted, which changed from pelleted to extruded feed. Together with adequate quantities of oxygen dissolved in the water and through the nitrification process, a decreased amount of TAN can be converted into nitrites and nitrates. Nitrates are less toxic to fish over a long time period throughout the nitrification process [24]. A trend of the reduction of undesirable physicochemical traits was clearly observed when the new feeding type was introduced in 2011. Based on a comparison of the pellet feed with the extrusion technology, the digestible level of the diet was shown to have improved. This extruded manufacture drastically reduces the amount of food not eaten by trout and increases the feed efficiency [25].

The property of the extruded feed, which floats more than the pellet feed, increased its availability to the trout for a longer time in the water column before falling to the bottom of the raceways [9,26]. Fish can quickly intercept food without its dispersion and waste production, which occurs more frequently when using pellet feed because it sinks and degrades more easily and quickly in the water column.

In particular, concerning the two different feed techniques, that is, the pellet feed and the extruded feed, the processing style changed from the first to the second in 2011 but the raw materials remained the same. The fish meal and fish oil feedstuff were the same and both were of a high quality. The owner of the farm where the study was conducted monitored the feed quality through periodic laboratory analysis to ensure that the composition of the feed remained unchanged.

During the decade of the study, no differences in feed quality were shown and for the most important nutrients, no raw material changes were reported. For this reason, in the current study, the specific chemical analysis of the feed was not taken into consideration in order to stress the topic of water quality.

NO_2-N represents an important stage in the oxidation of organic substances containing nitrogen. It is therefore a transitory form of nitrogen which is transformed into nitrates due to the bacterial activity in the presence of optimal quantities of dissolved oxygen in the water tanks.

Nitrates represent the final indicator of the degradation protein processes; the main sources are correlated with anthropic activity and in surface waters, the trend is usually seasonal.

Taking into consideration the TSS parameters detected in 2009 and comparing them with those of 2019, a significant reduction was clear, especially the trend observed during winter and Autumn.

According to a number of studies that have focused their attention on the relationship between feeding management and water quality [2,3,11,23,24], the significant improvement in water quality at this farm was due to the adoption of the modern type of feed based on the extrusion technique. In fact, compared to the pellet food, the new extruded feeding technique showed a greater stability in water and it allowed the food to be available to trout for a longer period of time. There are other positive elements to consider: The extruded feed had a higher fat absorption capacity and it was possible to know its specific weight. Therefore, it was easy to control the fish feeding and, as a consequence, the composition of the meat obtained from this fish. Our study supported the idea that extrusion is nowadays the best processing feed technique in aquaculture, as previously shown by other studies.

Welker [26] compared the extrusion technique with the pellet feed by analyzing the water stability, fecal durability and digestibility and found the best results with the use of extruded feed. Similar results were reported by Tyapkova [9].

Moreover, in aquaculture, the extrusion technique positively affected the water quality: Food waste due to dust, breaks and "leaking/leaching" was decreased, improving the availability of nutrients for fish and minimizing the environmental impact. Another element of the extruded feed is that it is characterized by a low sedimentation rate. To ensure feed suitable for trout, a low sedimentation rate is one of the main physical properties to consider because it means the extruded feed is available to the fish for a long time; consequently, the fish can rapidly intercept the food without its dispersion and waste production. On the other hand, the use of pellet food means that feed is crumbled, which involves the fragmentation of food cylinders. As a consequence, this could lead to a loss of food (even if decreased). The feed was pulverized and this waste could remain at the bottom of the raceway water, becoming a possible source of water degradation, as well as microbial contamination.

Based on data concerning the monitoring of the rearing waters and considering the loads of N and TP of extruded feed, which took place in the last 10 years, it is possible to say that the improvement of the feed, administered to rainbow trout, is proof of the excellent quality of the ingredients used in fish diets [27,28].

Our data are in accordance with previous studies [3–5,8] that estimated ammonia and phosphorus emissions in water tanks from dietary analysis. In particular, we found lower TAN values than those reported by Moraes [5] and Aubin [4], which is a sign of the high feed quality used on the farm involved in our study. Concerning the p values, our results are in line with those reported by Moraes [5] and are lower than the values reported by Aubin [4] and Dalsgaard [23].

In this study, the amount of TAN and TP present in the tanks that in the past had hosted trout that received pelleted feed was compared to current tanks that host trout fed with extruded feed. The significantly decreased TAN and TP loads are proof of an improvement in diet quality, with a consequent benefit for the environment and the health of the trout.

This study provides insights into the connection between fish feed and water quality, that is, the fish habitat, which has to be suitable for producing healthy fish and as waste water, which returns to natural water bodies, taking into consideration specific parameters to investigate the water quality. In particular, in our study, an improvement trend was mostly demonstrated by the TSS, confirming that feed manufactured by the extrusion technique was of high quality. In particular, a notable positive change in the general trend was demonstrated after the new extruded food processing style was introduced, which occurred in 2011. In fact, all parameters showed an important decrease after the adoption of the extruded feed. More specifically, in the last years of the analyzed decade, a more favorable feed conversion rate was shown with respect to when pellet feed was administrated, resulting in better growing performances exhibited by trout whilst saving approximately 40% of the feed. In fact, the feed conversion index and the protein efficiency index improve when rainbow trout are fed with a rationing level equal to 70% of the "ad libitum technique" [7].

Another important change, which could have contributed to improving the efficiency feed conversion, could be justified by the fact that, on the farm, only trout larger than 90 g are reared, which do not require meal or crumbled feed. The fragmentation of crumbled diets leads to a more pulverized feed. In this case, the uneaten extruded feed, over 4 mm in size, is easier to remove than the pelleted diet.

Regarding the feedstuffs of the extruded feed, the inclusion of vegetable sources (soybean, wheat and pea) from the owners' farm, located close to the fish plant, can be reported as an example of sustainability with positive effects on the production cost. On the fish farm, the fish stocking density was decreased (20 kg/m^3) with respect to the first years of the decade, when rainbow trout were stocked at double the biomass. The choice to limit the culture density is the basis of the application of the "multisite" rearing technique, which aims to increase biosecurity, preventing diseases due to a vertical transmission of pathogens [29]. The decision of this plant to rear only rainbow trout starting from pre-fattened fish showed more advantages in terms of the survival rate, which was more satisfactory in the last years.

Concerning, more specifically, the fish farming density, before the decade focused on in this study and in 2004, the owners decreased the fish density in their raceways to improve the welfare of the rainbow trout. From 2009 to 2019, the fish density remained the same due to the effective improvement in animal welfare.

As previously shown by Welker [11], the correct management of tanks is essential to ensure good water conditions for trout and the appropriate use of water oxygenation systems contributes to maintaining optimal living conditions for these fish. This aspect could be connected to the improvement of the outlet basin receiving water from the raceways before being introduced into the stream.

Similar to what was observed by Becke [14], the pH was always close to neutrality and it was kept as stable as possible. This helped to minimize environmental stress and allowed nitrifying bacteria to effectively remove the nitrogen that accumulated in the sediments. Conversely, in other studies, the pH values were closer to 6 [5] or between 6.5 and 7.5 [8], which means that the effluent water was more acidic than the water considered here.

On the farm, the workers carried out normal system control every day, in order to ensure an oxygen level in the water that never dropped below 85% of the saturation level and its concentration was never less than 5 mg/L. The water supply system used on the farm allowed at least one complete daily water change, which directly affected the water chemical–physical parameters. Furthermore, the receiving water body has benefitted from these modified techniques, in particular the outlet basin, which was populated by floating macrophyte duckweed (*Lemna* spp.) that are known to be suitable for wastewater treatment and for food for pigs and poultry [30–32]. The duckweed filters nutrients, with a reduction of the eutrophic load of the water; it is an excellent purifying plant and absorbs the nitrogen in the water by eliminating nitrates [30,31].

Another aspect to consider is that, by comparing the parameters of the inlet water with those of the outlet water, which are the focus of the current study and due to the fact that the inlet water showed the same values over the 10 years of the study, it was possible to affirm that there was an improvement in the farming water quality. In fact, as shown before, the values of TSS, BOD_5, COD, TAN, NO_2-N and NO_3-N were lower in 2019 than 10 years before.

Many papers have focused on the environmental impact of trout farms on the natural receiving waters in terms of uneaten food, fish metabolites and chemical treatments. For example, in a study carried out by Tahar et al., in 2018 [8], inlet and outlet concentrations of water parameters for four consecutive flow-through rainbow trout farms over a ten-year period were analyzed in Ireland to characterize the impact of each fish farm on the water quality as a function of their production and to identify any seasonal variability.

5. Conclusions

In the present trial, water parameters were investigated to determine whether the new feeding strategy based on extruded feed changed the water composition. The analyzed parameters (TSS, BOD_5, COD, NO_2-N, NO_3-N, TAN and TP) showed an important improvement from 2009 to 2019; the pH parameter did not show important variation during the studied period.

Water is effectively the habitat where rainbow trout live so it must be monitored and preserved as best as possible in terms of temperature and physico-chemical parameters and quality. All these aspects are the basis of ensuring high-quality rainbow trout farming. Sustainable management, together with a genetic program of rainbow trout specimens employed, based on selected fish showing the best performance in feed efficiency, will be the next challenge to further improve the fish performance and environment.

Considering what has been discussed, the feedback of these results should be considered a significant index of improvement in the quality of the environment and the adoption of the modern formulation certainly contributed to obtaining this result.

Author Contributions: Conceptualization, A.R. and E.F.; methodology, E.F. and A.R.; formal analysis, all authors; data elaboration, E.F.; writing, review and editing, all authors; funding acquisition, A.R. All authors have read and agreed to the published version of the manuscript.

Acknowledgments: We are grateful to the Erede Rossi Company for their valuable collaboration and technical assistance during the water sampling activities.

References

1. European Environment Agency. *European Waters—Assessment of Status and Pressures*; European Environment Agency: Copenhagen, Denmark, 2018; pp. 21–29. ISBN 978-92-9213-947-6.
2. Thorpe, J.E.; Cho, C.Y. Minimising waste through bioenergetically and behaviourally based feeding strategies. *Water Sci. Technol.* **1995**, *31*, 29–40. [CrossRef]
3. Boyd, C.E.; Queiroz, J.F. Nitrogen and phosphorus loads by system, USEPA should consider system variables in setting new effluent rules. *Advocate* **2001**, *1*, 85–86.
4. Aubin, J.; Tocqueville, A.; Kaushik, S.J. Characterisation of waste output from flow-through trout farms in France: Comparison of nutrient mass-balance modelling and hydrological methods. *Aquat. Living Resour.* **2011**, *24*, 63–70. [CrossRef]
5. Moraes, M.D.A.B.; Carmo, C.F.D.; Ishikawa, C.M.; Tabata, Y.A.; Mercante, C.T.J. Daily mass balance of phosphorus and nitrogen in effluents of production sectors of trout farming system. *Acta Limnol. Bras.* **2015**, *27*, 330–340. [CrossRef]
6. Parisi, G.; Terova, G.; Gasco, L.; Piccolo, G.; Roncarati, A.; Moretti, V.M.; Pais, A. Current status and future perspectives of Italian finfish aquaculture. *Fish Biol. Fish.* **2013**, *24*, 15–73. [CrossRef]
7. Bureau, D.P.; Hua, K.; Cho, C.Y. Effect of feeding level on growth and nutrient deposition in rainbow trout (Oncorhynchus mykiss Walbaum) growing from 150 to 600 g. *Aquac. Res.* **2006**, *37*, 1090–1098. [CrossRef]
8. Tahar, A.; Kennedy, A.M.; Fitzgerald, R.D.; Clifford, E.; Rowan, N. Longitudinal evaluation of the impact of traditional rainbow trout farming on receiving water quality in Ireland. *Peer. J.* **2018**, *6*, e5281. [CrossRef]
9. Tyapkova, O.; Osen, R.; Wagenstaller, M.; Baier, B.; Specht, F.; Zacherl, C. Replacing fishmeal with oilseed cakes in fish feed—A study on the influence of processing parameters on the extrusion behavior and quality properties of the feed pellets. *J. Food Eng.* **2016**, *191*, 28–36. [CrossRef]
10. Funk, D.; Rokey, G.; Russell, K.; Wenger Manufacturing. Laboratory analysis and lab scale demonstrations highlight extrusion processing conditions critical for high performance RAS feed production. *Aquafeed Adv. Process. Formul.* **2019**, *11*, 15–17.
11. Welker, T.L.; Overturf, K.; Abernathy, J. Effect of aeration and oxygenation on growth and survival of rainbow trout in a commercial serial-pass, flow-through raceway system. *Aquac. Rep.* **2019**, *14*, 100194. [CrossRef]
12. Burel, C.; Boujard, T.; Tulli, F.; Kaushik, S.J. Digestibility of extruded peas, extruded lupin, and rapeseed meal in rainbow trout (Oncorhynchus mykiss) and turbot (Psetta maxima). *Aquaculture* **2000**, *188*, 285–298. [CrossRef]
13. Alzieu, C. Water—The medium for culture. In *Aquaculture*; Barnabé, G., Ed.; Ellis Horwood Ltd.: London, UK, 1990; Volume 1, pp. 37–60.
14. Becke, C.; Schumann, M.; Steinhagen, D.; Rojas-Tirado, P.; Geist, J.; Brinker, A. Effects of unionized ammonia and suspended solids on rainbow trout (Oncorhynchus mykiss) in recirculating aquaculture systems. *Aquaculture* **2019**, *499*, 348–357. [CrossRef]
15. Alabaster, J.S.; Lloyd, R. *Water Quality Criteria for Freshwater Fish*, 2nd ed.; Elsevier: Amsterdam, The Netherlands, 1982; ISBN 9781483163116.
16. European Environment Agency. *Annual Report*; European Environment Agency: Copenhagen, Denmark, 2004; ISBN 92-9167-761-2, ISSN 1561–2120.
17. Melotti, P.; Roncarati, A.; Angellotti, A.; Dees, A.; Magi, G.E.; Mazzini, C.; Bianchi, C.; Casciano, R. Effects of rearing density on rainbow trout welfare, determined by plasmatic and tissue parameters. *Ital. J. Anim. Sci.* **2004**, *3*, 393–400. [CrossRef]
18. ISPRA. *Procedura di Misurazione Per La Determinazione Della Richiesta Chimica Di Ossigeno (COD) Mediante Test in Cuvetta: Metodo 5135. Manuali e Linee Guida 117/2004*; ISPRA: Lombardy, Italy, 2014; ISBN 978-88-448-0678-1.
19. APHA (American Public Health Association, American Water Works Association and Water Pollution Control Federation). *Standard Methods for the Examination of Water and Wastewater*; A.P.H.A.: Washington, DC, USA, 1995.

20. IRSA-CNR; APAT. *Metodi Analitici Per Le Acque. Manuali e Linee Guida 29/2003*; IRSA-CNR/APAT: Rome, Italy, 2003; Volume 1, pp. 161–170.
21. Bureau, D.P.; Harris, A.M.; Cho, C.Y. Apparent digestibility of rendered animal protein ingredients for rainbow trout (*Oncorhynchus mykiss*). *Aquaculture* **1999**, *180*, 345–358. [CrossRef]
22. IBM Corp. *IBM SPSS Statistics for Windows, Version 25.0*; IBM Corp: Armonk, NY, USA, 2017.
23. Dalsgaard, J.; Pedersen, P.B. Solid and suspended/dissolved waste (N, P, O) from rainbow trout (*Oncorynchus mykiss*). *Aquaculture* **2011**, *313*, 92–99. [CrossRef]
24. Galezan, F.H.; Bayati, M.R.; Safari, O.; Rohani, A. Modeling oxygen and organic matter concentration in the intensive rainbow trout (Oncorhynchus mykiss) rearing system. *Environ. Monit. Assess.* **2020**, *192*. [CrossRef]
25. Kaushik, S.J. Feed management and on-farm feeding practices of temperate fish with special reference to salmonids. In *On-Farm Feeding and Feed Management in Aquaculture*; FAO Fisheries and Aquaculture Technical Paper No. 583; Hasan, M.R., New, M.B., Eds.; FAO: Rome, Italy, 2013; pp. 519–551.
26. Welker, T.L.; Overturf, K.; Snyder, S.; Liu, K.; Abernathy, J.; Frost, J.; Barrows, F.T. Effects of feed processing method (extrusion and expansion-compression pelleting) on water quality and growth of rainbow trout in a commercial setting. *J. Appl. Aquac.* **2018**, *30*, 97–124. [CrossRef]
27. Islam, M.S. Nitrogen and phosphorus budget in coastal and marine cage aquaculture and impacts of effuent loading on ecosystem: Review and analysis towards model development. *Mar. Pollut. Bull.* **2005**, *50*, 48–61. [CrossRef]
28. Jia, B.; Tang, Y.; Tian, L.; Franz, L.; Alewell, C.; Huang, J.-H. Impact of Fish Farming on Phosphorus in Reservoir Sediments. *Sci. Rep.* **2015**, *5*, 16617. [CrossRef]
29. Delabbio, J.L.; Johnson, G.R.; Murphy, B.R.; Hallerman, E.; Woart, A.; McMullin, S.L. Fish Disease and Biosecurity: Attitudes, Beliefs, and Perceptions of Managers and Owners of Commercial Finfish Recirculating Facilities in the United States and Canada. *J. Aquat. Anim. Health* **2005**, *17*, 153–159. [CrossRef]
30. FAO. *Duckweed: A Tiny Aquatic Plant with Enormous Potential for Agriculture and Environment*; Food and Agricultural Organization of the United Nations: Geneva, Switzerland, 2010.
31. Stadtlander, T.; Forster, S.; Rosskothen, D.; Leiber, F. Slurry-grown duckweed (Spirodela polyrhiza) as a means to recycle nitrogen into feed for rainbow trout fry. *J. Clean. Prod.* **2019**, *228*, 86–93. [CrossRef]
32. Chakrabarti, R.; Clark, W.D.; Sharma, J.G.; Goswami, R.K.; Shrivastav, A.K.; Tocher, D.R. Mass Production of Lemna minor and Its Amino Acid and Fatty Acid Profiles. *Front. Chem.* **2018**, *6*, 479. [CrossRef] [PubMed]

14

Protein and Amino Acid Content in Four Brands of Commercial Table Eggs in Retail Markets in Relation to Human Requirements

Youssef A. Attia [1,*], Mohammed A. Al-Harthi [1], Mohamed A. Korish [1] and Mohamed H. Shiboob [2]

1 Arid Land Agriculture Department, Faculty of Meteorology, Environment, and Arid Land Agriculture, King Abdulaziz University, Jeddah 21589, Saudi Arabia; malharthi@kau.edu.sa (M.A.A.-H.); mmkorish@yahoo.com (M.A.K.)

2 Environmental Department, Faculty of Meteorology, Environment and Arid land Agriculture, King Abdulaziz University, P.O. Box 80208, Jeddah 21589, Saudi Arabia; mshiboob@kau.edu.sa

* Correspondence: yaattia@kau.edu.sa

Simple Summary: At present, great attention has been paid to the nutritional values of animal products to resilient humans against pathogens, boost their immunity, and cure diseases. Thus, this research investigated the nutritional value of four sources of commercial table eggs in the retail market in Jeddah, KSA, with the possible presence of raw protein, amino acid content, and protein quality indicators for different parts of eggs. The examined eggs showed a different percentage of essential and non-essential amino acids and antioxidant amino acids, suggesting a potential for enriching the nutritional values and prolonging the shelf life of the eggs by various nutritional strategic ways to enhance the antioxidant amino acids and the essential amino acid profile in eggs.

Abstract: Considering the common believe that all eggs in the retail market are nutritionally similar, four different commercial sources of eggs (A, B, C, and D) available in a retail market were collected to investigate the crude protein and amino acid content, as well as the protein quality in the whole edible part of eggs (albumen + yolk), egg albumen, and egg yolk, separately. Five egg samples per source were collected four times during the experimental period, which resulted in a total number of 20 samples that were pooled to finally present five samples per source of eggs. The results show that crude protein in albumen was significantly higher in A and B than that of C and D, but the difference was found among edible parts of eggs such as yolk > whole edible part > albumen. Essential amino acids (arginine, histidine, isoleucine, lysine, methionine, methionine + cysteine, phenylalanine, phenylalanine + tyrosine, threonine, and valine) of eggs significantly differed according to the source of eggs, but eggs from different sources could provide from 17.4–26.7% of recommended daily allowance (RDA) of amino acids for adults. Essential amino acids (EAAs) were higher ($p \leq 0.05$) in eggs from sources A and B than in source D, while source C exhibited intermediate values. Source B had greater ($p \leq 0.05$) non-essential amino acids (NEAAs) than did sources C and D in whole edible egg, while source A displayed intermediate values. The phenylalanine + tyrosine, histidine, and lysine were the 1st, 2nd, and 3rd limiting amino acids in all sources of eggs. In conclusion, the investigated eggs showed different EAAs/NEAAs ratio and antioxidant amino acids, indicating a potential for enhancing nutritional values and extending the shelf life of eggs by different nutritional additions.

Keywords: egg sources; amino acids ratios; antioxidant; recommended daily allowance

---

1. Introduction

Eggs are a portion of essential food for human consumption and contain most of the necessary nutrients, based on the daily need [1,2]. Eggs have all the crucial nutrients for life and are a valuable source of protein\amino acids, antioxidants, and bioactive components [3,4]. Furthermore, there has been considerable progress in the egg production industry, including the improvement of the genetic makeup of layers and production capacity, egg quality, and layers management [5,6].

Protein and amino acids are primary components of eggs and play critical roles in egg consumption and nutrition as they present the main part of the muscle, body function, hormones, enzymes and body fluids [7,8]. The total amino acids (TAAs) in eggs were 10.0 and 10.1 mg/g in the dry yolk of eggs in corn- and wheat-based diets, respectively [4]. According to the same authors, lysine was found to be 929, 1182 and 760 mg/100 g in fresh whole eggs, yolk, and albumen, respectively; for methionine, the corresponding values were 400, 375, and 396 [9]. The essential amino acids (EAAs) on dry yolk accounted for about 46–47%, while aromatic amino acids represented about 11% of the TAAs; arginine was the most copious amino acid (about 1.5 mg/g) in all samples, followed by glutamic acid (1.2 mg/g) and lysine (0.9 mg/g). The authors added that yolk samples content of the TAAs decreased by different cooking procedures, except for the boiled yolk of hens that were fed a corn-based diet, and tyrosine and tryptophan were noticed to be the key providers to the antioxidant properties of eggs.

Eggs were also found to be a rich source of antioxidants [4,9] with the amino acid cysteine having the potential as an antioxidant amino acid, and the combination of strong, weak, and non-antioxidant amino acids increase the antioxidant capacity [9,10]. Besides, the amino acid content of eggs was found to be affected by poultry breeds [5] and species [6], preparation method [11,12] and components (whole, albumen, and yolk) of an egg [13]. The literature review indicates that there is room to improve the nutritional values of eggs by breeds and dietary and management practices due to the differences found in amino acids and antioxidant profiles of table eggs [1,5,6]. However, data for amino acids profile and amino acids ratio of eggs in the retail market are scarce, and there is a common belief that all eggs are nutritionally equal. The present research aims to investigate the crude protein and amino acid content and to evaluate the protein and amino acid ratio in commercial table eggs from four sources in the retail market in terms of recommended daily allowance (RDA) for adults.

2. Materials and Methods

2.1. Sample Collection

The Deanship of Scientific Research, King Abdulaziz University Saudi Arabia, approved the experimental procedures under protocol no. D-156-155-1438. It recommends animal rights, welfare, and minimal stress and did not cause any harm or suffering to animals, according to the Royal Decree M59 on 14/9/1431H.

Four commercial table egg sources, named A, B, C, and D in the retail market from white eggshell hybrids laying hens, originated from Single Comb White Leghorn commercially available in Saudi Arabia. The hens had 40–60 weeks of age. The hens were housed in cages in environmental control houses, and fed commercial layer diets contained 17% crude protein (CP), 11.60 MJ/kg, 3.5% Ca, and 0.35% none-phytate phosphorus kcal, in mash form and offered free access to feed and water and eliminated with 14:10 light-dark cycle. The medical care, vaccination, and husbandry practice were as suggested by the primary breeders and carried under the supervision of a veterinarian.

Fresh eggs of grade A classes, stored at 5 °C, were gathered randomly to represent various sources of eggs in retail markets in Jeddah city, Saudi Arabia. The eggs were clean, of normal eggshell index, and of medium size, 55–60 g, of grade A. Thirty eggs were collected from A, B, C, D sources four times during February, March, April, and May 2017. The eggs of each source per time were broken opened, and five samples of each albumen, yolk, and albumen + yolk were randomly collected. Thus, there were 20 egg samples per time per egg part. After the last egg collection in May 2017, at 60 weeks of age, the samples (20 samples) of each egg source per part were pooled over time to finally present five samples per source of eggs per egg part albumen, yolk, and albumen + yolk, and thus used for protein and amino acids analyses.

2.2. Measurements

2.2.1. Determination of Crude Protein

The protein content of eggs (yolk, albumen and yolk with albumen) was determined following method number 954.01, the principle of the Kjeldahl method [14]. One gram of sample was digested using 15 mL sulphuric acid (96%), utilizing an electrically heated block digester. The resultant digest was made alkaline by dilution with 50 mL of 40% sodium hydroxide. Thereafter the diluted sample was rapidly steam distilled for ammonia into 25 mL 4% boric acid. The sample was then manually titrated with 0.2 N hydrochloric acid. The protein content was estimated using 6.25 as a conversion factor of nitrogen to crude protein. Triplicate samples were analyzed, and the results are expressed as g/100 g dry matter basis of a sample [14].

2.2.2. Determination of Amino Acids

The egg samples were dried using method number 934.01, and defatted using method number 920.39, as per reference [14]. A part of each sample, 0.2 g was hydrolyzed with 6 N HCl (10 mL) in a sealed tube and then heated in an oven at 100 °C for 24 h [15]. For the analysis of methionine and cysteine, samples were oxidized by performic acid before hydrolysis [16]. The resulting solution was brought to 25 mL with de-ionized water. After filtration, 5 mL of hydrolysate was evaporated until free from HCl. Then, the residue was dissolved in a diluting citrate buffer. The amino acids were measured using an Automatic Amino Acid Analyzer model AAA400 (Ingos Ltd., K Nouzovu 2090, 14316 Prague 4, Czech). The column was filled with Resin material and Ninhydrin reagent. The separation of amino acids depends on using different gradient pH buffers. Acid hydrolysis of samples was carried out following [15,17]. Tryptophan was determined, as reported by [18] and modified by [19].

2.2.3. Predicted Protein Quality and Amino Acids Ratios

The predicted protein efficiency ratio (P-PER) of the different sources of eggs was estimated from their amino acid content, according to [20]:

$$\text{P-PER} = -0.468 + 0.454\ (\text{Leucine}) - 0.105\ (\text{Tyrosine})$$

2.2.4. Amino Acids Ratios of Different Egg Sources

The quality of the amino acids was determined by estimating the ratio of amino acids in the egg samples compared with the requirements stated as a ratio [21]. The Amino Acid Score (AAS) was then predicted by applying the formula in reference [22].

The essential amino acid to TAAs ratio and the cystine to sulfur amino acids ratio were, thus calculated. The EAAs/non-essential amino acids (NEAAs) ratio was also calculated. The total of the aromatic amino acids was also estimated, and the ratio of aromatic amino acids to TAAs was calculated.

Tyrosine + tryptophan to TAAs and total aromatic amino acids were also calculated as indices of the antioxidant property of eggs [4].

2.3. Statistical Analysis

The Data were evaluated using the one-way ANOVA of SAS software program [23] according to the following model:

$$Yij = \mu + Ai + eij,$$

where: μ = general mean; Ai: influence of the source of the egg; eij: random error. The same model was used for the evaluation of amino acid patterns of different parts of the eggs.

The percentage of data were normalized by applying transformations to arcsin before running the analysis. Means differences were compared using the Student-Newman Keuls test ($p \leq 0.05$).

3. Results and Discussion

3.1. Growth Performance Crude Protein and Amino Acids Pattern of Whole Egg

The content of the crude protein and amino acids in the whole edible part of the egg is presented in Table 1. Crude protein of whole eggs was not significantly different among different egg sources. The results are in agreement with those recently published by Secci et al. [24,25] Arginine, lysine, threonine, and proline were significantly greater in eggs from sources C and D than in those of A, B, but phenylalanine + tyrosine (aromatic amino acids along with histidine and tryptophan) showed the contrary trend. However, the difference between sources A and B in these amino acids was not significant. Tyrosine is conditional NEAAs and an important for photosynthesis and signal transduction processes [26]. Phenylketonuria is a genetic disease in which the tyrosine becomes EAAs because tyrosine cannot be synthesized from phenylalanine [26]. The EAAs/NEAAs were significantly greater for sources A, B, and then source D, and the latter was greater than source C.

Table 1. Crude protein and amino acid patterns of whole edible parts of eggs of different sources in the retail market.

Crude Protein (g/100 g)/Amino Acid (mg/100 g)	Source of Eggs				Statistical Analysis	
	A	B	C	D	RMSE	*p*-Value
Crude protein	14.03	14.02	13.77	13.88	0.395	0.721
Arginine	769.0 ^{b}	693.3 ^{b}	915.3 ^{a}	882.3 ^{a}	61.55	0.001
Histidine	296.8 ^{bc}	279.0 ^{c}	315.8 ^{ab}	331.8 ^{a}	12.94	0.001
Isoleucine	698.3 ^{b}	720.0 ^{a}	641.8 ^{c}	613.3 ^{d}	10.82	0.001
Leucine	1100.0	1081.3	1128.0	1057.8	36.19	0.097
Lysine	923.0 ^{b}	870.3 ^{c}	1019.3 ^{a}	1012.8 ^{a}	19.21	0.001
Methionine	394.3 ^{b}	434.3 ^{a}	301.8 ^{c}	285.0 ^{c}	16.87	0.001
Methionine + cysteine	554.5 ^{b}	627.5 ^{a}	513.5 ^{c}	490.0 ^{c}	30 01	0.002
Phenylalanine	679.8 ^{b}	744.5 ^{a}	537.8 ^{d}	618.3 ^{c}	32.51	0.001
Phenylalanine + tyrosine	1199.0 ^{a}	1237.8 ^{a}	1099.0 ^{b}	1137.8 ^{b}	30.01	0.002
Threonine	613.0 ^{b}	578.3 ^{b}	674.8 ^{a}	699.5 ^{a}	23.48	0.001
Tryptophan	147.0	145.5	132.0	148.0	9.04	0.094
Valine	781.8 ^{b}	815.3 ^{a}	707.8 ^{c}	673.8 ^{d}	15.91	0.001
EAAs	5634.0 ^{b}	5668.3 ^{a}	5458.0 ^{c}	5439.5 ^{c}	106.4	0.021
Alanine	713.0 ^{a}	737.3 ^{a}	651.0 ^{b}	558.0 ^{c}	20.89	0.001
Aspartic acid	1295.8 ^{a}	1310.5 ^{a}	1260.0 ^{ab}	1218.5 ^{b}	36.66	0.019
Glutamic acid	1687.0 ^{a}	1713.5 ^{a}	1642.8^{a}	1555.3 ^{b}	37.32	0.001
Glycine	426.0 ^{ab}	441.3 ^{ab}	381.8 ^{b}	473.8 ^{a}	32.42	0.014
Proline	507.0 ^{b}	493.3 ^{b}	523.8 ^{a}	535.8 ^{a}	10.75	0.001
Serine	956.3 ^{c}	881.5 ^{d}	1099.8 ^{a}	1055.3 ^{b}	23.54	0.001
NEAAs	7033.5	6957.5	7247.0	7001.8	167.9	0.129
EAAs/NEAAs	0.801 ^{a}	0.815 ^{a}	0.753 ^{c}	0.777 ^{b}	0.0214	0.011

$^{a–d}$ means with varying superscripts differ significantly ($p < 0.05$); RMSE, Root mean square error; *p* value, probability level; EAAs, total essential amino acids; NEAAs, total non-essential amino acids; EAAs/NEAAs, total essential amino acids/total non-essential amino acids ratio.

Glycine, am NEAA, was higher in the D source than in the C source. Isoleucine, methionine, and methionine + cystine were significantly higher in the B source than in other sources. In addition, source A showed greater isoleucine, methionine, and methionine + cysteine than did sources C and D, although source C exhibited higher isoleucine than did source D.

Phenylalanine, an EAAs for the biosynthesis of norepinephrine and epinephrine, and valine, which is important amino acids for maintaining muscles, as well as for the regulation of the immune system were significantly higher in source B than in other sources. The other sources exhibited significant differences, showing a trend of B > A > D > C for phenylalanine and B > A > C > D for valine. Valine, along with leucine and isoleucine, are branched-chain amino acids and represented about two-thirds of amino acids in the body protein [27]. In whole edible parts of eggs, branched-chain amino acids amounted to ~45% of EAAs. Alanine, the 4th principal NEAAs, was similar in sources A and B, which was significantly greater than in sources C and D. The latter sources showed significant differences between them, with source C having higher. A similar trend was found for aspartic acid, but the differences between source C and other sources were not significant.

Glutamic acid, the chief NEAA, was significantly greater in sources A, B, and C than in source D. Serine, which is important in biosynthesis of purine and pyrimidines, was significantly higher in source C than in the other sources, which showed significant differences among them in the following order: D > A > B. These findings reflected to some extent difference in amino acids in egg albumen and egg yolk and indicated that eggs from various sources have different amino acid patterns, which could positively affect the nutritional values and consumer preference [4,27].

3.2. Crude Protein and Amino Acid Patterns of Albumen

Table 2 reports the crude protein and amino acid composition of the egg albumen. It was observed that the differences among the egg sources were significant for crude protein with sources A and B had usually higher values than sources C and D. Egg source shows a significant effect on some amino acids except for arginine, histidine (precursor for histamine and carnosine synthesis), lysine, threonine, tryptophan (important EAAs along with other EAAs in biosynthesis of protein), valine, alanine, aspartic, serine, and EAAs/NEAAs.

Table 2. Crude protein and amino acid patterns of albumen of eggs of different sources in the retail market.

Crude Protein (g/100 g)/Amino Acid (mg/100 g)	Source of Eggs				Statistical Analysis	
	A	B	C	D	RMSE	*p*-Value
Crude protein	12.33 ^{a}	12.45 ^{a}	11.73 ^{b}	11.83 ^{b}	0.536	0.036
Arginine	543.5	544.5	516.5	497.8	38.16	0.293
Histidine	221.8	222.8	194.8	198.8	19.16	0.130
Isoleucine	566.5	567.5	567.0	503.3	31.44	0.035
Leucine	845.0 ^{a}	845.8 ^{a}	818.3 ^{ab}	767.8 ^{b}	35.79	0.033
Lysine	745.8	746.8	743.5	738.0	25.48	0.962
Methionine	342.5 ^{a}	343.5 ^{a}	315.3 ^{a}	262.5 ^{b}	17.28	0.001
Methionine + Cystine	497.3 ^{a}	499.3 ^{a}	443.5 ^{b}	381.5 ^{c}	25.30	0.001
Phenylalanine	583.8 ^{a}	584.8 ^{a}	556.8 ^{ab}	519.5 ^{b}	25.24	0.012
Phenyalanine + Tyrosine	972.6 ^{a}	974.3 ^{a}	918.0 ^{b}	868.0 ^{c}	30.22	0.001
Threonine	454.5	455.5	427.3	424.0	27.11	0.251
Tryptophan	118.0	118.8	117.8	132.0	15.53	0.354
Valine	613.0	588.5	585.8	564.8	62.73	0.758
EAAs	4489.8 ^{a}	4473.5 ^{a}	4320.8 ^{b}	4110.8 ^{c}	76.45	0.001
Alanine	577.8	578.5	552.0	581.3	47.12	0.799
Aspartic acid	1023.5	1024.5	996.3	1035.3	29.22	0.325
Glutamic acid	1336.5 ^{a}	1337.8 ^{a}	1309.8 ^{a}	1066.5 ^{b}	34.65	0.001
Glycine	347.8 ^{b}	348.8 ^{b}	321.0 ^{b}	598.0 ^{a}	22.65	0.001
Proline	388.5 ^{a}	389.5 ^{a}	361.3 ^{b}	328.8 ^{c}	14.57	0.003
Serine	689.8	690.8	663.0	662.3	18.41	0.077
NEAAs	5450.8 ^{a}	5459.7 ^{a}	5209.8 ^{b}	5137.5 ^{b}	130.8	0.009
EAAs/NEAAs	0.824	0.819	0.829	0.800	0.0261	0.423

$^{a–c}$ means with varying superscripts differ significantly ($p < 0.05$); RMSE, Root mean square error; p value, probability level; EAAs, total essential amino acids; NEAAs, total non-essential amino acids; EAAs/NEAAs, total essential amino acids/total non-essential amino acids ratio.

Leucine and phenylalanine were similar in sources A and B and markedly ($p \leq 0.05$) higher than in source D, while the C source did not significantly differ from the other sources. Methionine and glutamic acid were significantly greater of sources A, B, and C than that of source D. Methionine + cystine, phenylalanine + tyrosine, EAAs, and proline were similar in sources A and B and significantly lower in sources C and D. Source D exhibited significantly lower values than source C. Glycine was significantly higher in source D than in the other sources. Non-essential amino acids were similar in sources A and B and significantly lower in sources C and D, with no significant differences between the latter sources.

3.3. *Crude Protein and Amino Acid Patterns of Egg Yolk*

Table 3 shows the findings related to crude protein and amino acid profile of egg yolk. The difference in crude protein was not significant among various sources. The observed differences between the sources of eggs were significant ($p \leq 0.05$) for some amino acids except for arginine, lysine, phenylalanine + tyrosine, threonine, tryptophan, valine, serine, and EAAs/NEAAs ratio.

Table 3. Crude protein and amino acid patterns of the yolk of eggs of different sources in the retail market.

Crude Protein (g/100 g)/Amino Acid (mg/100 g)	Source of Eggs				Statistical Analysis	
	A	B	C	D	RMSE	*p*-Value
Crude protein	15.97	16.12	15.83	16.04	0.796	0.873
Arginine	1156.3	1150.0	1133.8	1131.3	24.78	0.440
Histidine	418.3 [a]	412.0 [a]	393.3 [ab]	375.5 [b]	16.68	0.016
Isoleucine	819.8 [a]	813.0 [a]	795.3 [a]	712.5 [b]	19.11	0.001
Leucine	1418.0 [a]	1411.8 [a]	1393.3 [a]	1317.5 [b]	15.49	0.001
Lysine	1284.3	1278.0	1236.8	1243.5	33.96	0.174
Methionine	401.0 [a]	394.8 [a]	353.5 [b]	348.8 [b]	19.27	0.004
Methionine + cysteine	531.8 [c]	678.8 [a]	598.8 [b]	624.5 [ab]	36.01	0.001
Phenylalanine	691.8 [a]	685.0 [a]	655.8 [b]	644.3 [b]	17.53	0.008
Phenylalanine + tyrosine	1412.5	1399.0	1363.0	1342.0	36.99	0.073
Threonine	860.3	854.0	822.3	888.0	31.52	0.082
Tryptophan	185.5	218.5	263.8	208.8	35.89	0.059
Valine	860.8	854.3	865.8	878.0	58.07	0.947
EAAs	6939.8 [a]	6921.5 [a]	6768.0 [ab]	6628.5 [b]	112.7	0.008
Alanine	831.0 [a]	824.8 [a]	788.5 [ab]	761.3 [b]	22.40	0.003
Aspartic acid	1580.8 [a]	1574.3 [a]	1547.8 [a]	1492.3 [b]	33.99	0.014
Glutamic acid	2051.8 [a]	2045.0 [a]	2024.3 [a]	1958.0 [b]	27.41	0.002
Glycine	499.8 [b]	493.3 [b]	464.8 [c]	537.8 [a]	14.36	0.002
Proline	674.3 [a]	668.0 [a]	632.5 [b]	626.8 [b]	20.74	0.015
Serine	1343.5	1377.0	1335.8	1315.3	38.71	0.212
NEAAs	8988.5 [ab]	9131.0 [a]	8864.5 [b]	8811.5 [b]	115.1	0.011
EAAs/NEAAs	0.772	0.758	0.763	0.752	0.0169	0.481

[a–c] Means with varying superscripts differ significantly ($p < 0.05$); RMSE, Root mean square error; *p* value, probability level; EAAs, total essential amino acids; NEAAs, total non-essential amino acids; EAAs/NEAAs, total essential amino acids/total non-essential amino acids ratio.

Isoleucine, leucine, aspartic acid, and glutamic acid were similar in sources A, B, and C, but significantly lower in source D. Methionine, phenylalanine, and proline were similar in eggs from sources A and B and in sources C and D, although differences between the former and the latter groups were remarkable ($p \leq 0.05$) in favor of the former group.

Methionine + cystine was significantly higher in source B than in sources A and C, while the C source had higher values than source A. Source D also had higher values than that of source A but did not significantly differ from sources B and C. Essential amino acids and alanine were significantly lower in source D than in sources A and B, while source C had intermediate values. In contrast, glycine was significantly higher in source D than in other sources. In addition, sources A and B showed similar values for glycine, which were greater than in source C. Glycine is a precursor to proteins and for the biosynthesis of collagen and considered the simplest amino acid that can be formed from serine and this reaction is reversible [28].

Essential amino acids were higher ($p \leq 0.05$) in egg yolk from sources A and B than in source D, while source C exhibited intermediate values. Source B had greater ($p \leq 0.05$) NEAAs than did sources C and D, while source A had intermediate values.

In general, the results indicated that EAAs were the highest in source A and B of both egg albumen and egg yolk. However, the EAAs/NEAAs ratio was not different among various sources of eggs. Leucine, followed by lysine and arginine, were the most abundant EAAs. Glutamic acid and aspartic acids were the predominant NEAAs in the whole edible parts of eggs and the egg yolk. The principal amino acids in egg albumen were leucine, lysine, and valine, with the glutamic acid and aspartic acid were the chief NEAAs. These results are similar to the results reported by [9], the sum of phenylalanine and tyrosine was found to have the highest values, whereas tryptophan displayed the lowest values in different parts of eggs. According to [3,5,29], animal protein is better digested and contains more EAAs and greater available AAs than vegetable protein. Besides, the protein quality of animal protein was higher than that of leguminous seeds [30,31].

3.4. *Crude Protein and Amino Acid Patterns of Different Eggs Parts*

Table 4. displays the crude protein and amino acid profile of different egg parts compared to the recommended daily allowances. Results showed variations ($p \leq 0.05$) among different parts, e.g., the whole edible parts of eggs (albumen + yolk), albumen, and yolk. The yolk had a greater ($p \leq 0.05$) concentration of crude protein and amino acids than the whole edible parts of eggs, which in turn had greater values ($p \leq 0.05$) than the albumen. However, for the EAAs/NEAAs, the trend was opposite, being highest in the albumen and lowest in the yolk. The values for the whole edible parts of the eggs were intermediate.

Table 4. Amino acid patterns of different egg parts in the retail market compared to recommended daily allowance standard Food and Agriculture Organization (FAO)/World Health Organization (WHO) [22] values of adults (70 kg).

Crude Protein (g/100 g)/Amino Acid (mg/100 g)	Egg Part			Statistical Analysis		RDA, mg/kg Day
	Whole Eggs	Albumen	Yolk	RMSE	p-Value	
Crude protein	13.95 ^{b} (26.6%)	12.42 ^{c} (23.7%)	16.00 ^{a} (30.5%)	3.32	0.054	0.75 g/kg per day
Arginine	814.9 ^{b}	525.6 ^{c}	1142.8 ^{a}	43.93	0.001	–
Histidine	312.3^{b} (44.6%)	209.5^{c} (29.9%)	399.8 ^{a} (57.1%)	16.33	0.001	10 (700 mg/day)
Isoleucine	668.3 ^{b} (47.7%)	551.1 ^{c} (39.4%)	785.1 ^{a} (56.1%)	22.63	0.001	20 (1400 mg/day)
Leucine	1091.8 ^{b} (40.0%)	819.2 ^{c} (30.0%)	1385.1 ^{a} (70.7%)	31.42	0.001	39 (2730 mg/day)
Lysine	956.3 ^{b} (45.5%)	743.5 ^{c} (35.4%)	1260.6 ^{a} (60.0%)	27.13	0.001	30 (2100 mg/day)
Methionine	353.8 ^{b} (48.6%)	315.9 ^{c} (43.4)	374.5 ^{a} (51.4)	18.94	0.001	10.4 (728 mg/day)
Methionine + cysteine	546.4 ^{b} (52.0%)	455.4 ^{c} (43.4%)	608.4 ^{a} (57.9%)	29.99	0.001	15 (1050 mg/day)
Phenylalanine	645.1 ^{b}	516.2 ^{c}	669.2 ^{a}+	27.64	0.001	–
Phenyalanine + tyrosine	1168.4 ^{b} (66.8%)	933.1 ^{c} (53.3%)	1379.1 ^{a} (78.8%)	34.73	0.001	25 (1750 mg/day)
Threonine	614.4 ^{b} (58.5%)	440.3 ^{c} (41.9%)	856.1 ^{a} (81.5%)	27.87	0.001	15 (1050 mg/day)
Tryptophan	143.1 ^{b} (51.1%)	120.1 ^{c} (42.9%)	219.1 ^{a} (78.3%)	23.13	0.001	4 (280 mg/day)
Valine	744.6 ^{b} (40.9%)	588.0 ^{c} (32.3%)	864.7 ^{a} (47.5%)	49.06	0.001	26 (1820 mg/day)
EAAs	5549.9 ^{b}	4348.7 ^{c}	6814.4 ^{a}	104.2	0.001	–
Alanine	664.8 ^{b}	572.4 ^{c}	801.4 ^{a}	31.78	0.001	–
Aspartic acid	1271.2 ^{b}	1019.9 ^{c}	1548.8 ^{a}	33.79	0.001	–
Glutamic acid	1649.6 ^{b}	1262.6 ^{c}	2019.8 ^{a}	33.52	0.001	–
Glycine	430.7 ^{b}	378.8 ^{c}	498.9 ^{a}	23.81	0.001	–
Proline	514.9 ^{b}	367.1 ^{c}	650.4 ^{a}	15.51	0.001	–
Serine	998.2 ^{b}	676.4 ^{c}	1342.9 ^{a}	28.39	0.001	–
NEAAS	7059.9 ^{b}	5314.4 ^{c}	8948.9 ^{a}	136.2	0.001	–
EAAs/NEAAs	0.786 ^{b}	0.818 ^{a}	0.761 ^{c}	0.022	0.001	–

$^{a–c}$ Means with varying superscripts differ significantly ($p < 0.05$); RMSE, Root mean square error; p value, probability level; FAO/WHO, Food and Agriculture Organization/World Health Organization; RDA, recommended daily allowance; EAAs, total essential amino acids; NEAAs, total non-essential amino acids; EAAs/NEAAs, total essential amino acids/total non-essential amino acids ratio.

Eggs are well-known animal proteins with high protein/amino acid quality that fulfil the human RDA better than vegetable protein sources, and EAAs deficiency can depress growth and lead to many health problems [2,26]. In this regard, different egg sources showed different patterns in phenylalanine, methionine, lysine, isoleucine, valine, and threonine in the whole edible parts of eggs. Differences in these amino acids were 38.5, 52.5, 16.4, 17.5, 20.9, and 21.1%, respectively. Similar trends were found in the albumen and yolk of eggs. These results agree with those reported by [6,11], who reported that raw regular Kampung eggs (eggs that produced by village hens that allowed to roam in a building, room or open area that includes nest space and perches, but they did have access to the outdoors) and nutrient-enriched eggs have different amino acid patterns, with lysine, leucine, phenylalanine, and valine having the highest concentrations in eggs. The difference in protein and AA patterns of various eggs could be attributed to the impact of the diet composition on the level of CP and amino acids. Similar results were reported by [1,5–7].

Results indicate that the yolk fulfils the highest percentage of the RDA for adults (70 kg) between 19–30 years of age, followed by the whole egg and albumen. For the egg yolk, the highest percentages were for threonine (81.5%), phenylalanine + tyrosine (78.8%), and tryptophan (78.3%) while the lowest was for methionine (~51.4%). The corresponding values for the whole edible parts of eggs and albumen were 58.5% and 41.9% for threonine, respectively, 66.8 and 53.3% phenylalanine + tyrosine, respectively, and 51.1% and 42.9% for tryptophan, respectively.

From the results obtained for amino acid scores, phenylalanine + tyrosine was the most limiting, the amino acids that were deficient in protein and when supplemented resulted in the highest response, the amino acid in all sources of eggs, followed by histidine and lysine [32]. Limiting EAAs for humans that can't be synthesized at an adequate amount by human cells are phenylalanine, methionine, lysine, leucine, isoleucine, valine, threonine, and tryptophan [11,33]. In addition, histidine, arginine, cystine and tyrosine are regarded as EAAs for infants and growing children [34].

From a biochemical point of view, differences in amino acid patterns in egg proteins/amino acids could be attributed to the type and percentage of protein in the albumen and yolk, e.g., ovalbumin, ovotransferrin, lysozyme, ovomucoid, ovomucin, and immunoglobulin Y [2,35]. The ovalbumin is the major protein that amounts to greater than 50% of the protein in the albumen of eggs [36]. In this regard, lysine (509 mg/g) was the greatest amino acid in eggs, whereas cysteine (128 mg/g) was the lowest [37]. Arginine, serine, cysteine and iso-leucine amino acids in eggs were higher than those in soy protein, beef, casein, wheat flour, and egg white [38].

3.5. Amino Acids Ratios

Table 5 shows the data for amino acid scores, aromatic amino acids, and essential amino acids to TAAs, antioxidant amino acid protein of eggs from different sources. The results indicate that the amino acid scores for histidine and threonine were significantly different among egg sources, with the highest values from source D and the lowest from source B. Differences in amino acid scores were also significant for the amino acids valine, isoleucine, leucine lysine, methionine + cystine, and phenylalanine + tyrosine.

The present results show that source B had greater ($p \leq 0.05$) amino acid scores for valine, isoleucine, and methionine + cystine than did sources C and D. In addition, source C had greater scores ($p \leq 0.05$) for leucine, lysine, phenylalanine + tyrosine than did sources B and D, A and B and B, respectively. The results indicate that source C had the lowest quality of protein in terms of total aromatic amino acids (TAAAs) indices, but when tyrosine + tryptophan was considered as the main contributor to the antioxidant properties of eggs [4], source C had the highest value. Differences between the highest and lowest values for TAAAs and tyrosine + tryptophan were 7.4 and 8.5%, respectively.

The EAAs/TAAs and cystine/sulfur amino acids were also higher in source C and D compared to the resources A and B; the differences in these criteria reached 4.7 and 45%, respectively, showing higher variability in cystine/sulfur amino acids as indices. The results also show greater amino acid scores in sources A and B, particularly for valine, isoleucine, and methionine + cystine. In contrast, the P-PER based on the content of leucine and tyrosine showed insignificant differences among different egg sources, indicating that this measurement is not a suitable parameter for protein quality evaluation of eggs, as the amino acids involved in the equation of calculation did not reflect the abundant amino acids= in eggs, suggesting a need for precise equation.

The present results are in line with those cited by Adeyeye [6], who observed a TAAs of 10.0 and 10.1 mg/g of dry yolk in corn- and wheat-based diets, respectively. Moreover, differences in glutamic acid, glutamine, glycine, arginine, isoleucine, and ornithine were significant among dietary compositions. Similar to the present findings, EAAs accounted for about 46–47%, while TAAAs accounted for about 11% of the TAAs; arginine was the most copious amino acid, followed by glutamic acid and lysine [1,4]. Differences in the ratios of the amino acids of eggs from various sources could be due to the strains/breeds of layers and poultry species [5,6,39], dietary composition, cooking method, storage time, environment, [11,40,41] and egg parts [12,42,43].

The total aromatic amino acids, (TAAAs)/TAAs, as the index of antioxidant property of eggs [6], and EAAs/TAAs as an index of protein quality [42,43] were significantly different among various sources of eggs, with sources A and B having greater ($p \leq 0.05$) values than source C. In addition, TAAAs/TAAs were significantly greater in source D than in source C. Tyrosine + tryptophan, as an absolute value or relative to TAAAs or TAAs, was significantly different among different egg sources. The C sources had greater values ($p \leq 0.05$) than B sources. In addition, sources A and B showed similar values as source C except for tyrosine + tryptophan/TAAAs, which was higher ($p \leq 0.05$) in source C. No significant ($p \geq 0.05$) variations in the predicated protein efficiency ratio (P-PER) ratio among various egg sources. The cystine/sulfur amino acids ratio was significantly higher for eggs of source C.

Table 5. Amino acids score (%), aromatic amino acids, essential amino acids to the total amino acid ratio. Antioxidant amino acids of the whole edible egg parts (albumen + yolk) of different sources compared to standard Food and Agriculture Organization (FAO)/World Health Organization (WHO) [22] values.

Amino Acid (mg/100 g)	Source of Eggs				FAO/WHO, 2007 (g/100 g Protein)	Statistical Analysis	
	A	B	C	D		RMSE	*p*-Value
			Amino Acid Score				
Histidine	95.3 ^{ab}	89.5 ^{b}	101.4 ^{ab}	106.5 ^{a}	1.5	3.26	0.001
Valine	136.2 ^{ab}	141.9 ^{a}	123.3 ^{b}	117.4 ^{b}	3.9	4.16	0.025
Isoleucine	152.0 ^{a}	156.8 ^{a}	139.8 ^{b}	133.5 ^{b}	3.0	5.14	0.004
Leucine	101.6 ^{a}	99.9 ^{b}	104.2 ^{a}	97.7 ^{b}	5.9	2.93	0.005
Lysine	97.0 ^{b}	91.5 ^{b}	107.2 ^{a}	106.5 ^{a}	4.5	3.87	0.002
Methionine + cysteine	292.4 ^{ab}	301.9 ^{a}	268.3 ^{b}	277.6 ^{b}	2.2	5.29	0.001
Phenylalanine + tyrosine	50.2 ^{ab}	47.7 ^{b}	54.3 ^{a}	50.3 ^{ab}	3.8	1.56	0.001
Threonine	109.9 ^{b}	105.5 ^{b}	121.1 ^{a}	125.5 ^{a}	2.3	4.56	0.003
			Protein quality indices				
TAAAs	1643 ^{a}	1662 ^{a}	1547 ^{b}	1618 ^{ab}	—	46.5	0.024
TAAAs/TAAs ratio	0.1297 ^{a}	0.1317 ^{a}	0.1217 ^{b}	0.1300 ^{a}	—	0.003	0.001
Tyrosine + tryptophan	666.3 ^{ab}	638.8 ^{b}	693.3 ^{a}	667.5 ^{ab}	—	22.66	0.042
Tyrosine + tryptophan/TAAs ratio	0.0526 ^{a}	0.0506 ^{b}	0.0546 ^{a}	0.0536 ^{a}	—	0.001	0.001
Tyrosine + tryptophan/TAAAs ratio	0.406 ^{b}	0.384c	0.448 ^{a}	0.413 ^{b}	—	0.067	0.003
EAAs/TAAs ratio	0.445 ^{a}	0.449 ^{a}	0.429 ^{b}	0.437 ^{ab}	—	0.007	0.009
P-PER	3.15	3.10	3.21	2.99	—	1.53	0.196
Cystine/sulfur amino acids ratio	0.289 ^{b}	0.308 ^{b}	0.412 ^{a}	0.419 ^{a}	—	0.015	0.001

$^{a-c}$ means with varying superscripts differ significantly ($p < 0.05$); RMSE, Root mean square error; *p*-value, probability level; FAO/WHO, Food and Agriculture Organization/World Health Organization; EAAs, total essential amino acids; TAAAs, Total aromatic amino acids, TAAAs/TAAs, Total aromatic amino acids/total amino acids, EAAs/TAAs, essential amino acids/total amino acids, P-PER, predicated protein efficiency ratio.

4. Conclusions

In conclusion, the investigated eggs showed different EAAs/NEAAs ratio and antioxidant amino acids, indicating a potential for enhancing the nutritional values and extending the shelf life of eggs by different strategic nutritional approaches such as increasing antioxidants, and essential amino acid content of eggs.

Author Contributions: Y.A.A. conducted the experimental design, data collection, and first draft of the manuscript; M.A.A.-H. carried out the data collection and contribution to the experimental set-up, M.A.K. and M.H.S., lab work, and reading the manuscript; and Y.A.A. undertook the statistical analyses and proofreading of the manuscript. All authors have read and agreed to the published version of the manuscript.

Acknowledgments: The authors express their sincere gratitude and appreciation to DSR, King Abdulaziz University for technical and finical support.

References

1. Li, F.; Yang, Y.; Yang, X.; Shan, M.; Gao, X.; Zhang, Y.; Hu, J.; Shan, A. Dietary intake of broiler breeder hens during the laying period affects amino acid and fatty acid profiles in eggs. *Rev. Bras. Zootec.* **2019**, *48*, e20180292. [CrossRef]
2. Bradley, F.A.; King, A. Egg Basics for the Consumer: Packaging, Storage and Nutritional Information. 2000. Available online: http://anrcatalog.ucdavis.edu/pdf/8154.pdf (accessed on 12 January 2018).
3. Al-Harthi, M.A.; El-Deek, A.A.; Attia, Y.A. Impacts of dried whole eggs on productive performance, quality of fresh and stored eggs, reproductive organs and lipid metabolism of laying hens. *Br. Poult. Sci.* **2011**, *52*, 333–344. [CrossRef] [PubMed]
4. Nimalaratne, C.D.; Lopes-Lutz, A.; Schieber, A.; Wu, J. Free aromatic amino acids in egg yolk show antioxidant properties. *Food Chem.* **2011**, *129*, 155–161. [CrossRef]
5. Mori, H.; Takaya, M.; Nishimura, K.; Goto, T. Breed and feed affect amino acid contents of eggyolk and eggshell color in chickens. *Poult. Sci.* **2020**, *99*, 172–178. [CrossRef] [PubMed]
6. Adeyeye, E.I. The comparison of the amino acids patterns of whole eggs of duck, francolin and turkey consumed in Nigeria. *Glob. J. Sci. Front. Res. Chem.* **2013**, *13*. [CrossRef]
7. Attia, Y.A.; Al-Harthi, M.A.; Shiboob, M.M. Evaluation of quality and nutrient contents of table eggs from different sources in the retail market. *Ital. J. Anim. Sci.* **2014**, *13*, 369–376. [CrossRef]
8. Carvalho, T.S.M.; Sousa, L.S.; Nogueira, F.A.; Vaz, D.P.; Saldanha, M.M.; Triginelli, M.V.; Pinto, M.F.V.S.; Baiao, N.C.; Lara, L.J.C. Digestible methionine + cysteine in the diet of commercial layers and its influence on the performance, quality, and amino acid profile of eggs and economic evaluation. *Poult. Sci.* **2018**, *97*, 2044–2052. [CrossRef]
9. Meza-Espinoza, L.; Sáyago-Ayerdi, S.G.; de Lourdes García-Magaña, M.; Tovar-Pérez, E.G.; Yahia, E.M.; Vallejo-Cordoba, B.; González-Córdova, A.F.; Hernández-Mendoza, A.; Montalvo-González, E. Antioxidant capacity of egg, milk and soy protein hydrolysates and biopeptides produced by *Bromelia penguin* and *Bromelia karatas*-derived proteases. *Emir. J. Food Agric.* **2018**, *30*, 122–130.
10. Liu, J.; Zhang, D.; Zhu, Y.; Wang, Y.; He, S.; Zhang, T. Enhancing the in vitro antioxidants capacities via the interaction of amino acids. *Emir. J. Food Agric.* **2018**, *30*, 224–231.
11. Ismail, M.; Mariod, A.; Pin, S.S. Effects of preparation methods on protein and amino acid contents of various eggs available in Malaysian local markets. *Acta Sci. Pol. Technol. Aliment.* **2013**, *12*, 21.
12. Msarah, M.; Ahmed, A.A. Protein digestibility and amino acid content of Malaysian local egg protein prepared by different methods. *Environ. Ecosyst. Sci.* **2018**, *2*, 7–9. [CrossRef]
13. Roe, M.; Pinchen, H.; Church, S.; Finglas, P. *Nutrient Analyses of Eggs*; Analytical Report Revised Version; Institute of Food Research: Norwich, UK, 2013.
14. Association of Official Analytical Chemists (AOAC). *Official Methods of Analyses*, 16th ed.; Association of Official Analytical Chemists: Washington, DC, USA, 1999.
15. Moore, S.; Spackman, D.H.; Stein, W.H. Chromatography of amino acids on sulfonated polystyrene resins. *Anal. Chem.* **1958**, *30*, 1185–1190. [CrossRef]
16. Moore, S. On the determination of cysteine as cystine acid. *J. Biol. Chem.* **1963**, *238*, 235–237.
17. Csomos, E.; Simon-Sarkadi, L. Characterisation of tokaj wines based on free amino acid and biogenic amine using ion-exchange chromatography. *Chromatographia* **2002**, *56*, 185–188. [CrossRef]
18. Concon, J.M. Rapid and simple method for the determination of tryptophan in cereal grains. *Anal. Biochem.* **1975**, *67*, 206–219. [CrossRef]
19. Ogunsua, A.O. Amino acid determination in conophor nut by gas-liquid chromatography. *Food Chem.* **1988**, *28*, 287–298. [CrossRef]
20. Alsmeyer, R.H.; Cunigham, A.E.; Hapich, M.I. Equations predicted PER from amino acid analysis. *Food Technol.* **1974**, *28*, 34–38.
21. Oshodi, A.A.; Esuoso, K.O.; Akintayo, E.T. Proximate and amino acid composition of some under-utilized Nigerian legume flour and protein concentrate. *Riv. Ital. Sustanze Grasse* **1998**, *75*, 409–412.
22. Food and Agriculture Organization (FAO); World Health Organization (WHO); United Nations University (UNU). *Protein and Amino Acid Requirements in Human Nutrition*; Report of a Joint WHO/FAO/UNU Expert Consultation; WHO Technical Report Series No. 935; World Health Organization (WHO): Geneva, Switzerland, 2007.

23. SAS Institute. *User's Guide*; SAS institute Inc.: Cary, NC, USA, 2009.
24. Secci, G.; Bovera, F.; Parisi, G.; Moniello, G. Quality of eggs and albumen technological properties as affected by *Hermetia illucens* larvae meal in hens' diet and hen age. *Animals* **2020**, *10*, 81. [CrossRef]
25. Secci, G.; Bovera, F.; Nizza, S.; Baronti, N.; Gasco, L.; Conte, G.; Serra, A.; Bonelli, A.; Parisi, G. Quality of eggs from Lohmann Brown Classic laying hens fed black soldier fly meal as substitute for soya bean. *Animal* **2018**, *12*, 2191–2197. [CrossRef]
26. Whitney, E.; Rolfes, S.R. *Understanding Nutrition*, 14th ed.; Cengage Learning: Boston, MA, USA, 2015.
27. King, M.W. Amino Acids and Proteins: Introduction to Amino Acid Metabolism. The Medical Biochemistry Page. 2015. Available online: http://themedicalbiochemistrypage.org/amino-acid-metabolism.php (accessed on 20 March 2017).
28. Nelson, D.L.; Cox, M.M. *Principles of Biochemistry*, 4th ed.; Nelson, D.L., Cox, M.C., Freeman, W.H., Eds.; Freeman and Company: New York, NY, USA, 2004; p. 127.
29. Sakanaka, S.; Kitahata, K.; Mitsuya, T.; Gutierrez, M.A. Protein quality determination of delipidated egg-yolk. *J. Food Comp. Anal.* **2000**, *13*, 773–781. [CrossRef]
30. Mahan, L.K.; Escott-Stump, S. *Food Nutrition and Diet Therapy*; Saunders Publication Co.: Philadelphia, PA, USA, 2004.
31. Baudoin, J.P.; Maquet, A. Improvement of protein and amino acid contents in seeds of food legumes: A case study in Phaseolus. *Biotechnol. Agron. Soc. Environ.* **1999**, *3*, 220–224.
32. Harper, A.E.; Yoshimura, N.N. Protein quality, amino balance, utilization and evaluation of diets containing amino acids astherapeutic agents. *Nutrition* **1993**, *9*, 460–469. [PubMed]
33. Young, V.R. Adult amino acid requirements, the case for a major revision in current recommendations. *J. Nutr.* **1994**, *124*, 1517–1523. [CrossRef] [PubMed]
34. Imura, K.; Okada, A. Amino acid metabolism in pediatric patients. *Nutrition* **1998**, *4*, 143–148. [CrossRef]
35. Belitz, H.D.; Grosch, W.; Schieberle, P. Eggs. In *Food Chemistry*; Springer Publishing Co.: New York, NY, USA, 2009; Chapter 11; pp. 546–561.
36. McWilliams, M. Eggs. In *Foods, Experimental Perspectives*; Prentice Hall: Upper Saddle River, NJ, USA, 2001; p. 357.
37. Sosulski, F.W.; Imafidon, G.I. Amino acid composition and nitrogen-to-protein conversion factors of animal and plant foods. *J. Agric. Food Chem.* **1990**, *38*, 1351–1356. [CrossRef]
38. Sarwar, G.; Christiansen, D.A.; Finlayson, A.T.; Friedman, M.; Hackler, L.R.; Mckenzie, S.L.; Pellet, P.L.; Tkachuk, R. Intra-and inter-laboratory variation in amino acids analyses. *J. Food Sci.* **1983**, *48*, 526–531. [CrossRef]
39. Cook, F.; Briggs, G.M. Nutritive value of eggs. In *Eggs Science and Technology*; Avi Publishing Co: Westport, CT, USA, 1997; p. 92.
40. Tee, E.S.; Noor, M.I.; Azudin, M.N.; Idris, K. *Nutrient Composition of Malaysian Foods*; Malaysian Food Composition Database Programme C/O; Institute for Medical Research: Kuala Lumpur, Malaysia, 1997.
41. Taş, N.G.; Gokmen, V. Profiling of the contents of amino acids, water-soluble vitamins, minerals, sugars and organic acids in Turkish hazelunt varieties. *Pol. J. Food Nutr. Sci.* **2018**, *68*, 223–234. [CrossRef]
42. National Institute of Health-Office of Dietary Supplements. *Dietary Reference Intakes for Energy, Carbohydrate: Fiber, Fat, Fatty Acids, Cholesterol, Protein, and Amino Acids RDA*; National Institute of Health-Office of Dietary Supplements: Bethesda, MD, USA, 2005.
43. Hester, K.L. *Eggs for Young America*; University Press of New England: Lebanon, NH, USA, 1997; Chapter 1.

15

Effects of Propylene Glycol on Negative Energy Balance of Postpartum Dairy Cows

Fan Zhang [1,2], Xuemei Nan [1], Hui Wang [1], Yiguang Zhao [1], Yuming Guo [2] and Benhai Xiong [1,*]

[1] State Key Laboratory of Animal Nutrition, Institute of Animal Science, Chinese Academy of Agricultural Sciences, Beijing 100193, China; 18813015831@139.com (F.Z.); xuemeinan@126.com (X.N.); wanghui_lunwen@163.com (H.W.); zhaoyiguang@caas.cn (Y.Z.)

[2] State Key Laboratory of Animal Nutrition, College of Animal Science and Technology, China Agricultural University, Beijing 100193, China; guoyum@cau.edu.cn

* Correspondence: xiongbenhai@caas.cn.

Simple Summary: After calving, the milk production of dairy cows increases rapidly, but the nutrient intake cannot meet the demand for milk production, forming a negative energy balance. Dairy cows in a negative energy balance have an increased risk of developing clinical or subclinical ketosis. The ketosis in dairy cows has a negative impact on milk production, dry matter intake, health, immunity, and reproductive performance. Propylene glycol can be used as an important gluconeogenesis in ruminants and can effectively inhibit the formation of ketones. Supplementary propylene glycol to dairy cows during perinatal is an effective method to alleviate the negative energy balance. This review summarizes the reasons and consequences of negative energy balance as well as the mechanism and effects of propylene glycol in inhibiting a negative energy balance in dairy cows. In addition, the feeding levels and methods of using propylene glycol to alleviate negative energy balance are also discussed.

Abstract: With the improvement in the intense genetic selection of dairy cows, advanced management strategies, and improved feed quality and disease control, milk production level has been greatly improved. However, the negative energy balance (NEB) is increasingly serious at the postpartum stage because the intake of nutrients cannot meet the demand of quickly improved milk production. The NEB leads to a large amount of body fat mobilization and consequently the elevated production of ketones, which causes metabolic diseases such as ketosis and fatty liver. The high milk production of dairy cows in early lactation aggravates NEB. The metabolic diseases lead to metabolic disorders, a decrease in reproductive performance, and lactation performance decline, seriously affecting the health and production of cows. Propylene glycol (PG) can alleviate NEB through gluconeogenesis and inhibit the synthesis of ketone bodies. In addition, PG improves milk yield, reproduction, and immune performance by improving plasma glucose and liver function in ketosis cows, and reduces milk fat percentage. However, a large dose of PG (above 500 g/d) has toxic and side effects in cows. The feeding method used was an oral drench. The combination of PG with some other additives can improve the effects in preventing ketosis. Overall, the present review summarizes the recent research progress in the impacts of NEB in dairy cows and the properties of PG in alleviating NEB and reducing the risk of ketosis.

Keywords: propylene glycol; dairy cows; postpartum; negative energy balance; ketosis

---

1. Introduction

The transition period from late pregnancy to early lactation is well known as the critical time for the production of cows [1]. During this period, the higher energy demand for milk production coupled with the reduction in dry matter intake (DMI) around calving means that a large number of dairy cows are in a state of negative energy balance (NEB). To support the energy requirement, the body fat and protein of dairy cows are mobilized for hepatic gluconeogenesis, which leads to the increase of non-esterified fatty acids (NEFA), β-hydroxybutyrate (BHBA), and ammonia in plasma [2]. Invariably, the dairy cows with high-producing ability have the risk of subclinical ketosis (SCK, hyperketonemia without clinical signs) or clinical ketosis (CK, hyperketonemia with clinical signs). Cows with ketosis have a greater risk of several diseases including displaced abomasum, infections of the reproductive tract, mastitis, cystic ovarian disease, leg problems, and diseases of the digit and foot [3]. Due to the rapid increase in energy demand for milk after calving, the NEB usually has an adverse impact on health and thus decreases animal welfare, production, and profitability [4].

Earlier treatment of ketosis is important to reduce future economic losses in modern high-yield dairy farms. The goal of ketosis treatment is to stimulate gluconeogenesis, increase plasma glucose, and decrease lipolysis [5]. Propionate is the major precursor for gluconeogenesis. However, the limited feed intake during early lactation restricts ruminal propionate supply to the liver, raising the requirement for alternative gluconeogenic precursors [6]. Propylene glycol (PG) is a precursor of ruminal propionate that can be rapidly absorbed from the rumen for gluconeogenesis in the liver [7]. It has long been used as a treatment against ketosis [8]. Experimental studies have shown that an oral drench of PG can be effective in increasing glucose and decreasing BHBA and NEFA in plasma [9]. Therefore, this paper reviews the effects of NEB in dairy cows, and the research progress about the properties of PG in alleviating NEB and reducing the risk of ketosis during postpartum in dairy cows.

2. The Formation of NEB in Dairy Cows

During the transition period, dairy cows experience a dramatically increased demand for nutrients for the growing fetus and the initiation of lactation [9]. In the postpartum period, the nutrient requirements for milk yield, milk fat, milk protein, and milk lactose increase rapidly and exceed the supplies from DMI [10]. In addition, the diet of dairy cows after calving changes sharply from being mainly forage-based to concentrate-rich [11]. The sudden energy requirement for milk production and the DMI lags behind, inducing the negative nutrient balance, especially the NEB. The NEB symptoms appear in postpartum, but the dynamic changes of the physiological and metabolic status are verified from the prepartum period [12]. Responding to the NEB, the cow physiologically triggers the lipomobilization of body fat reserves, thereby amounts of NEFA are released into the blood circulation [12]. The NEFA are used as a fuel source by muscle, incorporated into milk fat, and taken up by the liver in proportion to their supply [13]. However, the excessive mobilization of body fat reserves leads to the accumulation of triglycerides in the liver or conversion to ketone bodies (i.e., BHBA, acetone, and acetoacetate) and leads to the onset of ketosis, which has adverse effects on the health, milk productivity, and reproduction in dairy cows. It is well known that dairy cows already go into a period of NEB a few days before calving and that the NEB extends to a few weeks after calving, with the nadir of NEB during the first week of lactation [14]. The feed intake increases slowly at the beginning of lactation [15] and the NEB switches to a positive range after approximately 45 d of lactation [16]. Other diseases such as rumen distention, abomasum displacement, metritis, mastitis, and so on also lead to insufficient nutrient supply and trigger secondary ketosis in the early lactation period. Therefore, methods of decreasing the release of NEFA from adipose tissue are important to alleviate the NEB of dairy cows in early lactation.

3. The Effects of NEB in Dairy Cows

3.1. *Increasing the Incidence of Metabolic Diseases*

During the transition period, in order to meet the energy requirement for improved milk production, the rate of lipolysis excesses the lipogenesis, which results in the increasing serum NEFA of dairy cows. When the NEFA could not be completely oxidized to carbon dioxide, it will partly be oxidized to ketone bodies or be stored in the liver as triglycerides [9]. During the period of high metabolic demands, the increased hepatic oxidation of NEFA induces satiety, decreases feed intake, and increases fat mobilization [17]. The blood BHBA level of above 1.2 mmol/L or 1.4 mmol/L is related to impaired health and performance, and is a common and costly metabolic disease, which is called hyperketonemia [18]. The high concentrations of NEFA and BHBA have negative effects on the health and productivity of dairy cows. Therefore, the physiological conditions associated with insufficient energy supply predispose dairy cows to metabolic and microbial diseases such as milk fever, endometritis, ketosis, displaced abomasum, and retained placenta [11].

The epidemiology of ketosis in dairy cows in early lactation increases the risk of displaced abomasum. Each 0.1 mmol/L increases in BHBA at the first SCK-positive test increased the risk of developing a displaced abomasum by a factor of 1.1 [95% confidence interval (CI) = 1.0 to 1.2] and increased the risk of removal from the herd by a factor of 1.4 (95% CI = 1.1 to 1.8) [19]. Raboisson et al. [20] summarized the association between SCK and displaced abomasum in 38 models from 10 publications and found that the risk (95% CI) of left displaced abomasum in cows with SCK were 3.55 (2.60–4.25). Fatty liver is also a metabolic disorder of dairy cows relating to NEB in early lactation. Fatty liver develops when the hepatic uptake of lipids exceeds the oxidation and secretion of lipids by the liver and thereby causes accumulation of triacylglycerol (TAG) in the liver [21]. The severe fatty liver causes metabolic dysfunction, which will reduce the hepatic metabolism, defense function, and insulin sensitivity [12]. The results of Fiore et al. [22] showed that fatty liver already developed before parturition, and increased from moderate to severe in 10 days after calving and then progressively disappeared. Therefore, methods of preventing hepatic lipidosis should be applied during this period.

3.2. *Decreasing the Milk Productivity Performance of Dairy Cows*

McArt et al. [19] concluded that each 0.1 mmol/L increase in BHBA at the first SCK-positive test was associated with a decrease in milk production by 0.5 kg/d for the first 30 days in milk (DIM). The cows with CK were lower in milk production and milk protein content, but milk fat content was higher than healthy cows [23]. A high percentage of fat and a low percentage of protein in the milk were associated with significant increases in the risk of SCK [24]. The mean fat to protein percentage ratio (FPR) and the frequency of FPR > 1.5 were higher in ketosis cows than healthy cows [25]. Therefore, the FPR of milk in early lactation is negatively correlated with energy balance and has been used as an indication of ketosis. The optimal FPR values are 1.05 to 1.18, while FPR values higher than 1.3 or 1.5 suggest a severe NEB and SCK [26]. In the cows with ketosis, the plasma NEFA from fat mobilization provided the precursor for milk fat synthesis in the mammary gland. The results [23] of in vivo and in vitro data indicated that NEFA could induce cell death-inducing DNA fragmentation factor-α-like effector A (CIDEA) expression in bovine mammary epithelial cells, leading to upregulation of de novo fatty acid synthesis enzymes (fatty acid synthase and acetyl-CoA carboxylase 1) and milk lipid secretion proteins (butyrophilin and xanthine dehydrogenase), thereby contributing to an increase in milk fat content in CK cows. The decrease in milk protein percentage might be related to the increased amino acid requirements for gluconeogenesis in ketosis cows, and the spared would be limited for protein synthesis in mammary gland.

3.3. *Decreasing the Reproductive Performance*

The reproductive performance of cows is one of the most important factors affecting the economic benefits of dairy production. The duration and severity of dairy cows' NEB in early postpartum are also related to reproductive performance. Extensive mobilization of fat has detrimental effects on liver function due to the accumulation of TAG, impairing the detoxification of ammonia into urea [27]. The NEB of dairy cows in early lactation will also increase the mobilization of protein, which will increase the metabolic residues of ammonia and urea. Ammonia is believed to play a role in starting before ovulation, whereas urea mainly interferes negatively after fertilization [28]. The first ovulation of a dairy cow is retarded by decreasing the luteinizing hormone (LH) pulsatility because of the low blood glucose [29].

The calving-to-first-service interval was 8 d longer and the calving-to-conception interval was 16 to 22 d longer in cows with SCK than in healthy cows [20]. Rutherford et al. [30] established that the SCK cows prolonged calving to the first estrus, calving to first insemination and calving to pregnancy intervals, and the first insemination was 4.3 times less likely to be successful compared to non-SCK cows. The high BHBA values, before, after, or before and after artificial insemination were reported associated with a six to 14% reduction in the pregnancy per artificial insemination compared with cows with low BHBA values [31]. Najm et al. [32] also showed the activity of healthy cows exceeded the ketosis cows by an average of 52.6% in 4–70 DIM and the mean motion activity on the day of estrus was also higher in healthy cows. The activity level of the cow will also affect the effective monitoring of estrus, which may be a factor decreasing the reproductive performance, especially detecting estrus by automated surveillance systems. The uterine inflammation could also be exacerbated by the elevated circulating concentration of BHBA or NEFA in early postpartum [33], which would delay uterine involution and successful conception. Therefore, severe NEB will reduce the reproductive performance of cows by delaying uterine recovery, prolonging calving to the first estrus, and reducing estrus activity and successful conception rate.

3.4. Inducing Immunosuppression

During the period from late pregnancy to early lactation, the NEB of dairy cows increases the risk of metabolic and infectious diseases. The metabolic status of early-lactating cows is known to affect the immune response to pathogens and impose immune challenges [34]. In this period, the NEB decreases the efficiency in pathogen clearance and increases the magnitude and duration of inflammation [35]. As a consequence, cows are more susceptible to several economically important disease such as metritis and mastitis [36]. In ketosis cows, the inflammation biomarkers of serum amyloid A, haptoglobin, and lipopolysaccharide binding protein are increased when compared with healthy counterparts [37]. The increase in circulating NEFA impairs peripheral blood mononuclear cells and polymorphonuclear leukocytes function, along with a weakening of those cells' phagocytosis capacity and a decrease in their ability to fight bacteria [38]. The inflammatory state in early lactation may disrupt normal nutrient partitioning and decrease the productivity of dairy cows [39].

Greater concentrations of both NEFA and BHBA have been associated with impaired immune functions and mastitis in dairy cows [40]. Glucose is considered as the preferred substrate for the immune system [41], and the activation of an immune response requires energy [42]. Serum glucose levels in cows with severe NEB are significantly reduced during early lactation, affecting the energy supply of the immune system. The increase in lipid infiltration in the liver also decreases the immune response. Additionally, cows with ketosis (blood BHBA > 3 mmol/L) have higher serum concentrations of proinflammatory cytokines interleukin (IL) 18, tumor necrosis factor (TNF)-α, and IL1B, and lower concentration of anti-inflammatory cytokine IL-10 [43]. The somatic cell count (SCC) in milk is closely related to the immune status of dairy cows. Van Straten et al. [44] concluded that the odds of an event of SCC > 250,000 cells/mL or SCC > 400,000 cells/mL were 44% and 33% greater for cows with ketosis when compared with cows without, respectively. In the results of Abuajamieh et al. [37] ketosis cows had increased circulating markers (serum amyloid A, haptoglobin, and lipopolysaccharide binding protein) of inflammation pre- and post-calving and before the clinical signs of ketosis. The higher prepartum NEFA increases the risk for metritis [45]. The infection and inflammation noticeably redirect resources toward the immune system and away from the utilization and synthesis of economically relevant products [41]. Therefore, severe NEB will lead to fatty liver and high serum NEFA in cows, which also contributes to immunosuppression and increases the risk of infections in the postpartum period. Inflammation postpartum upregulated immune gene expression and mitochondrial uncoupling further increase energy requirements, which exacerbates severe NEB status in cows [11].

4. The Anti-Ketogenic Properties of PG and the Mechanism of Inhibiting NEB

During early lactation, glucose synthesis should be increased to accommodate mammary demands [46]. To avoid the occurrence of dairy cow ketosis, it is important to provide extra gluconeogenesis for dairy cows. In 1954, PG was observed to be an effective treatment of ketosis in dairy cows [8]. PG supplementation appears to increase milk yield with a slight decrease in milk fat and an increase in milk lactose percentage [47]. Propylene glycol (1, 2-propanediol; $C_3H_8O_2$) is a sweet, hygroscopic, viscous liquid that has a gluconeogenic property and is routinely used because of its therapeutic effects on cows suffering from ketosis, based on the premise that it rapidly increases blood glucose [48]. As gluconeogenic precursors, it has been proven that PG is more effective at increasing plasma glucose concentration than glycerol, since 300 mL PG is at least as effective as 600 mL of glycerol [49]. Plasma concentrations of glucose and insulin are known to increase in response to dietary PG [50,51]. Propionate is the main product of PG fermentation, which can be rapidly metabolized with short lag time [52]. This is beneficial for cows to alleviate the NEB and anti-ketogenic. After oral administration, the majority of PG escapes from the rumen wall or gastrointestinal tract and is converted to glucose by the liver [53]. However, the other mechanism of the effects of PG involves the successive production of propionate together with propanal and with the latter being converted to propanol in the rumen, which in turn is converted to propionate and thereafter glucose in the liver [54]. The main effect of PG is to increase the glucogenic status, and as a consequence, the concentration of plasma BHBA is reduced and the cows have decreased risk of developing ketosis [55]. PG is metabolized to lactate, acetate, and pyruvate in the liver. Lactate enters gluconeogenesis via pyruvate, which can be converted to oxaloacetate. The concentration of oxaloacetate is the key metabolite in determining if the acetyl-CoA enters the tricarboxylic acid (TCA)-cycle or ketogenesis [7]. When the oxaloacetate is insufficient for citrate synthase to combine with acetyl-CoA, the excessive acetyl-CoA is then partitioned toward ketone synthesis [37]. The anti-ketogenic properties of PG are partly due to increasing the oxidation of acetyl-CoA into the TCA-cycle and the supply of gluconeogenic glucose [7]. The detailed anti-ketogenic pathways of PG are shown in Figure 1.

In addition, insulin resistance is an adaptation to the very high glucose requirements for lactation, thereby conserving glucose for lactation by limiting its use by insulin-sensitive tissues (muscle, adipose tissue etc.) [56]. The insulin resistance can hence be attributed to a decrease in insulin responsiveness and a decrease in insulin sensitivity [57]. The greater extent of insulin resistance in peripartal dairy cows can contribute to excessive adipose tissue lipolysis and thus greater metabolic disease risk [58]. Chalmeh et al. [59] confirmed that the supplementary feeding with PG reduced the insulin resistance in dairy cows during the transition period by the intravenous glucose tolerance test. The decrease in insulin resistance will inhibit lipolysis and decrease the metabolic disease risk in periparturient dairy cattle. Some researchers have found a negative effect of NEFA in the insulin sensitivity of dairy cattle [57]. Therefore, the effect of PG in decreasing the insulin resistance may be related to the property of PG as a main precursor of glucose and decreasing circulating NEFA.

The energy value of PG is 5.66 Mcal/kg, and according to the assumed PG metabolizable energy utilization efficiency for lactation (80%) of Miyoshi et al. [50], the net energy for lactation (NE_L) of PG was calculated to be 4.53 Mcal/kg. Due to the higher NE_L of PG, it can supply more energy intake than other concentrates for dairy cows in early lactation and reduce the incidence of ketosis.

Therefore, the effects of PG on alleviating NEB in dairy cows are mainly by improving the precursor for hepatic gluconeogenesis and increasing the oxidation of acetyl-CoA into the TCA-cycle. The high energy content of PG can increase the energy density of the diet for dairy cows. The fatty liver and ketone bodies of dairy cows will be inhibited with the increase in liver glucose synthesis.

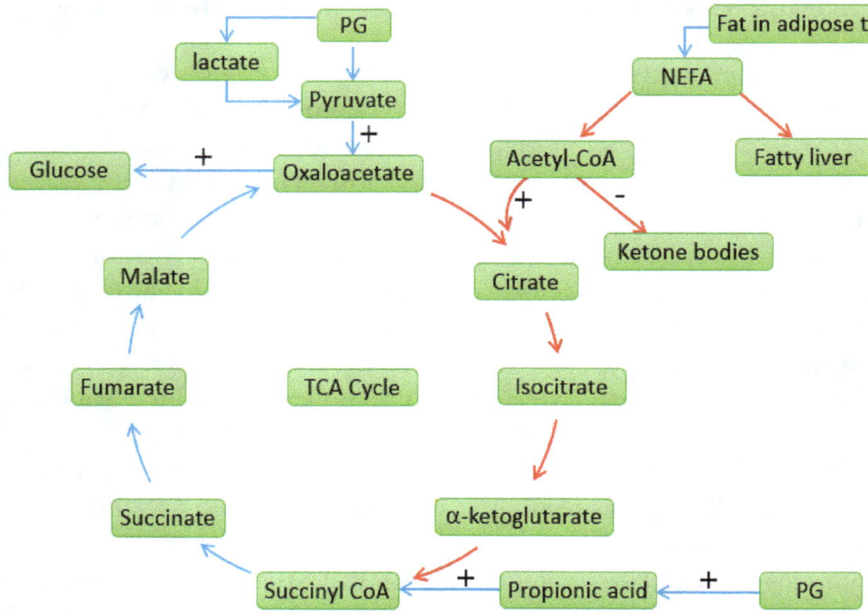

Figure 1. The anti-ketogenic pathways of propylene glycol (PG) in dairy cows [7]. The blue lines are to describe the gluconeogenic pathways of increasing the glucose to preventing ketosis. The red lines are the ways of increasing the oxidation of acetyl-CoA (coenzyme A) in the tricarboxylic acid cycle (TCA-cycle) and the supply of glucose by increasing the production of oxaloacetate, which will prevent acetyl-CoA convert to ketone bodies (β-hydroxybutyrate, acetone, and acetoacetate). PG can also reduce the triacylglycerol (TG) accumulation in liver.

5. Effects of PG on Alleviating NEB in Dairy Cows

5.1. The Effects of PG on DMI and Rumen Fermentation

The DMI at the postparturient stage has an essential effect on the NEB metabolism status of cows. The DMI of cows with CK is lower than healthy cows [23]. PG is considered unpalatable and usually reduces feed intake if not mixed thoroughly with other feed components or drenched [7]. Moallem et al. [60] reported the daily average DMI and NE_L intake from calving until 100 DIM was higher for cows supplemented with 500 g/d PG per cow until 21 DIM than the control group. The rumen fill score and body condition score (BCS) are also direct tools to evaluate the feed intake and energy balance status. The PG treatment improved the rumen fill score and lowed BCS loss in the dairy cows, which were diagnosed with ketosis in the results of Jeong et al. [61]. The increasing DMI and rumen fill scores are beneficial for decreasing the adverse impact of NEB.

The results of Kristensen and Raun [54] showed that infusion of PG did not affect ruminal pH or the total concentration of ruminal volatile fatty acids (VFA), but decreased the molar proportion of ruminal acetate and increased ruminal concentrations of PG, propanol, and propanal as well as the molar proportion of propionate. Chung et al. [62] also found the PG administration appeared to shift ruminal VFA patterns by producing more glucogenic VFA such as propionate and valerate at the expense of lipogenic VFA such as acetate. The increase of propionate and valerate can provide carbon sources for glucose biosynthesis, which is beneficial for dairy cows to alleviate NEB in early lactation. Acetate is the major source for milk fat synthesis in dairy cows [63]. Therefore, the decrease in acetate concentration may explain the decreased milk fat with PG supplementation.

5.2. The Effect of PG on Metabolic Index

It is widely accepted that PG has a glucogenic effect. The glucogenic status of the cows have effects on the liver metabolism of NEFA and, thereby, the regulation of ketogenesis [55]. The effects of PG treatment on SCK or CK have been explained by reduced adipose tissue mobilization, which leads to the decrease of NEFA in the liver and the reduction in the formation of ketone bodies [7]. As Sun et al. [12] reported, the dynamic changes of the physiological and metabolic status were from the prepartum period, so feeding of PG from prepartim is also a good method to alleviate NEB of dairy cows in postpartum. Juchem et al. [64] pointed out that prepartum PG administration increased concentrations of glucose and insulin, and decreased BHBA and NEFA in plasma. It has been validated that the dairy cows suffered deficiency of energy before calving, so the nutritional strategies should be implemented at the start of the prepartum period [12]. The supplement of PG to dairy cows before calving is effective in inhibiting the occurrence of cow ketosis. Therefore, prepartum PG administration has a glucogenic effect for dairy cows in postpartum.

Supplement PG to dairy cows in early lactation is also an important way to avoid energy metabolic diseases. The study of Butler et al. [48] found that drenching 500 mL PG to the dairy cow diet had significant beneficial effect on energy balance and increased plasma insulin and glucose, while the plasma NEFA was decreased. Kristensen and Raun [54] confirmed that when cows were dosed with PG, the plasma concentrations of PG, ethanol, propanol, propanal, glucose, L-lactate, propionate, and insulin were increased. Therefore, PG regulates the metabolism of cows by increasing the supply of L-lactate and propionate to gluconeogenesis and reducing insulin resistance. Insulin is a key hormone in the regulation of lipolysis in adipocytes. The increase of insulin is also useful for alleviating NEB for dairy cows. Bjerre-Harpøth et al. [55] observed 4-week postpartum PG allocation enhanced glucogenic status, which decreased plasma concentration of BHBA and increased plasma concentration of glucose, but had limited effect on adipose tissue mobilization. Although the metabolic changes in Simmental cows in the periparturient period were not as significant as in the case of Holstein-Friesian cows, the application of PG also resulted in higher milk yield, BCS, and serum glucose content [65]. Therefore, PG can enhance glucogenic status, and decrease the plasma NEFA and BHBA concentrations.

Displaced abomasum, fatty liver, and ketosis are common nutritional metabolic diseases of cows in the postpartum period. PG, as a glucogenic precursor of ruminants, plays an important role in inhibiting metabolic diseases caused by NEB in dairy cows. The results of McArt et al. [66] showed the cows with SCK were 1.6 times more likely to develop displaced abomasum and 2.1 times more likely to be dead or sold than SCK cows treated with PG within the first 30 DIM. The reasons for PG administration decreasing displaced abomasum and the ratio to be removed from the herd are that PG contributes to prevent ketosis and improve milk production. The results of Rukkwamsuk et al. [14] indicated dairy cows drenched with PG from seven days prepartum to seven days postpartum could reduce the risk of fatty liver. This is in accordance with PG decreasing the NEFA in plasma, which will subsequently reduce the TAG accumulation in the liver. Fatty liver is a major metabolic disease of dairy cows in early lactation. The main indicators of hepatic lesions and alterations of its function are the enzymes aspartate transaminase (AST), gamma-glutamyl transferase (GGT), and the blood metabolites glucose, cholesterol, and albumin [67]. The study of Hussein et al. [68] found that PG supplementation has the ability to reduce the enzyme activities of AST and GGT and improve serum glucose, but had no effect on the serum concentrations of total cholesterol and albumin. The PG treatment can thus reduce liver lesions. Stokes and Goff [69] reported offering PG at calving had the effects of lowering the health disorder risk (retained placenta, ketosis, hypocalcemia, displaced abomasum, and metritis etc.) in dairy cows. Feeding PG to SCK cows can effectively prevent the formation of ketone bodies, which will prevent SCK cows developing into CK cows. McArt et al. [70] showed that 300 mL/d of PG treated cows were 1.50 times more likely to resolve their SCK ($1.2 \leq$ BHBA < 3.0 mM/L) and 0.54 times less likely to develop CK (BHBA ≥ 3.0 mM/L) than the control cows. The reduction in plasma NEFA and increase in plasma glucose are related to the anti-ketogenic property of PG. The use of PG is likely to produce more propionate as the main precursor of glucose; therefore, it can reduce the NEB and insulin resistance [59]. Therefore, the PG supplement to dairy cows can decrease nutritional metabolic diseases in early lactation. The cows drenched with PG could improve the molar proportion of ruminal propionate and hepatic gluconeogenesis, which results in an elevation in serum glucose and a decrease in serum NEFA and BHBA. Drenching of PG during the transition period is therefore beneficial for dairy cows to alleviate NEB in the postpartum period.

5.3. The Effects of PG on Milk Production

In the study of Lomander et al. [9], cows supplemented with 300 g of liquid PG daily in the first 21 DIM trended to yield more milk (0.94 kg, 95% CI = −0.03–1.91) compared with control cows during the first 90 DIM, but no difference was found in energy-corrected milk. In the trial of Østergaard et al. [71] based on milk spectra analyses, the results of the treatment with PG (500 mL for 5 d) showed only few benefits in early lactation for physiologically imbalanced cows. The study of Juchem et al. [64] showed that prepartum PG administration had no effect on milk production during the first nine-weeks postpartum. Butler et al. [48] observed that when multiparous Holstein cows received 500 mL oral drench PG from d 10 before expected parturition to d 25 postpartum could increase milk lactose and tended to reduce milk fat content, but there was no difference in milk yield and milk protein percentage. McArt et al. [70] concluded that an oral dose of PG improved milk yield during early lactation in cows with SCK. Stokes and Goff [69] determined that the cows received PG within 4 h of calving and again 24 h post-calving had 3.1 kg/d greater milk production. Therefore, it is conducive to the improvement of postpartum milk performance when cows received PG after calving as soon as possible. PG can provide enough energy to support the increase of milk yield, especially to SCK cows. However, some reports showed there was no difference in milk yield. This may be because the dosages used of PG reduced the feed intake or it was used in the cows without ketosis. The reduced milk fat content affected by PG could be due to the decrease of plasma NEFA and the lower acetate in the rumen [62]. PG had no effect in milk protein possibly because there was no shortage of amino acids for milk protein synthesis. Glucose is necessary for dairy cows to synthesize milk lactose. When PG can supply enough energy and be converted to enough glucose, milk lactose will increase. Therefore, PG tends to increase milk yield and milk lactose, reduce the milk fat of ketosis cows, but has little effect on milk protein.

5.4. The Effects of PG on Reproductive Performance

The insufficient energy intake can result in poor reproductive performance such as prolonging postpartum anestrus, decreasing progesterone production by the corpus luteum, and reducing rates of conception [50]. As PG can alleviate the NEB, it therefore also effectively prevents the degradation of reproductive performance.

The results of Gamarra et al. [56] indicated that short-term dietary PG supplementation affected circulating concentrations of metabolites and metabolic hormones, increased progesterone concentrations, and the number of small follicles. The embryo losses were related to the reduced progesterone and the increase in progesterone stimulates and sustains endometrial functions essential for embryonic survival, implantation, and growth [72]. The increase in the number of small follicles is beneficial for early estrus and conception. McArt et al. [66] confirmed that oral administration of PG to SCK cows were 1.3 times more likely to conceive at first insemination than untreated cows. Insulin is necessary for maximal steroidogenesis in both follicular and luteal cells. Miyoshi et al. [50] suggested 500 mL/d of PG administration to NEB dairy cows was able to improve ovarian function in early lactation, which is due to PG induced insulin spike. PG can improve plasma glucose and stimulate insulin secretion. Thus, the increased plasma insulin in PG treated cows has effects on follicular development and LH secretion, leading to earlier ovulation [50]. However, the results of Castañeda-Gutiérrez et al. [73] showed that there was no difference in days to first ovulation in multiparous cows after daily topdressing with PG from last 21 d before expected calving to 21 DIM. Butler et al. [48] observed drenching of PG had no effect on the number of cows with follicles ovulating, undergoing atresia, or becoming cystic, but the day of maximum follicle diameter was earlier for PG treatment. PG treatment advanced the day of maximum follicle diameter, indicating that it promotes follicular growth and is conducive to early estrus.

From the above reports, it can be found that the results for the PG treatment are inconsistent. This may be due to some studies not determining the ketosis status of dairy cows or feeding the PG in diet instead of oral drench. The transient elevations in insulin and glucose, decreases in NEFA, and modest improvement in energy balance are insufficient to adequately stimulate the hypothalamic–pituitary–ovarian axis [48], especially to the dairy cows that do not have a ketosis status. However, to those cows with ketosis, PG can improve plasma glucose and decrease NEFA and therefore effectively improve reproductive performance.

5.5. The Effects of PG on Immune Performance

During the peripartum period, dairy cows experience the state of ketosis and fatty liver, which reduces the liver function coupled with increased inflammation and oxidative stress [74]. The cows drenched with PG have a remarkable reduction in TAG accumulation in the liver [14]. There are few direct studies of PG treatment on cow immunity in early lactation. However, metritis and milk SCC can indirectly reflect the immune status of cows. The cows that received PG at calving had significantly lower incidence of metritis [69]. Formigoni et al. [75] observed the mean linear SCC in the first 13 weeks of lactation period was reduced by PG administration (300 g/d from 10 d prior to expected calving until parturition and 300 g/d on days 0, 3, 6, 9, 12 d postpartum). PG is beneficial to increase the serum glucose concentration of postpartum cows and effectively inhibit the risk of fatty liver through gluconeogenesis. Therefore, feeding PG to high-ketone cows can improve the liver function by reducing the accumulation of liver fat, thus improving the immune function of cows.

6. The Toxicity and Side Effects of PG

Although PG can prevent ketosis, large doses (>500 g/d) have toxic and side effects on dairy cows due to the toxic compounds of PG during metabolism processes [12]. The clinical signs of PG in toxic doses include depression, ataxia, and excessive salivation as well as abnormal, malodorous, and foul breath and feces in dairy cows [76]. Farmers and veterinarians in Denmark found that some cows expressed the toxicity and side effects of PG [7]. PG toxicity causes oxygen saturation of arterial blood hemoglobin and the oxygen pressure in arterial blood decreases, along with the appearance of dyspnea and ruminal atony upon intake of concentrate containing PG [77]. The sulfur-containing gases produced during PG fermentation in rumen contribute to the toxic effects in rumen when high doses are administered for therapeutic purposes [76]. Hydrogen sulfide is an important signaling substance in hypoxic vasoconstriction, which can explain the link between PG application to the rumen and the dyspnea [77]. The toxicity and side effects of PG limit the maximum dosage in dairy cows to reduce the risk of ketosis. However, the side effects of PG are related to individual cow susceptibility, so it is important to consider the signs of toxicity in the administering of PG, especially at dosages above 500 g per day [7].

7. The Feeding Level and Method of PG

The cows drenched with PG could improve the molar proportion of ruminal propionate and hepatic gluconeogenesis, which results in an elevation of serum glucose and decrease in serum NEFA and BHBA. Therefore, the drenching of PG during the transition period is beneficial for the dairy cows to alleviate NEB postpartum. However, the feeding level and method may affect the effects of PG administration.

Gordon et al. [5] pointed out that feeding 300 g of PG daily to ketonic animals should be considered as the base of ketosis treatment. Due to the toxicity and side effects of PG, the maximum feeding level of PG is 500 g per day. The different parity cows have different milk production abilities, so the risk of ketosis is different, which will also affect the PG application strategy. The first lactation heifers had a 47% reduction in the risk of excessive NEB compared to the older cows because of the lower milk yields and lighter bodyweights [78]. Therefore, the use of PG to alleviate NEB should be primarily applied to multiple parity dairy cows.

PG usually exists in liquid form, which is not convenient for feeding in routine by oral drenching to dairy cows. Chung et al. [79] verified feeding PG as a dry product (65% PG and 35% silicon dioxide as the carrier) in total mixed rations (TMR) also reduced plasma BHBA concentration. However, the same amount of PG as a top dress (500 g/d of cookie meal (dried bakery by-product) mixed with dry PG) was more efficient than incorporating it into TMR. The administration method seems to be of importance for the metabolic response of PG in cows because the response of allocating PG as an oral drench or in a separately fed concentrate, is better than mixing into TMR [55]. When PG is added to TMR, the chronic delivery of PG alters the environment in the rumen and inhibits more propionate production, which would decrease the feed intake, increase fat mobilization, and perpetuate the problem of ketosis [5]. Thus, PG is best administered as an oral drench. Meanwhile, feeding PG (mixed with other carriers) in dry product is beneficial to decreasing labor.

Identifying the minimum effective durations for the treatment of ketosis is important for giving PG orally. Gordon et al. [80] observed that extended PG treatment from 3 d to 5 d increased milk production by 3.4 kg/d among the dairy cows that had low blood glucose (<2.2 mmol/L) and 1.7 times more likely to cure with blood BHBA > 2.4 mmol/L, but had no significant effect on milk production in blood glucose ≥ 2.2 mmol/L and cure risk with blood BHBA between 1.2 to 2.4 mmol/L. The additional treatment times to the lower blood BHBA concentration cows might have had little benefit with the stress of increasing labor. McArt et al. [81] confirmed that testing cows 2 days per week from 3 to 9 DIM and treating all positive cows with 300 mL of oral PG for 5 d were the most cost-effective strategy for herds with hyperketonemia incidences between 15% and 50%. For those above 50%, treating all fresh cows with 5 d of PG was the most cost-effective strategy. Therefore, the incidence and degree of hyperketonemia also influence the PG administration durations for treatment. As 75% of cows that developed SCK could be tested as positive within 1 week postpartum (with a peak at 5 DIM) [19], the test of ketones, and treatment of PG should be mainly done in this period. The economic benefit is also a factor influencing the application of PG in preventing ketosis in dairy cows. EI-Kasrawy et al. [82] found that continuous drenching of PG with 300 mL/d for long durations during the transition of dairy cows had a higher net return (1908.52 US$/cow) than drenching for a short duration in 400 mL/d (1171.34 US$/cow) and the control group (1440.21 US$/cow). The economic benefit is affected by the improved production and reproduction performance and the decreased treatment cost of diseases and incidence of early removal from herd. In general, 300 g of PG daily is the basic treatment and the maximum level is 500 g per day. The best feeding method of PG was administered as an oral drench from within one week postpartum and primarily applied to the multiple parities. However, the durations for the treatment of PG for ketosis need to consider the degree of hyperketonemia.

8. The Research of Combination Therapy

PG plays an important role in the treatment of ketosis in cows. With 5 d of PG therapy, the rate of cure from hyperketonemia was improved. However, approximately 40% of cows still remained hyperketonemia [66]. The feeding level of PG is restricted due to its potentially toxic effects [8]. Therefore, the use of PG alone still has a poor effect on the treatment of ketosis in some cows. So, the combination of PG with other additives may have a better effect in inhibiting ketosis in dairy cows. There has been a lot of research in this field, but different additives have different effects. The addition of dexamethasone [18] and glycerol [49] to PG showed no additional benefits and the fat appeared to blunt the metabolic response to PG administration of cows [13]. The combination of butaphosphan-cyanocobalamin [80], glucocorticoids [83], L-carnitine-methionine [61], and glucose [84] with PG to the cows of hyperketonemia are beneficial in improving the chances of the resolution of ketosis compared to PG only.

9. Conclusions

The NEB in early lactation reduces cow productivity and reproductive performance, and induces immunosuppression, increasing the chance of dairy cows being eliminated. As a precursor of gluconeogenesis, the addition of PG can provide energy and glucose for cows, thus preventing metabolic diseases such as ketosis and fatty liver as well as increasing milk yield and reducing milk fat percentage. PG can also increase the reproductive performance and immune function of cows due to glucose enhancement. However, due to the toxicity and side effects, PG is used in doses within 500 g/d per cow and offered in oral drench. To improve the effects in preventing ketosis, PG is better used with other additives.

Author Contributions: F.Z. wrote the manuscript; X.N., H.W., and Y.Z. amended the English grammar and designed the figures; Y.G. and B.X. discussed and revised the draft. All authors have read and agreed to the published version of the manuscript.

Acknowledgments: We thank the National Agricultural Information System, Chinese Academy of Agricultural Sciences (CAAS), Beijing, 100193, China for providing access to the literature during the COVID-19 pandemic.

References

1. Macrae, A.I.; Burrough, E.; Forrest, J.; Corbishley, A.; Russell, G.; Shaw, D.J. Prevalence of excessive negative energy balance in commercial united kingdom dairy herds. *Vet. J.* **2019**, *248*, 51–57. [CrossRef] [PubMed]
2. Moore, S.M.; DeVries, T.J. Effect of diet-induced negative energy balance on the feeding behavior of dairy cows. *J. Dairy Sci.* **2020**, *103*, 7288–7301. [CrossRef]
3. Lean, I.J. Non-infectious disease: Ketosis. In *Reference Module in Food Science*; Elsevier: Amsterdam, The Netherlands, 2020. [CrossRef]
4. Rousing, T.; Holm, J.R.; Krogh, M.A.; Ostergaard, S.; Gplus, E.C. Expert-based development of a generic haccp-based risk management system to prevent critical negative energy balance in dairy herds. *Prev. Vet. Med.* **2020**, *175*, 104849. [CrossRef] [PubMed]
5. Gordon, J.L.; LeBlanc, S.J.; Duffield, T.F. Ketosis treatment in lactating dairy cattle. *Vet. Clin. N. Am. Food Anim. Pract.* **2013**, *29*, 433–445. [CrossRef] [PubMed]
6. Chibisa, G.E.; Gozho, G.N.; Van Kessel, A.G.; Olkowski, A.A.; Mutsvangwa, T. Effects of peripartum propylene glycol supplementation on nitrogen metabolism, body composition, and gene expression for the major protein degradation pathways in skeletal muscle in dairy cows. *J. Dairy Sci.* **2008**, *91*, 3512–3527. [CrossRef]
7. Nielsen, N.I.; Ingvartsen, K.L. Propylene glycol for dairy cows—A review of the metabolism of propylene glycol and its effects on physiological parameters, feed intake, milk production and risk of ketosis. *Anim. Feed Sci. Technol.* **2004**, *115*, 191–213. [CrossRef]
8. Johnson, R.B. The treatment of ketosis with glycerol and propylene glycol. *Cornell Vet.* **1954**, *44*, 6–21.
9. Lomander, H.; Frossling, J.; Ingvartsen, K.L.; Gustafsson, H.; Svensson, C. Supplemental feeding with glycerol or propylene glycol of dairy cows in early lactation-effects on metabolic status, body condition, and milk yield. *J. Dairy Sci.* **2012**, *95*, 2397–2408. [CrossRef]
10. Bertoni, G.; Trevisi, E.; Lombardelli, R. Some new aspects of nutrition, health conditions and fertility of intensively reared dairy cows. *Ital. J. Anim. Sci.* **2009**, *8*, 491–518. [CrossRef]
11. Esposito, G.; Irons, P.C.; Webb, E.C.; Chapwanya, A. Interactions between negative energy balance, metabolic diseases, uterine health and immune response in transition dairy cows. *Anim. Reprod. Sci.* **2014**, *144*, 60–71. [CrossRef]
12. Sun, B.F.; Cao, Y.C.; Cai, C.J.; Yu, C.; Li, S.X.; Yao, J.H. Temporal dynamics of nutrient balance, plasma biochemical and immune traits, and liver function in transition dairy cows. *J. Integr. Agric.* **2020**, *19*, 820–837. [CrossRef]
13. Pickett, M.M.; Piepenbrink, M.S.; Overton, T.R. Effects of propylene glycol or fat drench on plasma metabolites, liver composition, and production of dairy cows during the periparturient period1. *J. Dairy Sci.* **2003**, *86*, 2113–2121. [CrossRef]
14. Rukkwamsuk, T.; Rungruang, S.; Choothesa, A.; Wensing, T. Effect of propylene glycol on fatty liver development and hepatic fructose 1,6 bisphosphatase activity in periparturient dairy cows. *Livest. Prod. Sci.* **2005**, *95*, 95–102. [CrossRef]

15. Harder, I.; Stamer, E.; Junge, W.; Thaller, G. Lactation curves and model evaluation for feed intake and energy balance in dairy cows. *J. Dairy Sci.* **2019**, *102*, 7204–7216. [CrossRef] [PubMed]
16. Grummer, R.R.; Rastani, R.R. When should lactating dairy cows reach positive energy balance? *Prof. Anim. Sci.* **2003**, *19*, 197–203. [CrossRef]
17. Allen, M.S.; Bradford, B.J.; Oba, M. Board invited review: The hepatic oxidation theory of the control of feed intake and its application to ruminants. *J. Anim. Sci.* **2009**, *87*, 3317–3334. [CrossRef]
18. Tatone, E.H.; Duffield, T.F.; Capel, M.B.; DeVries, T.J.; LeBlanc, S.J.; Gordon, J.L. A randomized controlled trial of dexamethasone as an adjunctive therapy to propylene glycol for treatment of hyperketonemia in postpartum dairy cattle. *J. Dairy Sci.* **2016**, *99*, 8991–9000. [CrossRef]
19. McArt, J.A.A.; Nydam, D.V.; Oetzel, G.R. Epidemiology of subclinical ketosis in early lactation dairy cattle. *J. Dairy Sci.* **2012**, *95*, 5056–5066. [CrossRef]
20. Raboisson, D.; Mounié, M.; Maigné, E. Diseases, reproductive performance, and changes in milk production associated with subclinical ketosis in dairy cows: A meta-analysis and review. *J. Dairy Sci.* **2014**, *97*, 7547–7563. [CrossRef]
21. Bobe, G.; Young, J.W.; Beitz, D.C. Invited review: Pathology, etiology, prevention, and treatment of fatty liver in dairy cows. *J. Dairy Sci.* **2004**, *87*, 3105–3124. [CrossRef]
22. Fiore, E.; Piccione, G.; Perillo, L.; Barberio, A.; Manuali, E.; Morgante, M.; Gianesella, M. Hepatic lipidosis in high-yielding dairy cows during the transition period: Haematochemical and histopathological findings. *Anim. Prod. Sci.* **2017**, *57*, 74–80. [CrossRef]
23. Sun, X.; Wang, Y.; Loor, J.J.; Bucktrout, R.; Shu, X.; Jia, H.; Dong, J.; Zuo, R.; Liu, G.; Li, X.; et al. High expression of cell death-inducing dffa-like effector a (cidea) promotes milk fat content in dairy cows with clinical ketosis. *J. Dairy Sci.* **2019**, *102*, 1682–1692. [CrossRef] [PubMed]
24. Steensels, M.; Maltz, E.; Bahr, C.; Berckmans, D.; Antler, A.; Halachmi, I. Towards practical application of sensors for monitoring animal health; design and validation of a model to detect ketosis. *J. Dairy Res.* **2017**, *84*, 139–145. [CrossRef] [PubMed]
25. Koeck, A.; Miglior, F.; Jamrozik, J.; Kelton, D.F.; Schenkel, F.S. Genetic associations of ketosis and displaced abomasum with milk production traits in early first lactation of canadian holsteins. *J. Dairy Sci.* **2013**, *96*, 4688–4696. [CrossRef] [PubMed]
26. Duričić, D.; Ljubić, B.B.; Vince, S.; Turk, R.; Valpotić, H.; Žaja, I.Ž.; Maćešić, N.; Benić, M.; Getz, I.; Samardžija, M. Effects of dietary clinoptilolite supplementation on β-hydroxybutirate serum level and milk fat to protein ratio during early lactation in holstein-friesian cows. *Microporous Mesoporous Mater.* **2020**, *292*, 109766. [CrossRef]
27. Tamminga, S. The effect of the supply of rumen degradable protein and metabolisable protein on negative energy balance and fertility in dairy cows. *Anim. Reprod. Sci.* **2006**, *96*, 227–239. [CrossRef]
28. Albaaj, A.; Foucras, G.; Raboisson, D. Changes in milk urea around insemination are negatively associated with conception success in dairy cows. *J. Dairy Sci.* **2017**, *100*, 3257–3265. [CrossRef]
29. Jorritsma, R.; Wensing, T.; Kruip, T.A.M.; Vos, P.L.A.M.; Noordhuizen, J.P.T.M. Metabolic changes in early lactation and impaired reproductive performance in dairy cows. *Vet. Res.* **2003**, *34*, 11–26. [CrossRef]
30. Rutherford, A.J.; Oikonomou, G.; Smith, R.F. The effect of subclinical ketosis on activity at estrus and reproductive performance in dairy cattle. *J. Dairy Sci.* **2016**, *99*, 4808–4815. [CrossRef]
31. Albaaj, A.; Jattiot, M.; Manciaux, L.; Saille, S.; Julien, C.; Foucras, G.; Raboisson, D. Hyperketolactia occurrence before or after artificial insemination is associated with a decreased pregnancy per artificial insemination in dairy cows. *J. Dairy Sci.* **2019**, *102*, 8527–8536. [CrossRef]
32. Najm, N.-A.; Zimmermann, L.; Dietrich, O.; Rieger, A.; Martin, R.; Zerbe, H. Associations between motion activity, ketosis risk and estrus behavior in dairy cattle. *Prev. Vet. Med.* **2020**, *175*, 104857. [CrossRef] [PubMed]
33. Pascottini, O.B.; LeBlanc, S.J. Modulation of immune function in the bovine uterus peripartum. *Theriogenology* **2020**, *150*, 193–200. [CrossRef] [PubMed]
34. Gross, J.J.; Grossen-Rösti, L.; Wall, S.K.; Wellnitz, O.; Bruckmaier, R.M. Metabolic status is associated with the recovery of milk somatic cell count and milk secretion after lipopolysaccharide-induced mastitis in dairy cows. *J. Dairy Sci.* **2020**, *103*, 5604–5615. [CrossRef]

35. Mann, S.; Sipka, A.; Leal Yepes, F.A.; Nydam, D.V.; Overton, T.R.; Wakshlag, J.J. Nutrient-sensing kinase signaling in bovine immune cells is altered during the postpartum nutrient deficit: A possible role in transition cow inflammatory response. *J. Dairy Sci.* **2018**, *101*, 9360–9370. [CrossRef] [PubMed]
36. Sordillo, L.M.; Contreras, G.A.; Aitken, S.L. Metabolic factors affecting the inflammatory response of periparturient dairy cows. *Anim. Health Res. Rev.* **2009**, *10*, 53–63. [CrossRef]
37. Abuajamieh, M.; Kvidera, S.K.; Fernandez, M.V.S.; Nayeri, A.; Upah, N.C.; Nolan, E.A.; Lei, S.M.; DeFrain, J.M.; Green, H.B.; Schoenberg, K.M.; et al. Inflammatory biomarkers are associated with ketosis in periparturient holstein cows. *Res. Vet. Sci.* **2016**, *109*, 81–85. [CrossRef]
38. Lacasse, P.; Vanacker, N.; Ollier, S.; Ster, C. Innovative dairy cow management to improve resistance to metabolic and infectious diseases during the transition period. *Res. Vet. Sci.* **2018**, *116*, 40–46. [CrossRef]
39. Bertoni, G.; Trevisi, E.; Han, X.; Bionaz, M. Effects of inflammatory conditions on liver activity in puerperium period and consequences for performance in dairy cows. *J. Dairy Sci.* **2008**, *91*, 3300–3310. [CrossRef]
40. Burvenich, C.; Bannerman, D.D.; Lippolis, J.D.; Peelman, L.; Nonnecke, B.J.; Kehrli, M.E.; Paape, M.J. Cumulative physiological events influence the inflammatory response of the bovine udder to escherichia coli infections during the transition period1. *J. Dairy Sci.* **2007**, *90*, E39–E54. [CrossRef]
41. Kvidera, S.K.; Horst, E.A.; Abuajamieh, M.; Mayorga, E.J.; Fernandez, M.V.S.; Baumgard, L.H. Glucose requirements of an activated immune system in lactating holstein cows. *J. Dairy Sci.* **2017**, *100*, 2360–2374. [CrossRef]
42. Ingvartsen, K.L.; Moyes, K.M. Factors contributing to immunosuppression in the dairy cow during the periparturient period. *Jpn. J. Vet. Res.* **2015**, *63*, S15–S24. [PubMed]
43. Shen, T.; Li, X.; Loor, J.J.; Zhu, Y.; Du, X.; Wang, X.; Xing, D.; Shi, Z.; Fang, Z.; Li, X.; et al. Hepatic nuclear factor kappa b signaling pathway and nlr family pyrin domain containing 3 inflammasome is over-activated in ketotic dairy cows. *J. Dairy Sci.* **2019**, *102*, 10554–10563. [CrossRef] [PubMed]
44. Van Straten, M.; Friger, M.; Shpigel, N.Y. Events of elevated somatic cell counts in high-producing dairy cows are associated with daily body weight loss in early lactation. *J. Dairy Sci.* **2009**, *92*, 4386–4394. [CrossRef]
45. Giuliodori, M.J.; Magnasco, R.P.; Becu-Villalobos, D.; Lacau-Mengido, I.M.; Risco, C.A.; de la Sota, R.L. Metritis in dairy cows: Risk factors and reproductive performance. *J. Dairy Sci.* **2013**, *96*, 3621–3631. [CrossRef] [PubMed]
46. Fiore, E.; Perillo, L.; Piccione, G.; Gianesella, M.; Bedin, S.; Armato, L.; Giudice, E.; Morgante, M. Effect of combined acetylmethionine, cyanocobalamin and alpha-lipoic acid on hepatic metabolism in high-yielding dairy cow. *J. Dairy Res.* **2016**, *83*, 438–441. [CrossRef] [PubMed]
47. Fisher, L.J.; Erfle, J.D.; Lodge, G.A.; Sauer, F.D. Effects of propylene glycol or glycerol supplementation of the diet of dairy cows on feed intake, milk yield and composition, and incidence of ketosis. *Can. J. Anim. Sci.* **1973**, *53*, 289–296. [CrossRef]
48. Butler, S.T.; Pelton, S.H.; Butler, W.R. Energy balance, metabolic status, and the first postpartum ovarian follicle wave in cows administered propylene glycol. *J. Dairy Sci.* **2006**, *89*, 2938–2951. [CrossRef]
49. Piantoni, P.; Allen, M.S. Evaluation of propylene glycol and glycerol infusions as treatments for ketosis in dairy cows. *J. Dairy Sci.* **2015**, *98*, 5429–5439. [CrossRef]
50. Miyoshi, S.; Pate, J.L.; Palmquist, D.L. Effects of propylene glycol drenching on energy balance, plasma glucose, plasma insulin, ovarian function and conception in dairy cows. *Anim. Reprod. Sci.* **2001**, *68*, 29–43. [CrossRef]
51. Studer, V.A.; Grummer, R.R.; Bertics, S.J.; Reynolds, C.K. Effect of prepartum propylene glycol administration on periparturient fatty liver in dairy cows. *J. Dairy Sci.* **1993**, *76*, 2931–2939. [CrossRef]
52. Ferraro, S.M.; Mendoza, G.D.; Miranda, L.A.; Gutiérrez, C.G. In Vitro gas production and ruminal fermentation of glycerol, propylene glycol and molasses. *Anim. Feed Sci. Technol.* **2009**, *154*, 112–118. [CrossRef]
53. Rizos, D.; Kenny, D.A.; Griffin, W.; Quinn, K.M.; Duffy, P.; Mulligan, F.J.; Roche, J.F.; Boland, M.P.; Lonergan, P. The effect of feeding propylene glycol to dairy cows during the early postpartum period on follicular dynamics and on metabolic parameters related to fertility. *Theriogenology* **2008**, *69*, 688–699. [CrossRef] [PubMed]
54. Kristensen, N.B.; Raun, B.M. Ruminal and intermediary metabolism of propylene glycol in lactating holstein cows. *J. Dairy Sci.* **2007**, *90*, 4707–4717. [CrossRef] [PubMed]

55. Bjerre-Harpøth, V.; Storm, A.C.; Eslamizad, M.; Kuhla, B.; Larsen, M. Effect of propylene glycol on adipose tissue mobilization in postpartum over-conditioned holstein cows. *J. Dairy Sci.* **2015**, *98*, 8581–8596. [CrossRef] [PubMed]
56. Gamarra, G.; Ponsart, C.; Lacaze, S.; Le Guienne, B.; Deloche, M.C.; Monniaux, D.; Ponter, A.A. Short term dietary propylene glycol supplementation affects circulating metabolic hormones, progesterone concentrations and follicular growth in dairy heifers. *Livest. Sci.* **2014**, *162*, 240–251. [CrossRef]
57. De Koster, J.D.; Opsomer, G. Insulin resistance in dairy cows. *Vet. Clin. N. Am. Food Anim. Pract.* **2013**, *29*, 299–322. [CrossRef]
58. Rico, J.E.; Bandaru, V.V.; Dorskind, J.M.; Haughey, N.J.; McFadden, J.W. Plasma ceramides are elevated in overweight holstein dairy cows experiencing greater lipolysis and insulin resistance during the transition from late pregnancy to early lactation. *J. Dairy Sci.* **2015**, *98*, 7757–7770. [CrossRef]
59. Chalmeh, A.; Pourjafar, M.; Badiei, K.; Jalali, M.; Sebdani, M.M. The comparative effects of dietary monensin and propylene glycol on insulin resistance of transition dairy cows. *Trop. Anim. Health Prod.* **2020**, *52*, 1573–1582. [CrossRef]
60. Moallem, U.; Katz, M.; Arieli, A.; Lehrer, H. Effects of peripartum propylene glycol or fats differing in fatty acid profiles on feed intake, production, and plasma metabolites in dairy cows. *J. Dairy Sci.* **2007**, *90*, 3846–3856. [CrossRef]
61. Jeong, J.-K.; Choi, I.-S.; Moon, S.-H.; Lee, S.-C.; Kang, H.-G.; Jung, Y.-H.; Park, S.-B.; Kim, I.-H. Effect of two treatment protocols for ketosis on the resolution, postpartum health, milk yield, and reproductive outcomes of dairy cows. *Theriogenology* **2018**, *106*, 53–59. [CrossRef]
62. Chung, Y.H.; Martinez, C.M.; Brown, N.E.; Cassidy, T.W.; Varga, G.A. Ruminal and blood responses to propylene glycol during frequent feeding. *J. Dairy Sci.* **2009**, *92*, 4555–4564. [CrossRef] [PubMed]
63. Urrutia, N.; Bomberger, R.; Matamoros, C.; Harvatine, K.J. Effect of dietary supplementation of sodium acetate and calcium butyrate on milk fat synthesis in lactating dairy cows. *J. Dairy Sci.* **2019**, *102*, 5172–5181. [CrossRef] [PubMed]
64. Juchem, S.O.; Santos, F.A.P.; Imaizumi, H.; Pires, A.V.; Barnabé, E.C. Production and blood parameters of holstein cows treated prepartum with sodium monensin or propylene glycol1. *J. Dairy Sci.* **2004**, *87*, 680–689. [CrossRef]
65. Adamski, M.; Kupczynski, R.; Chladek, G.; Falta, D. Influence of propylene glycol and glycerin in simmental cows in periparturient period on milk yield and metabolic changes. *Arch. Tierz.* **2011**, *54*, 238–248. [CrossRef]
66. McArt, J.A.A.; Nydam, D.V.; Oetzel, G.R. A field trial on the effect of propylene glycol on displaced abomasum, removal from herd, and reproduction in fresh cows diagnosed with subclinical ketosis. *J. Dairy Sci.* **2012**, *95*, 2505–2512. [CrossRef]
67. Gonzalez, F.D.; Muino, R.; Pereira, V.; Campos, R.; Benedito, J.L. Relationship among blood indicators of lipomobilization and hepatic function during early lactation in high-yielding dairy cows. *J. Vet. Sci.* **2011**, *12*, 251–255. [CrossRef]
68. Hussein, H.A.; Abdel-Raheem, S.M.; Abd-Allah, M.; Senosy, W. Effects of propylene glycol on the metabolic status and milk production of dairy buffaloes. *Tierärztliche Prax. Großtiere* **2015**, *43*, 25–34.
69. Stokes, S.R.; Goff, J.P. Evaluation of calcium propionate and propylene glycol administered into the esophagus of dairy cattle at calving11supported in part by a gift from kemin industries, inc., des moines, ia 50301-0070. *Prof. Anim. Sci.* **2001**, *17*, 115–122. [CrossRef]
70. McArt, J.A.A.; Nydam, D.V.; Ospina, P.A.; Oetzel, G.R. A field trial on the effect of propylene glycol on milk yield and resolution of ketosis in fresh cows diagnosed with subclinical ketosis. *J. Dairy Sci.* **2011**, *94*, 6011–6020. [CrossRef]
71. Østergaard, S.; Krogh, M.A.; Oliveira, V.H.S.; Larsen, T.; Otten, N.D. Only few benefits from propylene glycol drench in early lactation for cows identified as physiologically imbalanced based on milk spectra analyses. *J. Dairy Sci.* **2020**, *103*, 1831–1842. [CrossRef]
72. Madoz, L.V.; Rabaglino, M.B.; Migliorisi, A.L.; Jaureguiberry, M.; Perez Wallace, S.; Lorenti, N.; Domínguez, G.; Giuliodori, M.J.; de la Sota, R.L. Association between progesterone concentration and endometrial gene expression in dairy cows. *Domest. Anim. Endocrinol.* **2020**, *74*, 106481. [CrossRef] [PubMed]

73. Castañeda-Gutiérrez, E.; Pelton, S.H.; Gilbert, R.O.; Butler, W.R. Effect of peripartum dietary energy supplementation of dairy cows on metabolites, liver function and reproductive variables. *Anim. Reprod. Sci.* **2009**, *112*, 301–315. [CrossRef] [PubMed]
74. Osorio, J.S.; Trevisi, E.; Ji, P.; Drackley, J.K.; Luchini, D.; Bertoni, G.; Loor, J.J. Biomarkers of inflammation, metabolism, and oxidative stress in blood, liver, and milk reveal a better immunometabolic status in peripartal cows supplemented with smartamine m or metasmart. *J. Dairy Sci.* **2014**, *97*, 7437–7450. [CrossRef] [PubMed]
75. Formigoni, A.; Cornil, M.C.; Prandi, A.; Mordenti, A.; Rossi, A.; Portetelle, D.; Renaville, R. Effect of propylene glycol supplementation around parturition on milk yield, reproduction performance and some hormonal and metabolic characteristics in dairy cows. *J. Dairy Res.* **1996**, *63*, 11–24. [CrossRef] [PubMed]
76. Trabue, S.; Scoggin, K.; Tjandrakusuma, S.; Rasmussen, M.A.; Reilly, P.J. Ruminal fermentation of propylene glycol and glycerol. *J. Agric. Food Chem.* **2007**, *55*, 7043–7051. [CrossRef] [PubMed]
77. Bertram, H.C.; Petersen, B.O.; Duus, J.O.; Larsen, M.; Raun, B.M.L.; Kristensen, N.B. Proton nuclear magnetic resonance spectroscopy based investigation on propylene glycol toxicosis in a holstein cow. *Acta Vet. Scand.* **2009**, *51*, 1–9. [CrossRef]
78. Macrae, A.I.; Burrough, E.; Forrest, J.; Corbishley, A.; Russell, G.; Shaw, D.J. Risk factors associated with excessive negative energy balance in commercial united kingdom dairy herds. *Vet. J.* **2019**, *250*, 15–23. [CrossRef]
79. Chung, Y.H.; Brown, N.E.; Martinez, C.M.; Cassidy, T.W.; Varga, G.A. Effects of rumen-protected choline and dry propylene glycol on feed intake and blood parameters for holstein dairy cows in early lactation. *J. Dairy Sci.* **2009**, *92*, 2729–2736. [CrossRef]
80. Gordon, J.L.; LeBlanc, S.J.; Kelton, D.F.; Herdt, T.H.; Neuder, L.; Duffield, T.F. Randomized clinical field trial on the effects of butaphosphan-cyanocobalamin and propylene glycol on ketosis resolution and milk production. *J. Dairy Sci.* **2017**, *100*, 3912–3921. [CrossRef]
81. McArt, J.A.; Nydam, D.V.; Oetzel, G.R.; Guard, C.L. An economic analysis of hyperketonemia testing and propylene glycol treatment strategies in early lactation dairy cattle. *Prev. Vet. Med.* **2014**, *117*, 170–179. [CrossRef]
82. El-Kasrawy, N.I.; Swelum, A.A.; Abdel-Latif, M.A.; Alsenosy, A.E.A.; Beder, N.A.; Alkahtani, S.; Abdel-Daim, M.M.; Abd El-Aziz, A.H. Efficacy of different drenching regimens of gluconeogenic precursors during transition period on body condition score, production, reproductive performance, subclinical ketosis and economics of dairy cows. *Animals* **2020**, *10*, 937. [CrossRef] [PubMed]
83. Van der Drift, S.G.A.; Houweling, M.; Bouman, M.; Koets, A.P.; Tielens, A.G.M.; Nielen, M.; Jorritsma, R. Effects of a single glucocorticoid injection on propylene glycol-treated cows with clinical ketosis. *Vet. J.* **2015**, *204*, 144–149. [CrossRef] [PubMed]
84. Mann, S.; Yepes, F.A.L.; Behling-Kelly, E.; McArt, J.A.A. The effect of different treatments for early-lactation hyperketonemia on blood β-hydroxybutyrate, plasma nonesterified fatty acids, glucose, insulin, and glucagon in dairy cattle. *J. Dairy Sci.* **2017**, *100*, 6470–6482. [CrossRef] [PubMed]

16

Multiple Amino Acid Supplementations to Low-Protein Diets: Effect on Performance, Carcass Yield, Meat Quality and Nitrogen Excretion of Finishing Broilers under Hot Climate Conditions

Youssef A. Attia [1,*], Fulvia Bovera [2,*], Jinquan Wang [3], Mohammed A. Al-Harthi [1] and Woo Kyun Kim [3,*]

1 Arid Land Agriculture Department, Faculty of Meteorology, Environment and Arid Land Agriculture, King Abdulaziz University, Jeddah 21589, Saudi Arabia; malharthi@kau.edu.sa

2 Department of Veterinary Medicine and Animal Production, University of Napoli Federico II, via F. Delpino1, 80137 Napoli, Italy

3 Department of Poultry science, University of Georgia, Athens, GA 30602, USA; jq.wang@uga.edu

* Correspondence: yaattia@kau.edu.sa (Y.A.A.); bovera@unina.it (F.B.); wkkim@uga.edu (W.K.K.)

Simple Summary: Crude protein is an essential nutrient in poultry feed. Reducing the use of crude protein not only reduces the feed cost, but also minimizes pollution during poultry production. Thus, finding the minimum protein requirement in broiler diet without compromising broiler growth is the objective in the present study. Supplementing essential amino acids including methionine and lysine to the low-protein diet showed comparable growth performance and carcass yield to the regular protein diet. Thus, reducing the crude protein level is possible if the essential amino acid balance is adequate for broiler growth.

Abstract: The objective of this study was to evaluate the effect of low-protein diets with amino acid supplementation on growth performance, carcass yield, meat quality and nitrogen excretion of broilers raised under hot climate conditions during the finisher period. In trial 1, broilers from 28 to 49 days of age were fed 18% crude protein (CP) as a positive control or 15% CP supplemented with (1) DL-methionine (Met) + L-lysine (Lys), (2) Met + Lys + L-Arginine (Arg), or (3) Met + Lys + L-Valine (Val). In trial 2, broilers from 30 to 45 days of age, were fed an 18% CP diet as a positive control or 15% CP supplemented with Met, Lys, Arg, Val, L-Isoleucine (Ile) or combination with glycine (Gly) and/or urea as nitrogen sources: (1) Met + Lys, (2) Met + Lys + Arg, (3) Met + Lys + Val, (4) Met + Lys + Ile, (5) Met + Lys + Arg +Val + Ile + Gly, and (6) Met+ Lys + Arg + Val + Ile + Gly + urea. Protein use was improved by feeding low-protein amino acid-supplemented diets as compared to the high-protein diet. Feeding 15% crude protein diet supplemented with only methionine and lysine had no negative effects on carcass yield, CP, total lipids and moisture% of breast meat while decreasing nitrogen excretion by 21%.

Keywords: broilers; amino acid; low crude protein; carcass yield; nitrogen excretion; hot climate

---

1. Introduction

In hot climates, the growth of broilers is usually slow, as high temperature adversely affects feed intake, and high-protein diets, used to sustain animal growth, may impair broiler tolerance to heat stress due to the high heat increment induced by protein metabolism [1,2]. Decreasing dietary protein levels of broiler diets could be a good strategy, but the amino acid requirements for broiler chicks must be satisfied by supplementing synthetic amino acids to the basal diets. In this way it is possible to achieve several goals: (1) minimize the excess of amino acid input and thus nitrogen pollution; (2) improve the bird tolerance to hot climate; (3) obtain satisfactory growth performance; all this can increase the profit for the poultry industry [3–5].

Environmental pollution due to nitrogen excretion is an important issue in poultry houses and results in negative impacts on the health of workers and birds, soil, and ground water [4,5]. Thus, reducing dietary protein levels with proper amino acid supplementation has been of great interest [2,4,6–14]. It is well recognized that methionine (Met) is typically the first limiting amino acid in practical broiler diets, whereas lysine (Lys) is the second in broiler diets. However, this depends on the ingredient composition of the diets [4,15,16]. There are several commercially available synthetic amino acids, such as methionine, lysine, threonine (Thr), tryptophan (Trp), and arginine (Arg) [4,12,17] that can be used to fulfill amino acid requirements in low-protein diets.

Fancher and Jensen [6], feeding female broilers on low-protein, corn/soybean meal diets supplemented with synthetic amino acids from 3 to 6 weeks of age, found worse performance in comparison to a higher-protein finisher diet. Liu et al. [18] reported that a decrease in protein level from 22.5% to 20% impaired body weight gain from 1 to 21 days of age. Along the same line, Holsheimer and Janssen [19] reported that Thr, Trp and Arg were limiting in diets containing 17% CP supplemented with Met and Lys when compared to 20% CP diet during 3–7 weeks of age. They also indicated that 0.77% Thr and 0.22% Trp were enough in finisher broiler diets during 3–7 weeks of age. Han et al. [7] observed that broiler chicks fed low-protein diets (16% CP) supplemented with Met, Lys, Arg, valine (Val), and glutamic acid during 3-6 weeks of age had high body weight and better feed conversion ratio (FCR) than those fed a 20% CP diet while maintaining similar body fat content. Moreover, Laudadio et al. [8] indicated that female broiler chicks fed 17% CP diet supplemented with Met plus cysteine (Cys) showed higher body weight than chickens fed 20% CP during 3–7 weeks of age. Lipstein et al. [20] and Lecercq et al. [21] reported that low-protein diets supplemented with Met and Lys prevented increased carcass fat deposition during the finisher period. However, severely low crude protein diets may increase abdominal fat deposition in carcass [22,23]. A 3% reduction in CP compromised body weight and feed conversion during 21–42 day of age but the supplementation of Val+Ile (isoleucine)+Arg+Gly was able to restore both [24].

Broiler chicks consume the highest proportion (~60%) of their feed during the finisher period, when they are more susceptible to heat stress because of their bigger body size. However, few studies have been performed to evaluate the effects of low-protein diets supplemented with multiple key synthetic amino acids on growth performance and carcass yield, under hot climate during finisher period. Thus, the objective of this study, organized in two different trials (Trial 1 and 2) was to evaluate the effect of low-protein diets supplemented with different levels of various essential amino acids on body weight gain, feed conversion ratio, nitrogen excretion, carcass yield, and inner organ weight of broilers during the finisher period and farmed under hot climate conditions.

2. Materials and Methods

2.1. Animal Husbandry:

Broiler chicks were reared in battery-brooders (35 cm length × 25 cm width × 30 cm height) with 23:1 light-dark cycle at day 1, gradually reduced to 18:6 light-dark cycle. Birds were fed from tube feeders and drank from automatic nipple drinkers. Mash diets and water were offered ad libitum throughout the trials. Birds were vaccinated against Newcastle disease virus with Hatchner (B1) at 7 days old and Lasota at 20, 30 and 45 days of age. The average indoor ambient temperature and relative humidity were (mean ± standard deviation): 34 ± 6 °C and 54 ± 9% in Trial 1 and 35 ± 5 °C and 57 ± 11% in Trial 2, respectively. The experimental protocol was approved by the Animal and Poultry Production Scientific and Ethics Committee of Damanhour University. The care and handling of the animals were performed so as to maintain their rights, ensure their welfare, and cause minimal stress, according to International Guidelines for research involving animals (Directive2010/63/EU).

2.2. Experimental Design and Diets

Trial 1 comprised 7 groups (each with 25 Hubbard male broilers, 5 replicates of 5 birds) which had the same feed ingredients and only differed in the protein source and the type of amino acid included. Two different CP diets were formulated based on the protein source: vegetable diets with corn soybean meal (groups 2, 3 and 4) or animal protein diets with fish and meat meals (groups 5, 6 and 7). From 28 to 49 days of age, the groups were submitted to the following dietary treatment: 18% CP (group 1); 15% CP supplemented with Met and Lys (group 2 and group 5); 15% CP supplemented with Met, Lys and Arg (group 3 and group 6); 15% CP supplemented with Met, Lys, Arg and Val (group 4 and group 7). In trial 1, the amounts of Met and Lys supplemented were 0.20% and 0.28% for vegetable diets (group 2, 3 and 4), and 0.19% and 0.26% for animal diets (groups 5, 6 and 7), respectively. The supplemented amounts of Arg, Val and Ile were 0.17%, 0.15% and 0.0% for the vegetable protein diet (group 2, 3 and 4), and 0.22%, 0.15% and 0.05% for the animal protein diet (groups 5, 6 and 7), respectively.

Trial 2 comprised 7 groups (25 male Hubbard broilers, 5 replicates of 5 birds per group) which had the same feed ingredients and only differed in the type of amino acid included. From 30 to 45 days of age, the groups were submitted to the following dietary treatments: group 1 as positive control with 18% CP; 15% CP supplemented with Met at 0.19%, Lys at 0.25% (group 2); 15% CP supplemented with Met at 0.19%, Lys at 0.25% and Arg 0.21% (group 3); 15% CP supplemented with Met at 0.19%, Lys at 0.25% and Val at 0.10% (group 4); 15% CP supplemented with Met at 0.19%, Lys at 0.25% and Ile (group 5); 15% CP supplemented with Met at 0.19%, Lys at 0.25%, Arg at 0.21% Val at 0.10%, Ile and Gly as an amino nitrogen source (group 6); and 15% CP supplemented with Met at 0.19%, Lys at 0.25%, Arg at 0.21% Val at 0.10%, and Ile (as a nitrogen source to equalize nitrogen content between the positive control diet and the 15% CP diet; group 7).

The diets (Trial 1 and 2) were formulated based on Nutrient Requirements of Poultry from National Research Council (NRC) [25] tabulated values for feedstuffs (Table 1). Met and Lys levels of the low protein 15% CP diets were equal to those of the positive control diet. Amino acids were supplemented and analyzed by Evonik (Degussa-Hüls AG, Frankfurt am Main, Germany). Amino acids were supplemented to meet NRC amino acid requirements [25] for broiler chicks.

Table 1. Composition, and calculated and analyzed nutrients of diets in trial 1 and 2.

Ingredients, %	Trial 1			Trial 2	
	18%	15% Animal	15% Plant	18%	15%
Yellow corn	71.63	73.00	73.00	66.86	72.40
Soybean meal	21.00	15.00	20.00	24.71	16.90
Fish meal (72% CP herring)	2.00	1.73	0.00	2.00	2.00
Meat meal	2.00	1.73	0.00	0.00	0.00
Soybean oil	0.50	1.45	1.70	1.65	1.56
Bone meal	1.40	1.68	2.38	1.70	1.88
Lime stone	0.82	0.72	0.63	0.75	0.72
Vit + Min premix [1]	0.25	0.25	0.25	0.25	0.25
NaCl	0.30	0.30	0.30	0.30	0.30
DL-methionine	0.10	0.19	0.20	0.11	0.19
L-lysine	0.00	0.26	0.28	0.06	0.25
Sand	0.00	3.69	1.26	1.61	3.55
Total	100.0	100.0	100.0	100.0	100.0

Table 1. *Cont.*

Ingredients, %	Trial 1			Trial 2	
	18%	15% Animal	15% Plant	18%	15%
		Calculated values, %			
ME kcal/ kg	3008	3000	3041	3000	3004
Tryptophan	0.23	0.18	0.19	0.24	0.19
Ca	1.01	1.01	1.01	0.91	0.91
Available P	0.41	0.41	0.41	0.37	0.37
		Determined values [2], %			
Crude protein	19.04	15.54	15.18	18.63	15.23
Methionine	0.30	0.25	0.24	0.31	0.26
TSAA	0.92	0.70	0.68	0.59	0.51
Lysine	0.92	0.70	0.72	0.96	0.74
Arg	1.02	0.84	0.81	1.03	0.86
Val	0.89	0.69	0.71	0.82	0.71
Ile	0.71	0.62	0.60	0.73	0.59

[1] Premix provides per kg of diet: vitamin A, 8000 international units (IU); vitamin E, 9 mg; menadione (as menadione sodium bisulfite), 150 IU; Vit. D3, 1,000 ICU; riboflavin, 4.0 mg; Ca pantothenate, 10 mg; nicotinic acid, 12 mg; choline chloride, 300 mg; vitamin B12,2 mg; vitamin B6, 1.2 mg; thiamine (as thiamine mononitrate), 2.0 mg; folic acid, 40 mg; d-biotin, 0.05 mg. Trace minerals (mg per kg of diet): Mn, 75; Zn, 40; Fe, 40; Cu, 3; Se, 0.15; iodine, 0.8 and 500 mg an antioxidant. [2] Calculated based on the analyzed amino acids of feed ingredients except for tryptophan, which was not determined.

2.3. *Growth Performance and Carcass Yield*

Chicks were weighed at the beginning, at 28 days of age in Trial 1 and at 30 days in Trial 2, and at the end at 49 days of age for Trial 1 and 45 days of age for Trial 2. Average initial body weights in Trial 1 and 2 were (mean ± standard deviation) 1,008 ± 10 g and 1,164 ± 12 g, respectively. In both trials there were no differences among groups for initial body weight and the Body Weight Gain was calculated as the difference between final and initial body weight. Additionally, for both trials, the amounts of administered and residual feed were weighed on the same day as the weighing of the birds, and total feed and protein intakes were calculated as the difference between administered and residual feed. Feed and protein conversion ratios were calculated by dividing feed and protein intake by body weight gain. In Trial 2, at 40 days of age, represented samples (500 g) of excreta (n = 5 per treatment) as one sample per replicate were collected and cleaned from feathers, feeds and scabies, and then moisture and nitrogen contents were determined according to the Association of Official Agricultural Chemists (AOAC) [26]. A total of 35 broilers at the end of Trials 1 and 2 were slaughtered for carcass evaluation. In Trial 1, moisture, crude protein, ether extract, and ash of the skinless boneless breast meat (pectorals major) from the slaughtered birds were further determined in pooled triplicate samples according to AOAC [26]. The crude protein, ether extract, and ash content were divided by dry weight for breast meat chemical characteristics.

2.4. *Statistical Analysis:*

Data from each trial were analyzed using the general linear model procedure of Statistical Analysis Software [27] using a one-way analysis of variance. The experimental unit was the replicate. Mean differences were tested using student Newman Keuls Test [27] to evaluate the differences among means at $P \leq 0.05$.

3. Results

In general, broilers in both trials showed signs of heat stress, including panting, lying down, and flapping their wings to dissipate heat, but differences in behavior responses were not quantitative among different treatments (data not shown).

In Trials 1 and 2, during the whole finishing period, there were no significant differences among groups for body weight, weight gain, feed intake and feed conversion ratio (Tables 2 and 3).

Table 2. Growth performance from 28 to 49 day of age (Trial 1).

Item	Groups							SEM [1]	*P* Value
	1	2	3	4	5	6	7		
Body weight at 28 d, g	1009	1006	1011	1005	1002	1018	1006	15.3	0.99
Body weight at 49 d, g	1964	1940	1970	1938	1947	1970	1853	28.7	0.88
Body weight gain, g	955	934	959	933	945	952	947	21.1	0.78
Feed intake, g	2212	2223	2257	2263	2223	2226	2226	12.7	0.15
Protein intake, g	409.2 [a]	333.4 [b]	338.5 [b]	339.4 [b]	333.5 [b]	333.9 [b]	333.9 [b]	1.96	0.01
Feed conversion ratio, g/g	2.32	2.38	2.35	2.43	2.35	2.34	2.35	0.11	0.74
Protein Conversion Ratio, g/g	0.428 [a]	0.357 [b]	0.353 [b]	0.364 [b]	0.353 [b]	0.351 [b]	0.353 [b]	0.009	0.01

[a,b]: Means within a row not sharing a common a superscript differ significantly $P \leq 0.05$, based on Duncan's test; [1] SEM, standard error of the mean.

Table 3. Growth performance from 30 to 45 day of age (Trial 2).

Item	Groups							SEM	*P* Value
	1	2	3	4	5	6	7		
Body weight 30 d, g	1155	1176	1167	1153	1171	1160	1166	25.1	0.99
Body weight 45 d, g	1959	1977	1992	1988	1981	1993	1971	37.2	0.99
Body weight gain,	804	801	824	835	810	834	806	25.8	0.95
Feed intake, g	1717	1756	1763	1782	1796	1763	1753	23.4	0.38
Protein intake, g	309 [a]	263 [b]	264 [c]	267 [b]	269 [b]	265 [b]	263 [b]	3.05	0.01
Feed conversion ratio, g/g	2.14	2.19	2.14	2.13	2.22	2.11	2.17	0085	0.98
Protein conversion ratio, g/g	0.384 [a]	0.329 [b]	0.321 [b]	0.320 [b]	0.333 [b]	0.317 [b]	0.326 [b]	0.014	0.01
Excreta nitrogen, %	4.03	3.18	3.25	3.47	3.79	3.71	3.90	0.54	0.17
Excreta dry matter, %	21.0	20.9	21.8	21.1	21.6	22.2	22.7	1.6	0.46

[a–c] Means within a row not sharing a common a superscript differ significantly $P \leq 0.05$, based on Duncan's test.

However, in both trials the 15% CP diet groups with amino acid supplementation showed lower ($P < 0.05$) protein intake and protein conversion ratio (PCR) compared to group 1 (Tables 2 and 3).

In Trial 1, there were no differences among groups for carcass yield, body organs and breast meat chemical compositions (Table 4). In Trial 2 (Table 5), the supplemented groups 2, 3 and 4 showed a significantly higher ($P < 0.05$) gizzard percentage compared to groups 1 and 7. Groups 2, 3 and 7 showed a lower ($P < 0.05$) spleen percentage compared to group 1 (Table 5). However, there were no differences among the other carcass traits.

Table 4. Carcass yield, body organs percentage and breast meat chemical composition (Trial 1).

Item	Groups							SEM	*P* Value
	1	2	3	4	5	6	7		
Carcass yield									
Dressing [1], %	62.2	61.9	62.2	62.0	63.2	63.4	62.5	0.97	0.17
Breast + wings, %	23.4	23.3	22.5	24.4	23.6	22.4	22.1	0.77	0.40
Thigh + legs, %	21.7	22.3	22.0	21.6	21.6	23.6	22.6	0.67	0.40
Abdominal fat, %	1.71	1.82	1.83	1.81	1.80	2.19	1.93	0.24	0.36
Liver, %	2.42	2.67	2.24	2.18	2.14	2.31	2.29	0.44	0.10
Heart, %	0.64	0.65	0.64	0.69	0.62	0.65	0.65	0.041	0.93
Pancreas, %	0.23	0.25	0.22	0.21	0.24	0.24	0.23	0.012	0.43
Spleen, %	0.131	0.125	0.108	0.161	0.125	0.134	0.149	0.013	0.14
Chemical characteristics of breast meat									
Moisture, %	74.2	74.4	73.8	75.1	74.9	74.6	74.9	2.42	0.58
Crude protein, %	82.7	81.4	81.4	82.0	81.9	82.0	82.0	1.73	0.74
Ether extract, %	13.1	13.4	13.8	13.0	13.2	13.6	13.5	0.74	0.68
Ash, %	4.2	5.2	4.8	5.0	4.9	4.4	4.5	0.12	0.39

[1] Head, giblets, feet and eviscerate were not included.

Table 5. Carcass yield of broiler chickens (Trial 2).

Item	Groups							SEM	*P* Value
	1	2	3	4	5	6	7		
Dressing [1], %	62.4	63.7	61.8	60.8	61.2	63.2	60.0	0.96	0.11
Breast + wings, %	30.5	31.1	30.2	29.2	30.4	32.4	29.6	0.71	0.07
Thigh + legs, %	32.9	34.8	32.1	32.1	31.7	31.8	31.4	0.89	0.16
Abdominal fat, %	1.71	1.80	2.10	1.84	1.45	1.33	1.62	0.25	0.39
Liver, %	1.84	1.95	2.22	2.09	2.02	1.93	1.99	0.13	0.49
Heart, %	0.44	0.46	0.51	0.47	0.42	0.44	0.45	0..03	0.43
Gizzard, %	1.54 ^{b}	2.00 ^{a}	1.90 ^{a}	1.88 ^{a}	1.83 ^{a,b}	1.76 ^{a,b}	1.53 ^{b}	0.10	<0.01
Pancreas, %	0.21	0.18	0.20	0.19	0.20	0.18	0.19	0.015	0.69
Spleen, %	0.129 ^{a}	0.083 ^{c}	0.087 ^{b,c}	0.124 ^{a,b}	0.115 $^{a\text{–}c}$	0.121 ^{a,b}	0.088 ^{b,c}	0.012	0.03

$^{a\text{–}c}$ Means within a row not sharing a common a superscript differ significantly $P \leq 0.05$, based on Duncan's test; [1] Head, giblets, feet and eviscerate were not included.

4. Discussion

4.1. Growth Performance, Feed/Protein Conversion Ratio and Nitrogen Excretion of Broiler Chicks

Our previous studies showed that broiler growth was low at high temperature (34 ± 6 – 35 ± 5 °C and 54 ± 9 and 57 ± 11% Relative Humidity) [1,2]. In the present study, the results from Trials 1 and 2 indicated that reducing the CP level from 18% to 15% with the supplementation of amino acids under hot climates and during the finisher period reduced the overall protein intake of broilers without detriment to growth performance. In both trials, the low final body weight compared to Attia et al. [28] could be attributed to the high ambient temperature and humidity (34 and 35 °C and 54% and 57%, respectively). Heat stress in tropical and subtropical regions not only causes economic effects in the form of retarded growth performance, but also induces animal health and welfare concerns [29]. The hot temperatures during the daytime retarded the growth of 7-week-old broilers under 2 kg in both trials. Extreme environmental conditions should be avoided in general through ventilation or reducing stock density; however, hot climate conditions are unavoidable in tropical areas. The results in the current study indicate that satisfactory broiler growth and feed conversion could be achieved with a reduction by 16.6% (from 18% to 15%) of the crude protein content of the diet in finishing broilers raised under hot climate conditions, but the low-protein diet must be supplemented with adequate amounts of Met and Lys, and the same results were also obtained with only vegetable or partially animal protein sources in the diet.

Several studies have suggested that reducing protein levels in diets with adequate supplementation of indispensable amino acids will not compromise broiler growth performance during the finisher period [10–14]. Recent research indicates that in broilers from 1 to 35 days of age, the CP of the diets can be reduced from 210 to 180 g/kg without detriment to broiler performance, but further reduction from 180 to 165 g/kg of diet CP worsened FCR, although the level of essential amino acids was maintained [30]. Additionally, van Harn et al. [31] observed that, in broilers from 1 to 35 days of age, during the finisher period, the crude protein of the diets could be reduced by 2.2–2.3 percentage points (from 19.8%) without negative effects on bird performance if the essential amino acids were correctly balanced, but further decrease could penalize some carcass traits. Chrystal et al. [30] attributed these results to a possible deficiency of non-essential amino acids such as glycine and serine at the lowest level of diet CP. As animal age increases in the present trial (28–49 and 30–45 day of age for Trials 1 and 2, respectively), the requirements of birds for non-essential amino acids could be satisfied even while reducing the protein content of the diet to 15%. The retarded growth performance under hot climate also contributed to the change in requirements. However, the current results could not distinguish the confounding effects of heat stress and reduction of protein. It is interesting to observe that, under hot climate conditions, reducing the CP level to 15% with key amino acid supplementation can maintain the same growth performance of finishing broilers as 18% CP.

Arginine or Valine supplementation in 15% CP animal/vegetable diets over Met and Lys did not show any additive impact in growth performance. In the literature, the availabilities of Val and Arg in fish meal were low (67.3% and 62.6%, respectively); thus, supplementation of Val and Arg was necessary to maintain optimum growth performance in the fish meal-containing diet [32,33]. Wide variations in amino acid digestibility have been reported in animal source proteins, particularly in terms of batch-to-batch differences [34]. The differences in responses to Arg or Val addition may be due to the difference in dietary protein sources, inclusion levels and amino acid requirements for broiler males and females [35–38]. It has been suggested that the protein and/or amino acid levels of the diets, strain, and the age of the birds may affect the response to the level of amino acid addition [7,39]. The supplementation of Met, Lys, Thr, Arg and Val was found to improve the growth of broilers [39], and the supplementation of Met, Lys, Arg, Val and Thr plus glutamic acid to 16% CP diet during 3–6 weeks of age also increased growth performance [36].

In the 14% CP soybean meal diet, the limiting amino acids were Met, Lys, Thr and Val [40], whereas Met, Lys, Thr, Arg, Val and Ile were limited in the 12% CP diet [41]. It has also been reported a low-protein diet (17%) with balanced Lys, Met, Thr and Trp levels had the same performance as a 19% CP diet during 22 to 56 days of age [42,43]. However, the present research indicates that 15% CP diet supplemented with Met + Lys showed no significant differences in growth and FCR compared to the positive control diet. Therefore, our trial suggests that limiting amino acids are related to CP percentage, environmental temperature and ingredient composition (availability of amino acids) of the experimental diets.

In the present study, soybean meal, a high-quality protein ingredient [34], has been used as primary protein source. The lack of response to Ile in Trial 2 indicated that this amino acid is not limiting in 15% CP diet under heat stress and/or there might be the antagonism between Ile and leucine, because both are branched-chain amino acids [17,44,45]. This inconsistency among many studies may be attributed to diet composition, amino acid content and availability in ingredients, age of birds, and environment conditions (temperature and humidity).

Glycine as a source of amino nitrogen or urea, as a nitrogenous source had no additive influence on the growth of broilers in Trial 2, indicating that the 15% CP diet supplemented with Met and Lys furnishes adequate amounts of dispensable amino acids for broiler growth during 30–45 days of age under a hot climate. Awad et al. [46] reported that the addition of glycine did not compensate the depressed growth performance in broilers under hot climate. The lack of responses to glycine as a nonessential amino acid on improving broiler performance is in contrast to the finding of Dean et al. [47]. Supplementation of glycine enhanced growth performance of broilers fed a low protein amino acid-supplemented diet because low-protein diets are low in nitrogen sources for making non-essential amino acids and other essential nitrogen compounds in the body [47]. Urea in non-ruminant animals had no nutrient values [48,49]. Thus, further studies are needed to evaluate the amino acid requirements under different conditions involving ingredients and rearing conditions.

The present results indicated that the diet containing 15% of CP supplemented with Met and Lys may satisfy all amino acid requirements for optimal broiler growth under hot climate. Some studies have shown that broilers eat more feed to satisfy their protein or amino acid needs under hot climate [9,50,51]. The feed conversion ratios in this study are similar to other studies showing that there were no significant differences in feed intake and efficiency of feed use among birds fed a low protein and a high-protein diets [8,51–53].

The present results indicated that protein intake was high, and protein/gain ratio was poorer in the 18% CP group than the groups fed low-protein amino acid-supplemented diets. Aletor et al. [50] indicated that a lower dietary protein level improves the protein use due to the lower protein turnover rate. There was a numerical 21.1% reduction in nitrogen excreted when 15% CP supplemented with Met and Lys in the present study. These results are in agreement with several studies reporting that low-protein amino acid-supplemented diets had no negative effect on broiler performance, while reducing nitrogen excretion by 33.6% [24,43,52,54,55]. The reduction of nitrogen excretion also influences ammonia volatilization in poultry houses: supplementing amino acids in low-protein diets reduced ammonia volatilization by 45% in poultry [4,5,56].

In general, due to the limits of the experimental conditions, it is not clear that the amino acid requirement change was triggered by heat stress or a decrease on performance or both; thus, further research is needed.

4.2. *Carcass Yield, Body Organs and Chemical Composition of Breast Meat*

Carcass yield results in the current study indicate that Met and Lys supplementation in 15% CP diet provides sufficient amount of amino acids for maintaining optimum growth and development of body organs compared to 18% or 18.5% CP diets under hot climate. Alleman and Leclercq [51] reported that reducing CP content from 20% to 16% during 23–44 days of age had no negative effect on breast muscle percentage. However, several researchers have concluded that lowering the protein level in diet, maintaining appropriate levels of essential amino acids (Arg, Ile, Val and Trp) allowed normal protein gain, but increased fat deposition in both fat and lean lines of broilers, revealing that protein or lipid deposition was potentially controlled independently [8,57,58].

In agreement with the present results, Aletor et al. [50] found that decreasing dietary CP percentage while satisfying the amino acid requirements of broilers had no detrimental effect on relative weights of breast or thigh, which is similar to the present findings. However, in contrast to the present results, there were significant increases in the relative weights of liver and abdominal fat which may indicate higher lipogenetic activity in the body. Other studies have shown a significant increase in abdominal fat deposition when broilers or ducks were fed low-CP diets [22,23,59]. However, according to the current study, under hot climates, broilers may use amino acid and protein from low-CP diets better than those under the normal temperature conditions.

The gizzard and spleen percentage differences were unexpected in the current study. Further studies are necessary to explain this change. The protein level in finisher diets for broiler chickens could be reduced from 18% to 15% without adverse effect on broiler growth performance during 28–49 days of age under hot climates if Met and Lys are supplemented to meet requirements while decreasing nitrogen excretion by 21%. Therefore, protein use and economic efficiency in birds fed 15% CP diet supplemented with Met and Lys were improved because dietary protein input was lower compared to the high-protein control diet. Furthermore, the non-significant influences on carcass yield, abdominal fats, and total lipids of meat indicated that these traits are not compromised feeding the low-protein diet supplemented with Met and Lys.

5. Conclusions

Under hot climate conditions and during the finisher period, feeding a low-protein diet fortified with essential amino acids improved protein use and deceased nitrogen excretion without negative effect on performance and carcass yield of broiler. Thus, in the above-mentioned environmental conditions, the protein level of the finishing broiler diets can be reduced from 18% to 15% with an adequate supplementation of only methionine and lysine.

Author Contributions: Conceptualization, Y.A.A. and F.B.; methodology and data collection, and formal analysis, M.A.A.-H.; writing—review and editing, W.K.K. and J.W.; resources, writing—original draft preparation, review and proofreading, Y.A.A., M.A.A.-H. and F.B. All authors have read and agreed to the published version of the manuscript.

Acknowledgments: This article was funded by the Deanship of Scientific Research (DSR), King Abdulaziz University, Jeddah. Therefore, the authors appreciate the technical and financial support of the DSR.

References

1. Attia, Y.A.; Böhmer, B.M.; Roth-Maier, D.A. Responses of broiler chicks raised under constant relatively high ambient temperature to enzymes, amino acid supplementations, or diet density. *Archive Für Geflügelkunde* **2006**, *70*, 80–91.
2. Attia, Y.A.; Hassan, R.A.; Tag El-Din, A.E.; Abou-Shehema, B.M. Effect of ascorbic acid or increasing metabolizable energy level with or without supplementation of some essential amino acids on productive and physiological traits of slow-growing chicks exposed to chronic heat stress. *J. Anim. Physiol. Anim. Nutr.* **2011**, *95*, 744–755. [CrossRef] [PubMed]
3. Dozier, W.A.; Kidd, M.T.; Corzo, A. Dietary amino acid responses of broiler chickens. *J. Appl. Poult. Res.* **2008**, *17*, 157–167. [CrossRef]
4. Kim, J.H.; Patterson, P.H.; Kim, W.K. Impact of dietary crude protein, synthetic amino acid and keto acid formulation on nitrogen excretion. *Int. J. Poult. Sci.* **2014**, *13*, 429–436. [CrossRef]
5. Chalova, V.I.; Kim, J.H.; Patterson, P.H.; Ricke, S.C.; Kim, W.K. Reduction of nitrogen excretion and emissions from poultry: A review for conventional poultry. *World's Poult. Sci. J.* **2016**, *72*, 509–520. [CrossRef]
6. Fancher, B.I.; Jensen, L.S. Dietary protein level and essential amino acid content: Influence upon female broiler performance during the grower period. *Poult. Sci.* **1989**, *68*, 897–908. [CrossRef]
7. Han, Y.; Suzuki, H.; Parsons, C.M.; Baker, H. Amino acid fortification of a low-protein corn and soybean meal diet for chicks. *Poult. Sci.* **1992**, *71*, 1168–1178. [CrossRef]
8. Laudadio, V.; Dambrosio, A.; Normanno, G.; Khan, R.U.; Naz, S.; Rowghani, E.; Tufarelli, V.; Bunchasak, C.U.; Tanaka, S.K.; Ohtani, S.; et al. The effect of supplementing methionine plus cystine to a low-protein diet on the growth performance and fat accumulation of growing broiler chicks. *Asian-Australas. J. Anim. Sci.* **1997**, *10*, 185–191.
9. Alleman, F.; Michel, J.; Chagneau, A.M.; Leclercq, B. The effects of dietary protein independent of essential amino acids on growth and body composition in genetically lean and fat chickens. *Br. Poult. Sci.* **2000**, *41*, 214–218. [CrossRef]
10. Aletor, V.A.; Hamid, I.I.; Niess, E.; Pfeffer, E. Low-protein amino acids-supplemented diets in broiler-chickens: Effects on performance, carcass characteristics whole-body–composition and efficiencies of nutrient utilization. *J. Sci. Food Agric.* **2000**, *80*, 547–554. [CrossRef]
11. Aletor, V.A.; Roth, F.X.; Paulicks, B.R.; Roth-Maier, D.A. Growth, body-fat deposition, nitrogen excretion and efficiencies of nutrients utilization in broiler-chicks fed low-protein diets supplemented with amino acids, conjugated linoleic acid or an -glucosidase inhibitor. *Archive Geflügelk* **2001**, *66*, 21–30.
12. Laudadio, V.; Dambrosio, A.; Normanno, G.; Khan, R.U.; Naz, S.; Rowghani, E.; Tufarelli, V. Effect of reducing dietary protein level on performance responses and some microbiological aspects of broiler chickens under summer environmental conditions. *Avian Biol. Res.* **2012**, *5*, 88–92. [CrossRef]
13. McGill, E.; Kamyab, A.; Firman, J.D. Low crude protein corn and soybean meal diets with amino acid supplementation for broilers in the starter period. 1. Effects of feeding 15% crude protein. *Int. J. Poult. Sci.* **2012**, *11*, 161–165.
14. McGill, E.; Kamyab, A.; Firman, J.D. Low crude protein corn and soybean meal diets with amino acid supplementation for broilers in the starter period. 2. Effects of feeding 13% crude protein. *Int. J. Poult. Sci.* **2012**, *11*, 166–171.
15. Scott, M.L.; Nesheim, M.C.; Young, R.J. *Nutrition of the Chicken*, 3rd ed.; M. L. Scott and Associates: Ithaca, NY, USA, 1982.
16. Fernandez, S.R.; Aoyagi, S.; Han, Y.; Parsons, C.; Baker, D.H. Limiting order of amino acids in corn and soybean meal for growth of the chick. *Poult. Sci.* **1994**, *73*, 1887–1896. [CrossRef]
17. Baker, D.M. Problems and pitfalls in animal experiments designed to establish dietary requirements for essential nutrients. *J. Nutr.* **1986**, *116*, 2339–2349. [CrossRef]
18. Liu, N.; Wang, J.Q.; Gu, K.T.; Deng, Q.Q.; Wang, J.P. Effects of dietary protein levels and multienzyme supplementation on growth performance and markers of gut health of broilers fed a miscellaneous meal based diet. *Anim. Feed Sci. Technol.* **2017**, *234*, 110–117. [CrossRef]
19. Holsheimer, J.P.; Janssen, W.M.M.A. Limiting amino acids in low protein maize-soybean meal diets fed to broiler chicks from 3 to 7 weeks of age. *Br. Poult. Sci.* **1991**, *32*, 151–158. [CrossRef]

20. Lipstein, B.; Bornstein, S.; Bartov, I. The replacement of some of the soybean meal by the first limiting amino acids in practical broiler diets. 3. Effects of protein concentration and amino acid supplementation in broiler finisher diets on fat deposition on the carcass. *Br. Poult. Sci.* **1975**, *16*, 627–635. [CrossRef]
21. Leclercq, B.; Chagneau, A.M.; Cochard, T.; Khoury, J. Comparative responses of genetically lean and fat chickens to lysine, arginine and non-essential amino acid supply. I. Growth and body composition. *Br. Poult. Sci.* **1994**, *35*, 687–696. [CrossRef]
22. Sterling, K.G.; Costa, E.F.; Henry, M.H.; Pesti, G.M.; Bakalli, R.I. Responses of broiler chickens to cottonseed and soybean meal-based diets at several protein levels. *Poult. Sci.* **2002**, *81*, 217–226. [CrossRef] [PubMed]
23. Xie, M.; Jiang, Y.; Tang, J.; Wen, Z.G.; Zhang, Q.; Huang, W.; Hou, S.S. Effects of low-protein diets on growth performance and carcass yield of growing White Pekin ducks. *Poult. Sci.* **2017**, *96*, 1370–1375. [CrossRef] [PubMed]
24. Ospina-Rojas, I.C.; Murakami, A.E.; Duarte, C.R.A.; Eyng, C.; Oliveira, A.L.; Janeiro, V. Valine, isoleucine, arginine and glycine supplementation of low-protein diets for broiler chickens during the starter and grower phases. *Br. Poult. Sci.* **2014**, *55*, 766–773. [CrossRef] [PubMed]
25. National Research Council (NRC). *Nutrient Requirements of Poultry*, 9th ed.; National Academy of Sciences, National Research Council: Washington, DC, USA, 1994.
26. Association Official Analytical Chemistry (AOAC). *Official Methods of Analysis*, 19th ed.; Association Official Analytical Chemistry: Washington, DC, USA, 2007.
27. SAS Institute. *SAS®User's Guide for Personal Computer*; SAS Institute Inc.: Cary, NC, USA, 2007.
28. Attia, Y.A.; Al-Harthi, M.A.; Hassan, S.S. Turmeric (Curcuma longa Linn.) as a phytogenic growth promoter alternative for antibiotic and comparable to mannan oligosaccharides for broiler chicks. *Rev. Mex. Cienc. Pecu.* **2017**, *8*, 11–21. [CrossRef]
29. Quinteiro-Filho, W.M.; Ribeiro, A.; Ferraz-de-Paula, V.; Pinheiro, M.L.; Sakai, M.; Sá, L.R.M.D.; Palermo-Neto, J. Heat stress impairs performance parameters, induces intestinal injury, and decreases macrophage activity in broiler chickens. *Poult. Sci.* **2010**, *89*, 1905–1914. [CrossRef]
30. Chrystal, P.V.; Moss, A.F.; Khoddami, A.; Naranjo, V.D.; Selle, P.H.; Liu, S.Y. Impacts of reduced-crude protein diets on key parameters in male broiler chickens offered maize-based diets. *Poult. Sci.* **2020**, *99*, 505–516. [CrossRef]
31. Van Harn, J.; Dijkslag, M.A.; van Krimpen, M.M. Effect of low protein diets supplemented with free amino acids on growth performance, slaughter yield, litter quality, and footpad lesions of male broilers. *Poult. Sci.* **2019**, *98*, 4868–4877. [CrossRef]
32. Smith, R.E. Assessment of the availability of the amino acid in fish meal, soybean meal and feather meal by chick growth assay. *Poult. Sci.* **1968**, *47*, 1624. [CrossRef]
33. Miller, D.; Kifer, R.R. Factors affecting protein evaluation of fish meal by chick bioassay. *Poult. Sci.* **1970**, *49*, 999–1004. [CrossRef]
34. Ravindran, V.; Hew, L.; Ravindran, G.; Bryden, W. Apparent ileal digestibility of amino acids in dietary ingredients for broiler chickens. *Anim. Sci.* **2005**, *81*, 85–97. [CrossRef]
35. Han, Y.; Baker, D.H. Effects of sex, heat stress, body weight and genetic strain on the lysine requirement of broiler chicks. *Poult. Sci.* **1993**, *72*, 701–708. [CrossRef] [PubMed]
36. Han, Y.; Baker, D.H. Lysine requirement of fast and slow growing broiler chicks. *Poult. Sci.* **1991**, *70*, 2108–2114. [CrossRef] [PubMed]
37. Aggoor, F.A.M.; El-Naggar, N.M.; Mehrez, A.Z.; Attia, Y.A.; Qota, E.M.A. Effect of different dietary protein and energy levels during roaster period on: 1-Performance and economic evaluation of broiler chicks. *Egypt. Poult. Sci.* **1997**, *17*, 81–105.
38. Costa, M.J.; Da Zaragoza-Santacruz, S.; Frost, T.J.; Halley, J.; Pesti, G.M. Straight-run vs. sex separate rearing for 2 broiler genetic lines Part 1: Live production parameters, carcass yield, and feeding behavior. *Poult. Sci.* **2017**, *96*, 2641–2661. [CrossRef]
39. Waguespack, A.M.; Powell, S.; Bidner, T.D.; Payne, R.L.; Southern, L.L. Effect of incremental levels of L-lysine and determination of the limiting amino acids in low crude protein corn-soybean meal diets for broilers. *Poult. Sci.* **2009**, *88*, 1216–1226. [CrossRef]
40. Warnick, R.E.; Anderson, J.O. Limiting essential amino acids in soybean meal for growing chickens and the effects of heat upon availability of the essential amino acids. *Poult. Sci.* **1968**, *47*, 281–287. [CrossRef]

41. Schwartz, R.W.; Bray, D.J. Limiting amino acids in 40:60 and 15:85 blends of corn: Soybean protein for the chicks. *Poult. Sci.* **1975**, *54*, 1814.
42. Uzu, G. Limit of reduction of the protein level in broiler feeds. *Poult. Sci.* **1982**, *61*, 1557–1558.
43. Shao, D.; Shen, Y.; Zhao, X.; Wang, Q.; Hu, Y.; Shi, S.; Tong, H. Low-protein diets with balanced amino acids reduce nitrogen excretion and foot pad dermatitis without affecting the growth performance and meat quality of free-range yellow broilers. *Ital. J. Anim. Sci.* **2018**, *17*, 698–705. [CrossRef]
44. Jensen, L.S.; Mendonca, C.X. Amino acid nutrition of broilers during the grower period. In *Georgia Nutrition Conference for the Feed Industry*; University of Georgia: Athens, GA, USA, 1988; pp. 76–83.
45. D'Mello, J.P.E. (Ed.) Amino acid imbalances, antagonisms and toxicities. In *Amino Acid in Farm Animal Nutrition*; CAB International: Wallingford, UK, 1994; pp. 63–98.
46. Awad, E.A.; Zulkifli, I.; Soleimani, A.F.; Loh, T.C. Individual non-essential amino acids fortification of a low-protein diet for broilers under hot and humid tropical climate. *Poult. Sci.* **2014**, *94*, 2772–2777. [CrossRef]
47. Dean, D.W.; Bidner, T.D.; Southern, L.L. Glycine supplementation to low protein, amino acid-supplemented diets supports optimal performance of broiler chicks. *Poult. Sci.* **2006**, *85*, 288–296. [CrossRef] [PubMed]
48. El-Boushy, A.R. Broiler growth response from practical low-protein diets supplemented with urea and diammonium hydrogen phosphate. *Neth. J. Agric. Sci.* **1980**, *28*, 147–155.
49. Kagan, A.; Balloun, S.L. Urea and aspartic acid supplementation of low-protein broiler diets. *Br. Poult. Sci.* **1976**, *17*, 403–413. [CrossRef] [PubMed]
50. Aletor, V.A.; Eder, K.; Becker, K.; Paulicks, B.R.; Roth, F.X.; Roth-Maier, D.A. The effects of conjugated linoleic acids or an alpha-glucosidase inhibitor on tissue lipid concentrations and fatty acid composition of broiler chicks fed a low-protein diet. *Poult. Sci.* **2003**, *82*, 796–804. [CrossRef]
51. Alleman, F.; Leclercq, B. Effect of dietary protein and environmental temperature on growth performance and water intake of male broiler chickens. *Poult. Sci.* **1997**, *38*, 607–610. [CrossRef]
52. Jacob, J.; Blair, R.; Ibrahim, S.; Scott, T.; Newberry, R. Using reduced protein diets to minimize nitrogen excretion of broilers. *Poult. Sci.* **1995**, *74* (Suppl. 1), 127.
53. Nahashon, S.M.; Bartlett, J.R.; Smith, E.J. Responses to dietary crude protein and energy levels by cross of chickens involving White Plymouth Rock. *Poult. Sci.* **1995**, *74* (Suppl. 1), 207.
54. Hegedüs, M. Dietary factors influencing protein utilization: A review. *Acta Vet. Hung.* **1992**, *40*, 133–143.
55. Moran, E.T.; Bushong, R.D.; Bilgili, S.F. Reducing dietary crude protein for broilers while satisfying amino acid requirements by least-cost formulation: Live performance, litter composition, and yield of fast-food carcass cute of six weeks. *Poult. Sci.* **1992**, *71*, 1687–1694. [CrossRef]
56. Keshavarz, K.; Austic, R.E. The use of low-protein, low-phosphorus, amino acid- and phytase-supplemented diets on laying hens performance and nitrogen and phosphorus excretion. *Poult. Sci.* **2004**, *83*, 75–83. [CrossRef]
57. Leclercq, B. The influence of dietary protein content on the performance of genetically lean or fat growing chickens. *Br. Poult. Sci.* **1983**, *24*, 581–587. [CrossRef] [PubMed]
58. Cahaner, A.; Leenstra, F. Effects of high temperature on growth and efficiency of male and female broilers from lines selected for high weight gain, favorable Feed conversion, and high or low fat content. *Poult. Sci.* **1992**, *71*, 1237–1250. [CrossRef] [PubMed]
59. Smith, E.R.; Pesti, G.M.; Bakalli, R.I.; Ware, G.O.; Menten, J.F.M. Further studies on the influence of genotype and dietary protein on the performance of broilers. *Poult. Sci.* **1998**, *77*, 1678–1687. [CrossRef] [PubMed]

Permissions

All chapters in this book were first published in MDPI; hereby published with permission under the Creative Commons Attribution License or equivalent. Every chapter published in this book has been scrutinized by our experts. Their significance has been extensively debated. The topics covered herein carry significant findings which will fuel the growth of the discipline. They may even be implemented as practical applications or may be referred to as a beginning point for another development.

The contributors of this book come from diverse backgrounds, making this book a truly international effort. This book will bring forth new frontiers with its revolutionizing research information and detailed analysis of the nascent developments around the world.

We would like to thank all the contributing authors for lending their expertise to make the book truly unique. They have played a crucial role in the development of this book. Without their invaluable contributions this book wouldn't have been possible. They have made vital efforts to compile up to date information on the varied aspects of this subject to make this book a valuable addition to the collection of many professionals and students.

This book was conceptualized with the vision of imparting up-to-date information and advanced data in this field. To ensure the same, a matchless editorial board was set up. Every individual on the board went through rigorous rounds of assessment to prove their worth. After which they invested a large part of their time researching and compiling the most relevant data for our readers.

The editorial board has been involved in producing this book since its inception. They have spent rigorous hours researching and exploring the diverse topics which have resulted in the successful publishing of this book. They have passed on their knowledge of decades through this book. To expedite this challenging task, the publisher supported the team at every step. A small team of assistant editors was also appointed to further simplify the editing procedure and attain best results for the readers.

Apart from the editorial board, the designing team has also invested a significant amount of their time in understanding the subject and creating the most relevant covers. They scrutinized every image to scout for the most suitable representation of the subject and create an appropriate cover for the book.

The publishing team has been an ardent support to the editorial, designing and production team. Their endless efforts to recruit the best for this project, has resulted in the accomplishment of this book. They are a veteran in the field of academics and their pool of knowledge is as vast as their experience in printing. Their expertise and guidance has proved useful at every step. Their uncompromising quality standards have made this book an exceptional effort. Their encouragement from time to time has been an inspiration for everyone.

The publisher and the editorial board hope that this book will prove to be a valuable piece of knowledge for researchers, students, practitioners and scholars across the globe.

List of Contributors

Raffaella Rossi, Sabrina Ratti and Carlo Corino
Dipartimento di Medicina Veterinaria, Università Degli Studi di Milano, Via dell'Università 6, 26900 Lodi, Italy

Francesco Vizzarri
Department of Agricultural and Environmental Science, Università di Bari Aldo Moro, Via G. Amendola 165/A, 70126 Bari, Italy

Marisa Palazzo and Donato Casamassima
Department of Agricultural, Environmental and Food Sciences, Università Degli Studi del Molise, Via F. De Sanctis 1, 86100 Campobasso, Italy

Janghan Choi and Woo Kyun Kim, Jinquan Wang
Department of Poultry Science, University of Georgia, Athens, GA 30602, USA

Youssef A. Attia and Mohammed A. Al-Harthi
Arid Land Agriculture Department, Faculty of Meteorology, Environment and Arid Land Agriculture, King Abdulaziz University, Jeddah 21589, Saudi Arabia

Ali S. El-Shafey
Animal and Poultry Production Department, Faculty of Agriculture, Damanhour University, Damanhour 22516, Egypt

Wen Yang Chuang, Chu Ling Liu, Chia Fen Tsai and Wei Chih Lin
Department of Animal Science, National Chung Hsing University, Taichung 402, Taiwan

Shen Chang Chang
Kaohsiung Animal Propagation Station, Livestock Research Institute, Council of Agriculture, Tainan 71246, Taiwan

Hsin Der Shih
Taiwan Agricultural Research Institute Council of Agriculture, Executive Yuan, Taichung City 41362, Taiwan

Yi Ming Shy
Hsinchu Branch, Taiwan Livestock Research Institute, Council of Agriculture, Executive Yuan, Tainan 71246, Taiwan

Tzu-Tai Lee
Department of Animal Science, National Chung Hsing University, Taichung 402, Taiwan
The iEGG and Animal Biotechnology Center, National Chung Hsing University, Taichung 402, Taiwan

Wuchen Du
The United Graduate School of Agricultural Sciences, Tottori University, Tottori 680-8550, Japan

Fujiang Hou
State Key Laboratory of Grassland Agro-Ecosystems, Key Laboratory of Grassland Livestock Industry Innovation, Ministry of Agriculture, College of Pastoral Agriculture Science and Technology, Lanzhou University, Lanzhou 730020, China

Atsushi Tsunekawa and Nobuyuki Kobayashi
Arid Land Research Center, Tottori University, Tottori 680-0001, Japan

Toshiyoshi Ichinohe
Faculty of Life and Environmental Science, Shimane University, Matsue 690-8504, Japan

Fei Peng
International Platform for Dryland Research and Education, Tottori University, Tottori 680-0001, Japan
Key Laboratory of Desert and Desertification, Northwest Institute of Eco-Environment and Resources, Chinese Academy of Sciences, Lanzhou 730000, China

Lukshman Jay. Ekanayake
Department of Animal Science, Faculty of Agriculture, University of Peradeniya, Peradeniya 20000, Sri Lanka

Rene Anne Corner-Thomas, Lydia Margaret Cranston, Paul Richard Kenyon, Stephen Todd Morris and Sarah Jean Pain
School of Agriculture and Environment, Massey University, Palmerston North 4442, New Zealand

Laurynas Vadopalas, Sarunas Badaras, Dovile Klupsaite and Sonata Sidlauskiene
Institute of Animal Rearing Technologies, Lithuanian University of Health Sciences, Tilžės g. 18, LT-47181 Kaunas, Lithuania

Agila Dauksiene
Institute of Animal Rearing Technologies, Lithuanian University of Health Sciences, Tilžės g. 18, LT-47181 Kaunas, Lithuania
Department of Physiology and Anatomy, Lithuanian University of Health Sciences, Tilžės g. 18, LT-47181 Kaunas, Lithuania

Modestas Ruzauskas
Department of Physiology and Anatomy, Lithuanian University of Health Sciences, Tilžės g. 18, LT-47181 Kaunas, Lithuania
Microbiology and Virology Institute, Lithuanian University of Health Sciences, Tilžės g. 18, LT-47181 Kaunas, Lithuania

Vita Lele, Vytaute Starkute, Paulina Zavistanaviciute, Egle Zokaityte, Erika Mozuriene and Elena Bartkiene
Institute of Animal Rearing Technologies, Lithuanian University of Health Sciences, Tilžės g. 18, LT-47181 Kaunas, Lithuania
Department of Food Safety and Quality, Lithuanian University of Health Sciences, Tilžės g. 18, LT-47181 Kaunas, Lithuania

Vadims Bartkevics
University of Latvia Faculty of Chemistry, Jelgavas iela 1, LV-1004 Riga, Latvia
Institute of Food Safety, Animal Health and Environment BIOR, Lejupesiela 3, LV-1076 Riga, Latvia

Romas Gruzauskas
Department of Food Science and Technology, Kaunas University of Technology, Radvilenu str. 19, LT-50254 Kaunas, Lithuania

Simone Mancini and Barbara Turchi
Department of Veterinary Sciences, University of Pisa, Viale delle Piagge 2, 56124 Pisa, Italy

Filippo Fratini and Gisella Paci
Department of Veterinary Sciences, University of Pisa, Viale delle Piagge 2, 56124 Pisa, Italy
Interdepartmental Research Center "Nutraceuticals and Food for Health", University of Pisa, Via del Borghetto 80, 56124 Pisa, Italy

Simona Mattioli and Alessandro Dal Bosco
Department of Agricultural, Food and Environmental Science, University of Perugia, Borgo XX Giugno 74, 06100 Perugia, Italy

Tiziano Tuccinardi
Department of Pharmacy, University of Pisa, Via Bonanno 6, 56124 Pisa, Italy

Sanjin Nozic
Department of Veterinary Sciences, University of Pisa, Viale delle Piagge 2, 56124 Pisa, Italy
Department of Pharmacy, University of Pisa, Via Bonanno 6, 56124 Pisa, Italy

Giulia Secci and Giuliana Parisi
Department of Agriculture, Food, Environment and Forestry, University of Firenze, via delle Cascine 5, 50144 Firenze, Italy

Fulvia Bovera
Department of Veterinary Medicine and Animal Production, University of Napoli Federico II, via Federico Delpino 1, 80137 Napoli, Italy

Giuseppe Moniello
Department of Veterinary Medicine, University of Sassari, via Vienna 2, 07100 Sassari, Italy

Jernej Poteko
ETH Zürich, Institute of Agricultural Sciences, 8092 Zurich, Switzerland
Agroscope, Ruminants Research Unit, 8356 Ettenhausen, Switzerland

Michael Kreuzer
ETH Zürich, Institute of Agricultural Sciences, 8092 Zurich, Switzerland

Sabine Schrade and Michael Zaehner
Agroscope, Ruminants Research Unit, 8356 Ettenhausen, Switzerland

Kerstin Zeyer and Joachim Mohn
Empa, Laboratory for Air Pollution/Environmental Technology, 8600 Duebendorf, Switzerland

Johanna O. Zeitz
ETH Zürich, Institute of Agricultural Sciences, 8092 Zurich, Switzerland
Laboklin, Klinische Labordiagnostik, 97688 Bad Kissingen, Germany

Angela Schwarm
ETH Zürich, Institute of Agricultural Sciences, 8092 Zurich, Switzerland
Department of Animal and Aquacultural Sciences, Norwegian University of Life Sciences, 1432 Ås, Norway

Shiqin Wang and Qiyu Diao
State Key Laboratory of Grassland Agro-Ecosystems, Key Laboratory of Grassland Livestock Industry Innovation, Ministry of Agriculture and Rural Affairs, College of Pastoral Agriculture Science and Technology, Lanzhou University, Lanzhou 730020, China
Feed Research Institute, Chinese Academy of Agricultural Sciences/Key Laboratory of Feed Biotechnology of the Ministry of Agriculture and Rural Affairs, Beijing 100081, China

Fadi Li
State Key Laboratory of Grassland Agro-Ecosystems, Key Laboratory of Grassland Livestock Industry Innovation, Ministry of Agriculture and Rural Affairs, College of Pastoral Agriculture Science and Technology, Lanzhou University, Lanzhou 730020, China

Tao Ma, Guohong Zhao, Naifeng Zhang, Yan Tu, Kai Cui, Yanliang Bi and Hongbiao Ding
Feed Research Institute, Chinese Academy of Agricultural Sciences/Key Laboratory of Feed Biotechnology of the Ministry of Agriculture and Rural Affairs, Beijing 100081, China

Soomin Shim, Seungsoo Kim, Sungil Ahn and Changsix Ra
Department of Animal Industry Convergence, College of Animal Life Sciences, Kangwon National University, Chuncheon 24341, Korea

Seunggun Won
Department of Animal Resources, College of Life and Environmental Science, Daegu University, Gyeongsan 38453, Korea

Arif Reza
Department of Animal Industry Convergence, College of Animal Life Sciences, Kangwon National University, Chuncheon 24341, Korea
Department of Environmental Science, College of Agricultural Sciences, IUBAT—International University of Business Agriculture and Technology, Dhaka 1230, Bangladesh

Baedong Jung and Byungil Yoon
Department of Veterinary Science, Kangwon National University, Chuncheon 24341, Korea

Elisa Fiordelmondo, Gian Enrico Magi, Francesca Mariotti and Alessandra Roncarati
School of Bioscience and Veterinary Medicine, University of Camerino, Circonvallazione 93-95, 62034 Matelica, Italy

Rigers Bakiu
Department of Aquaculture, University of Tirana, Place Mother Tereza 183, 695581 Tirana, Albania

Mohamed A. Korish
Arid Land Agriculture Department, Faculty of Meteorology, Environment, and Arid Land Agriculture, King Abdulaziz University, Jeddah 21589, Saudi Arabia

Mohamed H. Shiboob
Environmental Department, Faculty of Meteorology, Environment and Arid land Agriculture, King Abdulaziz University, Jeddah 21589, Saudi Arabia

Fan Zhang
State Key Laboratory of Animal Nutrition, Institute of Animal Science, Chinese Academy of Agricultural Sciences, Beijing 100193, China
State Key Laboratory of Animal Nutrition, College of Animal Science and Technology, China Agricultural University, Beijing 100193, China

Xuemei Nan, Hui Wang, Yiguang Zhao and Benhai Xiong
State Key Laboratory of Animal Nutrition, Institute of Animal Science, Chinese Academy of Agricultural Sciences, Beijing 100193, China

Yuming Guo
State Key Laboratory of Animal Nutrition, College of Animal Science and Technology, China Agricultural University, Beijing 100193, China

Index

Printed in the USA
CPSIA information can be obtained
at www.ICGtesting.com
LVHW080042211023
761596LV00010BA/1876

9 781641 168144